G. Hale Puckle

An elementary treatise on conic sections and algebraic geometry

With numerous examples and hints for their solution

G. Hale Puckle

An elementary treatise on conic sections and algebraic geometry
With numerous examples and hints for their solution

ISBN/EAN: 9783742892331

Manufactured in Europe, USA, Canada, Australia, Japa

Cover: Foto ©berggeist007 / pixelio.de

Manufactured and distributed by brebook publishing software
(www.brebook.com)

G. Hale Puckle

An elementary treatise on conic sections and algebraic geometry

AN

ELEMENTARY TREATISE

ON

CONIC SECTIONS

AND

ALGEBRAIC GEOMETRY,

WITH NUMEROUS EXAMPLES AND HINTS FOR THEIR SOLUTION.

Especially designed for the Use of Beginners.

BY

G. HALE PUCKLE, M.A.,

ST JOHN'S COLLEGE, CAMBRIDGE.

FIFTH EDITION, REVISED AND ENLARGED.

London:

MACMILLAN AND CO.

1884

PREFACE.

THE first edition of this book was published shortly after the appearance of Dr Salmon's admirable treatise, with the hope that I could write a short and easy work upon a similar plan, without losing the obvious advantages of his harmonious and consecutive arrangement of the subject. Hence the Reduction of the General Equation of the Second Degree will be found to precede the discussion of the properties of the curves, and the Parabola is treated of after the Ellipse and Hyperbola; but I have arranged the chapter on the Reduction of the General Equation, so that a very small part of it will suffice, when the subject is read for the first time.

My chief object was to write with special reference to those difficulties and misapprehensions, which I had found most common to beginners. In the later editions I have tried, without losing sight of my original purpose, to make the book suitable to the requirements of the time. The present edition has been carefully revised throughout. I have added a considerable amount of new matter, especially in the way of illustrative examples worked out. I have also changed the notation of the General Equation of the Second Order, to that used by Dr Salmon.

It would be quite beyond the scope of an elementary work to notice all the modern algebraic methods which have been applied to the Conic Sections. I have wished to make the treatise complete, as far as regards Cartesian Co-ordinates; and, as Abridged Notation naturally leads to Trilinear Co-ordinates, I have given a short account of the latter method also.

It is almost unnecessary to say, that I am very much indebted to Dr Salmon's *Conic Sections.* Indeed, I have found very little in other modern treatises, that has not first appeared in Dr Salmon's book. In preparing the later editions, I have received much valuable assistance from the Rev. D. J. Davies, late Fellow of Emmanuel College, Cambridge. Most of the alterations in the present edition have been made by the advice, very kindly given, of Mr J. S. Yeo, Fellow of St John's College, Cambridge.

G. HALE PUCKLE.

WINDERMERE,
August, 1881.

CONTENTS.

Students reading this subject for the first time may omit the Chapters, Articles, and Examples marked with an asterisk.

ALGEBRAIC GEOMETRY.

CHAPTER I.

Position of a Point on a Plane—Loci—Equations.

1. ALGEBRA is applied to Geometry, to investigate problems which concern the *magnitudes* of lines or areas, or to express the *position* of points and the *form* of curves. The Algebraic proofs of the propositions of the second book of Euclid are examples of the former application, and the use of positive and negative signs with the Trigonometrical ratios has introduced the reader to the latter. It is upon the latter branch of the subject that we are now about to enter; and we proceed to explain the system invented by Descartes, and hence called 'the Cartesian system,' in which the positions of points are determined by means of *Co-ordinates*.

2. In ordinary Algebra we have been in the habit of considering the symbols + and − as symbols of the reverse operations of addition and subtraction. In Algebraic Geometry, as in Trigonometry, we use these symbols to indicate contrariety of position. Thus, in any indefinite straight line XX', let us consider O as a fixed point from which distances are to be measured, and let us take two points M, M', equidistant from O and on opposite sides of it; then, if we

denote the distance OM by $+a$, we shall express the distance

X' M' O A M X

OM' by $-a$, *i.e.* we shall consider lines measured from left to right as positive, and lines measured from right to left as negative.

The propriety of this convention will best appear from an example. Let us suppose X the east and X' the west of O; then, if a man starting from O walk 6 miles to the east and then 2 miles to the west, the *magnitude* of his walk will be represented by 8, but his *position* with reference to his starting place by $+4$, or, in order to find his position, if we consider the distance he has walked east as positive, we must consider the distance he has walked west as negative. Similarly, if he walked 4 miles east and then 6 miles west, his *position* would be expressed by -2, or two miles to the left of O, but the *magnitude* of his walk by $+10$. This distinction must be carefully borne in mind, and it must be remembered that the convention established above has reference simply to the *position* of points on the line XX', *with reference to* O; hence if we take $OM = +4$ and $OM' = -4$, the student must be careful to avoid the error of assuming MM' to be represented by $4-4$ or 0, as this would be reasoning about the magnitude of a line upon assumptions which have only been made about the position of its extremities with regard to a fixed point.

3. A simple example will shew the great advantage of the preceding convention, in rendering generally true formulæ that would otherwise be true for a particular case only. Let O and A be two fixed points upon the line XX', and M a point moving upon the same line. In order to express the distance of M from the point O by means of the distances OA and AM, let

$$OM = x,\ OA = a,\ AM = x';$$

then, if the point M is placed to the right of A, we have

$$x = a + x' \quad \text{.....................} \quad (1);$$

if it is between O and A,

$$x = a - x' \quad \text{.....................} \quad (2);$$

if it is to the left of O,

$$x = x' - a \quad \text{.....................} \quad (3).$$

Here we have supposed x and x' to be the *magnitudes* of OM and AM, without regard to sign; and we see that three formulæ are necessary to express the distance of M from O, when we consider all the positions that it can occupy. But, if attention is paid to the *signs* of x and x', the formula first obtained is equally true for the other two cases. For, if we suppose M to pass from the right to the left of A, x' becomes $-x'$, and formula (1) becomes the same as (2), as long as x' is less than a or M lies between O and A. If x' is greater than a, M will fall to the left of O, and x becomes negative; in this case formula (1) becomes

$$-x = a - x',$$

which is the same as formula (3).

4. *To determine the position of a point on a plane.*

Let us suppose that we know the position of two straight

lelogram $OMPN$, P will be the point whose position we wished to determine.

5. The line PM is usually denoted by the letter y, and is called the *ordinate** of the point; OM, which $=PN$, is denoted by the letter x, and is called the *abscissa* of the point; the two lines are called the *co-ordinates* of P. The lines xx', yy' are called the *axes of co-ordinates*, and their point of intersection O is called the *origin*: xx' is called the *axis of x*, and yy' the *axis of y*.

The point P is said to be determined when the values of x and y are given, as by the two equations $x=a$, $y=b$; as, for example, if it were given that $x=3$ feet, $y=2$ feet, we should determine the point of which x and y are the co-ordinates by measuring 3 feet along Ox and 2 feet along Oy, and completing the parallelogram of which these two lines formed the adjacent sides. The corner of the parallelogram opposite to O would be the position of the point. The point whose position is defined by the equations $x=a$, $y=b$ is commonly spoken of as the point (ab). The axes are said to be rectangular or oblique, according as angle yOx is or is not a right angle.

6. We have supposed hitherto that x and y, the co-ordinates of the point, are positive quantities, and have measured the distances along the lines Ox and Oy. If x or y be negative, it will indicate, according to the convention established above, that we must measure along Ox' or Oy', in order to find the position of the point. For

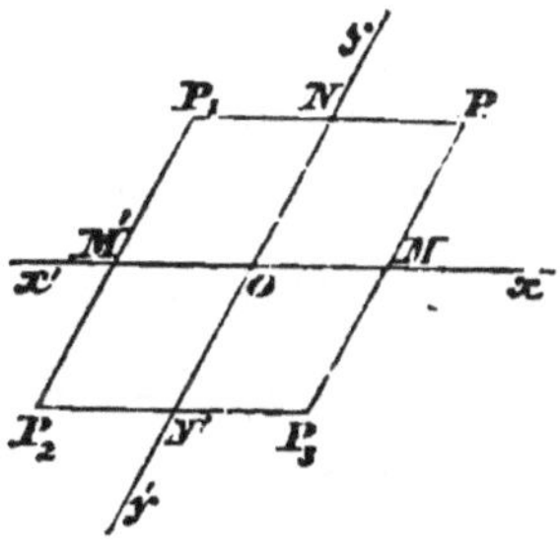

* The lines PM, &c. drawn parallel to one another from a series of points, were called by Newton 'lineæ ordinatim applicatæ,' and the abscissæ OM, &c. were the distances *cut off* by these lines from a fixed line as Ox.

example, if P, P_1, P_2, P_3 be points, situated in the four angles made by the axes, whose co-ordinates are of the same *magnitude, i.e.*

$$PM = P_1M' = P_2M' = P_3M = b,$$

and $OM = OM' = a$, these points will be represented by the following equations:

$$P\begin{cases}x = a,\\ y = b;\end{cases} \qquad P_1\begin{cases}x = -a,\\ y = b;\end{cases}$$

$$P_2\begin{cases}x = -a,\\ y = -b;\end{cases} \qquad P_3\begin{cases}x = a,\\ y = -b.\end{cases}$$

The point represented by $x = 0$, $y = 0$ is the origin O; by $x = 0$, $y = b$ is the point N on the axis of y; by $x = a$, $y = 0$ is the point M on the axis of x, and so on.

7. *To find the distance between two points P, R, whose co-ordinates are known with reference to axes inclined to each other at a given angle.*

Let angle $yOx = \omega$, and let the co-ordinates of P be PM $(=y')$, OM $(=x')$, and of R, RN $(=y'')$, ON $(=x'')$;

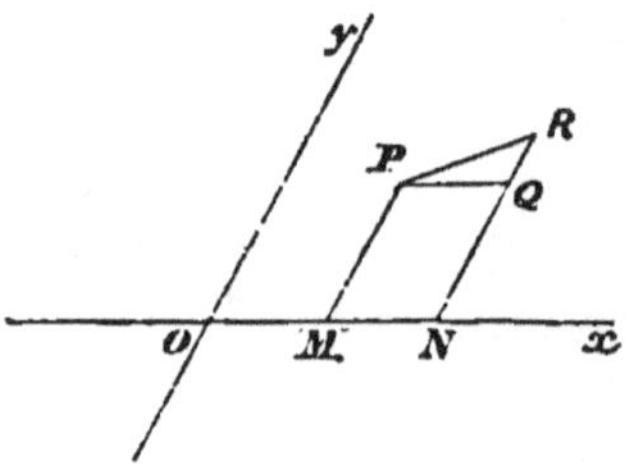

draw PQ parallel to Ox, then

$$PQ = ON - OM = x'' - x',$$

$$RQ = RN - PM = y'' - y',$$

and $\angle\, PQR = \pi - \omega$; hence

$$PR^2 = PQ^2 + RQ^2 - 2PQ\,.\,RQ \cos PQR$$
$$= (x'' - x')^2 + (y'' - y')^2 + 2(x'' - x')(y'' - y') \cos\omega.$$

If one of the points as P were the origin, so that $x' = 0$, $y' = 0$, we should have

$$PR^2 = x''^2 + y''^2 + 2x''y'' \cos\omega.$$

Cor. These formulæ become much more simple when $\omega = 90°$, or the axes are rectangular; in that case, since $\cos\omega = 0$,

$$PR^2 = (x'' - x')^2 + (y'' - y')^2,$$

or, if P be at the origin,

$$PR^2 = x''^2 + y''^2.$$

8. In using these formulæ, attention must be paid to the *signs* of the co-ordinates. If the point P, for instance, be in the angle xOy', the sign of its ordinate (y') will be negative, and we must write $y'' + y'$ instead of $y'' - y'$ in the formulæ; this may be seen to agree with the figure, as RQ will now $= RN + PM$.

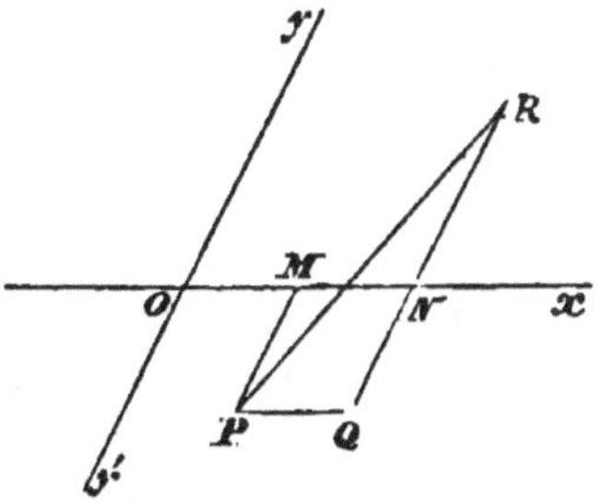

The reader should draw figures, placing P and R in different compartments and in other varieties of position, that he may assure himself of the universal truth of the expressions obtained for the distance PR.

Ex. *To find the distance between two points whose co-ordinates are $x = 2$, $y = -3$ and $x = -5$, $y = 6$, the axes being inclined at an angle of $60°$.*

Here $$x'' - x' = -5 - 2 = -7,\ y'' - y' = 6 + 3 = 9,$$

and $\cos\omega = \frac{1}{2}$: hence, if d be the distance,

$$d^2 = 49 + 81 - 2 . 7 . 9 . \tfrac{1}{2} = 67,$$

$$d = \sqrt{67}.$$

9. We have said that the student should, by different figures, convince himself of the universal truth of the formula of Art. 7; but a careful consideration of a few propositions in this manner will shew him that this is not necessary, and that the formula first obtained will, as in Art. 3, adapt itself to the changes of the figure. Thus, in the fig. of Art. 7, QR is the *difference* between the actual magnitudes of RN and

PN, and must remain so, as long as P and Q are above Ox, that is, as long as both the ordinates are positive. If P falls below Ox, while R remains above it, QR is the *sum* of PM and RN; and this change is provided for by the change of the sign of PM. Hence, in Algebraic Geometry, we need not examine every modification of the figure, as in Euc. II. 13; for any general figure, that we draw consistent with the conditions of a problem, will lead us to a result which will be true for all possible cases comprehended in it.

10. *To find the co-ordinates of a point (hk), where the straight line joining two given points $(x'y')$, $(x''y'')$ is cut in a given ratio.*

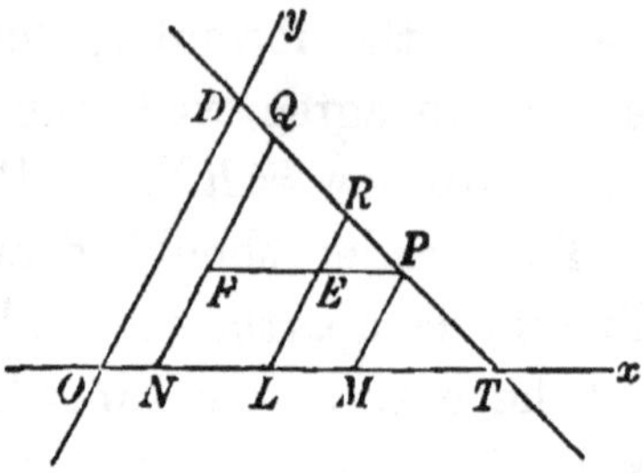

Let P, Q be the two given points $(x'y')$, $(x''y'')$, R the point (hk) whose co-ordinates are sought, and let

$$PR : RQ :: m : n.$$

Draw the ordinates PM, RL, QN, and the line PEF parallel to OT; then

$$\frac{PR}{RQ} = \frac{PE}{EF} = \frac{ML}{LN},$$

or
$$\frac{m}{n} = \frac{x' - h}{h - x''},$$

whence
$$h = \frac{mx'' + nx'}{m + n}.$$

Similarly
$$k = \frac{my'' + ny'}{m + n}.$$

If $m = n$, or PQ is bisected in R,

$$h = \frac{x'' + x'}{2}, \quad k = \frac{y'' + y'}{2},$$

a result which is very frequently of use. It should be verified for different figures, as in Art. 8.

11. *Polar Co-ordinates.*

Besides the method of expressing the position of a point, that we have hitherto made use of, there is another which can often be employed with advantage. If a fixed point O be given, and a fixed line OA through it, we shall evidently know the position of any point P, if we know the length OP and the angle POA. The line OP is called the *radius vector*, the fixed point is called the *pole*, the line OA the *initial line*; and this method is called the method of *polar co-ordinates*. We shall, for the sake of brevity, call the point whose polar co-ordinates are ρ and θ, 'the point $(\rho\theta)$.'

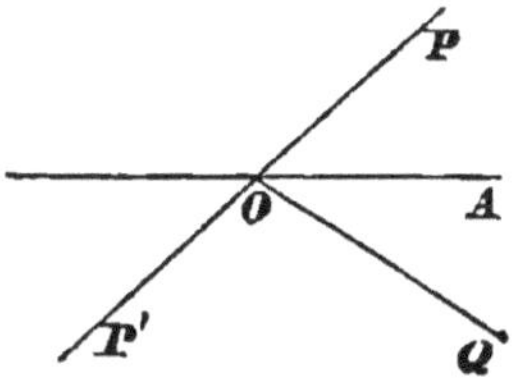

12. The sign $-$ is applied to polar co-ordinates on exactly the same principles as those already explained in the case of rectangular co-ordinates. Thus, if ρ represent any distance measured from O towards P, $-\rho$ represents an equal distance measured from O towards P'; and, if θ represent any angle measured from A towards P, $-\theta$ will represent an equal angle measured from A towards Q. We shall define the *positive* direction of ρ as that part of the line which marks the boundary of the angle θ. A few examples will make this clearer.

Let a be any distance OP, measured from O towards P, θ being the angle which OP makes with OA; then

Fig. 1.

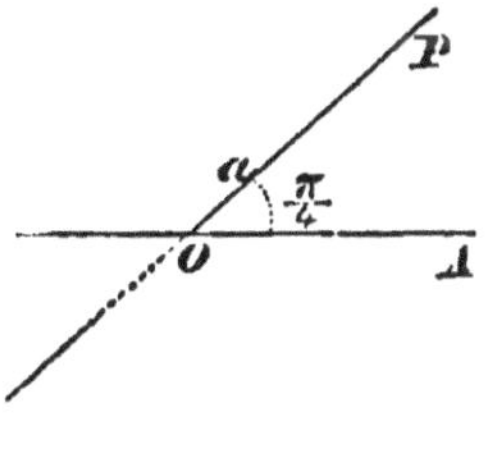

Fig. 2.

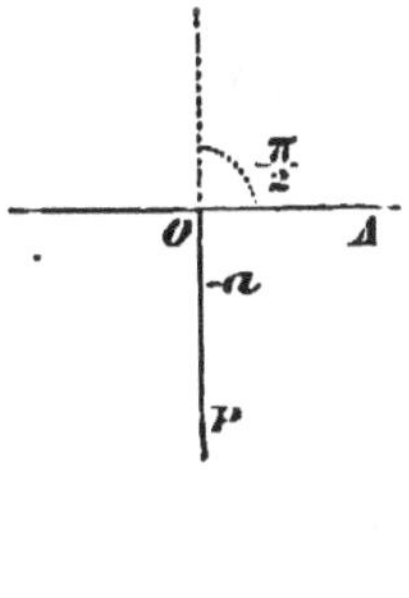

Fig. 3.

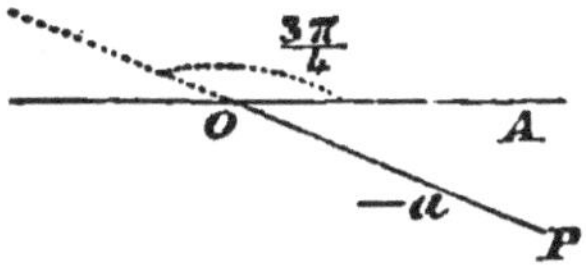

Fig. 4.

Fig. 5.

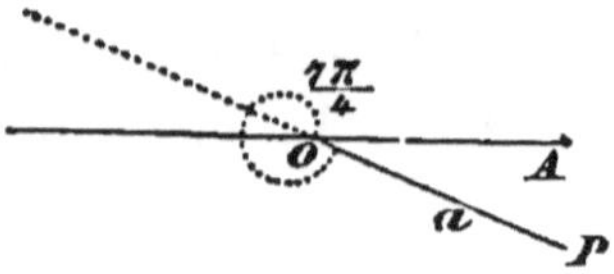

$\theta = \frac{\pi}{4}$, $\rho = a$ represents P in Fig. 1;

$\theta = \frac{\pi}{2}$, $\rho = -a$.................. Fig. 2;

$\theta = \frac{3\pi}{4}$, $\rho = -a$...... .... Fig. 3;

$\theta = \pi$, $\rho = -a$.................. Fig. 4;

$\theta = \frac{7\pi}{4}$, $\rho = a$ .... Fig. 5.

It is important to observe that the direction in which ρ is measured depends not only on its sign, but also on the value of θ; thus, when $\theta = \frac{3\pi}{4}$ and $\rho = -a$, ρ must be measured from O to P as in fig. 3; and when $\theta = \frac{7\pi}{4}$, $\rho = a$, ρ must be measured in exactly the same direction. Again, when $\theta = 0$, $\rho = a$, and when $\theta = \pi$, $\rho = -a$, ρ must in both cases be measured from O towards A.

13. *To transform polar co-ordinates into rectangular, or rectangular into polar.*

Suppose P to be the point whose polar co-ordinates OP $(= \rho)$, and angle POA $(= \theta)$ are known. Take the pole O as origin of rectangular co-ordinates, OA and a perpendicular through O, as axes of x and y, and let OM $(= x)$ and PM $(= y)$ be the rectangular co-ordinates of P.

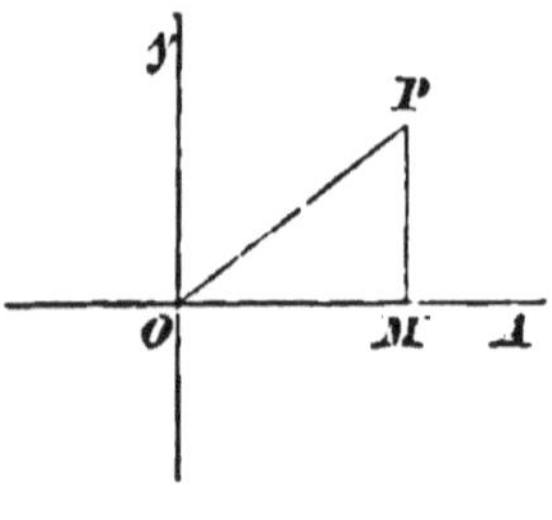

Hence we have

$$y = \rho \sin \theta \qquad x = \rho \cos \theta,$$

$$x^2 + y^2 = \rho^2, \qquad \tan \theta = \frac{y}{x}.$$

These equations will enable us to transform any equation from polar co-ordinates to rectangular, and *vice versâ*.

Ex. *To transform the equation*

$$a^2\rho \sin^2\theta = 2ab^2 \cos \theta - b^2\rho \cos^2 \theta$$

to rectangular co-ordinates.

Multiplying by ρ we have

$$a^2\rho^2 \sin^2\theta = 2ab^2\rho \cos \theta - b^2\rho^2 \cos^2 \theta,$$

or $$a^2y^2 = 2ab^2x - b^2x^2,$$

the initial line being now the axis of x, and the pole the origin.

14. *To determine the distance between two points whose polar co-ordinates are given.*

Let P, P' be the two points, $OP = \rho$, angle $POA = \theta$, $OP' = \rho'$, angle $P'OA = \theta'$; then from the triangle POP' we have

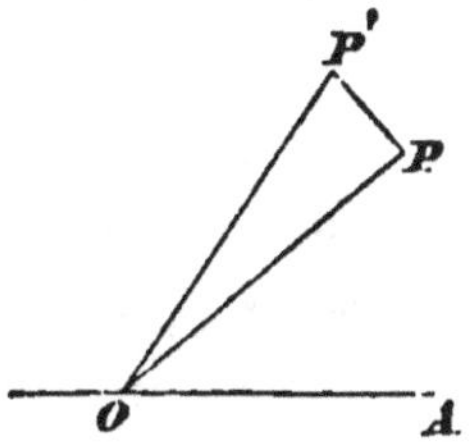

$$PP'^2 = OP'^2 + OP^2 - 2OP' . OP \cos POP',$$

or $$PP'^2 = \rho'^2 + \rho^2 - 2\rho'\rho \cos (\theta' - \theta),$$

which gives the required distance.

15. We shall now shew how the method of co-ordinates may be applied to determine the position of straight lines and the form of curves, and we will begin with a few of the simplest cases.

We have seen that the position of a point P is completely determined by two equations $x = a$, $y = b$. Suppose we have *one* only of these equations, $x = a$, given us; then, evidently, if we draw a straight line PM parallel to Oy so that $OM = a$, the equation $x = a$ is satisfied by every point in that line produced every way indefinitely. Hence the equation $x = a$, instead of representing a point, represents a straight line parallel to the axis of y. In like manner the equation $y = b$ represents a straight line parallel to the axis of x. Hence the point (ab) may be considered as the point of intersection of the two straight lines represented by the equations $x = a$, $y = b$, for at that point both equations are satisfied.

Again, the equation $x = 0$ is evidently satisfied by every point in the axis of y, and may therefore be said to represent that axis; and similarly, the equation $y = 0$ represents the axis of x. The intersection of these straight lines is the point where both the equations $x = 0$, $y = 0$ are satisfied, or the origin.

16. If, instead of two equations to determine x and y, we have a single equation, expressing a relation between them, (as, for instance, $2x + 3y + 9 = 0$,) the position of the point P will be indeterminate; for we may assign any values we please, OM, OM_1, OM_2, &c., to x, and the equation will furnish us with corresponding values of y, PM, P_1M_1, P_2M_2, &c., where we draw the ordinate downwards, if the value which we assign to x gives us a negative value of y; hence we may obtain any number of

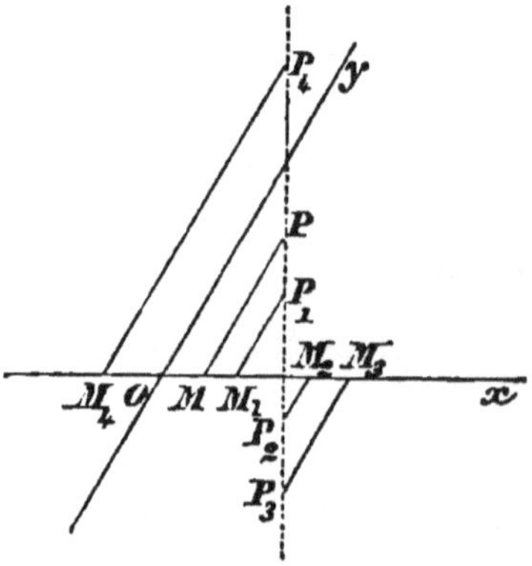

points P, P_1, P_2, &c., whose co-ordinates satisfy the equation. We may take the points M, M_1, M_2, &c. as near as we please, and so get an assemblage of points P, P_1, P_2, &c. as near to one another as we please. If the points are brought indefinitely near to one another, they will form a continuous line; hence we may consider the equation to represent the line which passes through *all* the points whose co-ordinates satisfy it.

Ex. Suppose the equation to be $y=8-3x$; then, the unit of measure being known, if we take $OM=1$, we have $PM=5$, and P is a point in the line represented; if $OM_1=2$, $P_1M_1=2$, and P_1 is another point; by taking values of x between 1 and 2 we may determine any number of points between P and P_1. If $OM_2=3$, $P_2M_2=-1$, and must therefore be drawn downwards; if $OM_3=4$, $P_3M_3=-4$, and must also be drawn downwards; by taking values of x between 2 and 3, or 3 and 4, we may obtain any number of points between P_1 and P_2 or P_2 and P_3. If we take a negative value of x, or suppose $OM_4=-1$, we have $P_4M_4=11$, and so on: hence the line represented by the equation will be that which passes through P, P_1, &c. and all other points which satisfy the equation.

17. An equation between x and y may give us more than one value of y for each assumed value of x, or *vice versâ*; for example, the equation $y^2=6x$ will give, for every value of x, two equal values of y with opposite signs, so that, for every position of M, we shall have two positions of P, and two lines will be traced out at the same time.

When a point is restricted by conditions of any kind, to occupy any of a particular series of positions, that series of positions is called the *locus of the point*, or the *locus of the equation* that expresses the conditions; and the equation is called the *equation to the locus*, or the *equation to the curve* which passes through all the positions of the point.

The curve then, or locus, represented by an equation between x and y, is the assemblage of all those points whose co-ordinates satisfy the equation; hence, if x', y' be quantities which, when substituted for x and y, satisfy the equation, the point $(x'\ y')$ is a point on the locus.

18. Remarks similar to the above (Arts. 15—17) will apply to polar co-ordinates; for, if ρ and θ are connected by an equation, every value of θ will, according to the degree of the equation, give one or more values of ρ, which must be interpreted according to the rules of Art. 12.

19. The following simple examples will shew the method of representing loci by means of equations.

(i) *To determine the locus of the equation $y=x-4$, the axes being rectangular.*

Let Ox, Oy be the axes; now, when $x=0$ in the equation, $y=-4$; hence, if we take $OD=4$ on the negative part of the axis of y, D will be a point in the locus (Art. 6). Again, when $y=0$, $x=4$; hence, if we take $OT=4$, on the positive part of the axis of x, T will be a point in the locus. Draw the straight line DT, and produce it indefinitely both ways; this will be the locus of the equation; for take any point P in this line and let its co-ordinates be $OM=x$, $PM=y$; then, evidently, $PM=MT$, since the line makes an angle of 45^0 with Ox, and therefore

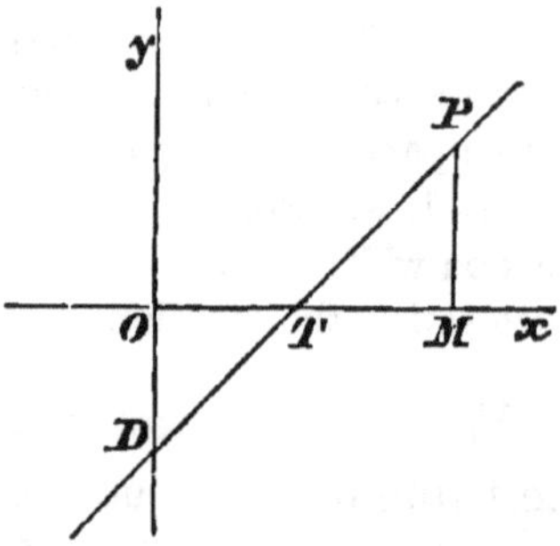

$$PM=OM-OT,$$

$$\text{or } y=x-4.$$

Hence the co-ordinates of any point in the line DT satisfy the equation, and DT is the locus required.

(ii) *To determine the loci of the equations $y=x$ and $y=-x$.*

It is easily seen that, whatever be the inclination of the axes, the former equation is satisfied by every point in a straight line bisecting the angle yOx, and the latter by a straight line perpendicular to it, bisecting the angle DOx.

(iii) *To determine the locus of the equation $x^2+y^2=9$, the axes being rectangular.*

This equation asserts (Art. 7) that the distance of the point (xy) from the origin is constant and $=3$. Hence the locus of the point is a circle whose centre is the origin and radius $=3$.

(iv) *To determine the locus of the equation, $\rho=p\sec\theta$.*

Let O be the pole, OA the initial line. Take $OD=p$, and draw DP perpendicular to OA; then, if P is any point on the line, $OP=OD\sec POD$, or, if $OP=\rho$, and the angle $POD=\theta$, we have $\rho=p\sec\theta$ for any point on DP; hence DP is the locus.

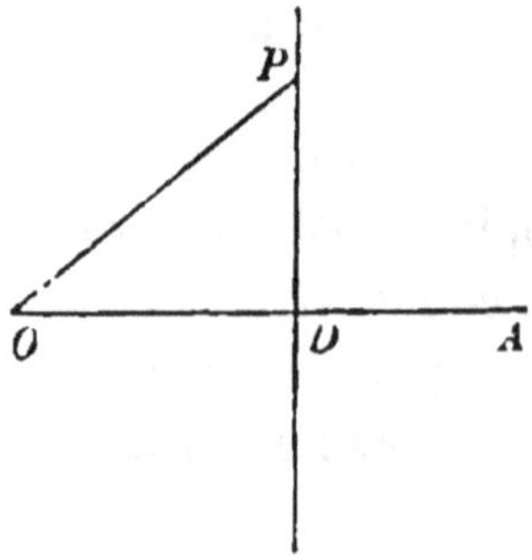

(v) *To determine the loci of the equations ρ = a constant, θ = a constant.*

The first is evidently a circle with the pole for centre. The second is a straight line passing through the pole, and is the initial line, if $\theta=0$.

20. As examples of determining the equation to a locus from some known property it possesses, we may take the following.

(i) *To find the equation to the locus, of which every ordinate (PM) is greater than the corresponding abscissa (OM) by a given distance (b).*

The condition is evidently expressed by the equation $y=x+b$, and, if we take $OD=OT=b$, as in the figure, the line DT will plainly be the locus of the equation; for, whatever be the position of P on the line, $PM=MT$, or $y=x+b$.

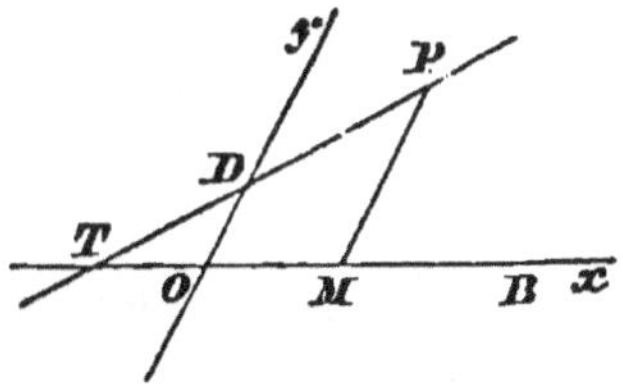

(ii) *To find the equation to the locus, of which every ordinate (PM) is a mean proportional between MO and MB, B being a point in Ox, and the axes being rectangular.*

Let $OB=2r$ and the angle yOx be supposed $=90°$; then since $PM^2=OM.MB$, the equation is

$$y^2=x(2r-x) \text{ or } y^2=2rx-x^2.$$

21. It is evident then, that the examination of any equation between two unknown quantities must generally give rise to an assemblage of points, or a geömetrical locus; and, conversely, if we can, by knowing some property of a locus or the circumstances of its description, determine a relation between the co-ordinates of any point taken arbitrarily on the locus, that relation is the equation to the locus.

If a problem relate to the position of a single point, and the data be sufficient to determine the position of that point, the problem is determinate; but, if one or more of the conditions be omitted, the data which remain may be sufficient to determine more than one point, each of which satisfies the conditions of the problem; the problem is then indeterminate. Such problems will in general result in indeterminate equations representing straight lines or curves, upon which the required points are situated. If, for example, the hypotenuse and one side of a right-angled triangle be given, the position of the vertex is fixed, for not more than one triangle can be described on the same side of the hypotenuse, with these data; but, if the hypotenuse only be given, the vertex may lie anywhere in the circumference of the semicircle described on the hypotenuse as diameter; the problem of finding the vertex is then indeterminate, and the semicircle is said to be the *locus* of all the vertices of all right-angled triangles described on one side of the given hypotenuse.

22. In the following pages we purpose to investigate those lines only, which are represented by equations of the first and second degree between two variables (x and y). An equation of the first degree is an equation that involves no power of either of the variables higher than the first, nor their product; its most general form is

$$Ax + By + C = 0.$$

An equation of the second degree contains no term in which the sum of the indices of the variables is greater than two; its most general form is

$$Ax^2 + 2Hxy + By^2 + 2Gx + 2Fy + C = 0.$$

This peculiar arrangement of coefficients is adopted, because formulæ derived from the equation thus become simpler and more easy to be remembered. It is obtained from the symmetrical equation

$$Ax^2 + By^2 + Cz^2 + 2Fyz + 2Gzx + 2Hxy = 0,$$

by making $z = 1$.

EXAMPLES I.

1. FIND the points whose co-ordinates are $(0, 1)$, $(-2, 1)$, $(-5, 0)$, $(-2, -3)$.

2. Draw a triangle, the co-ordinates of whose angular points are $(0, 0)$, $(2, -3)$, $(-1, 0)$, and find the co-ordinates of the middle points of its sides.

3. A straight line cuts the positive part of the axis of y at a distance 4, and the negative part of the axis of x at a distance 3 from the origin: find the co-ordinates of the point where the part intercepted by the axes is cut in the ratio of $3 : 1$, the smaller segment being adjacent to the axis of x.

4. There are two points P $(7, 8)$ and Q $(4, 4)$: find the distance PQ, (i) with rectangular axes, and (ii) with axes inclined at an angle of 60°.

5. Work Ex. 4 when P is $(-2, 0)$, Q $(-5, -3)$.

6. The co-ordinates of P are $x=2$, $y=3$, and of Q, $x=3$, $y=4$; find the co-ordinates of R, so that $PR : RQ :: 3 : 4$.

7. The polar co-ordinates of P are $\rho=5$, $\theta=75^\circ$, and of Q, $\rho=4$, $\theta=15^\circ$; find the distance PQ.

8. Find the polar co-ordinates of the points whose rectangular co-ordinates are

(1) $x=\sqrt{3}$, $y=1$; (2) $x=-\sqrt{3}$, $y=1$; (3) $x=-1$, $y=1$;

and draw a figure in each case.

9. Find the rectangular co-ordinates of the points whose polar co-ordinates are

(1) $\rho=5$, $\theta=\frac{\pi}{4}$; (2) $\rho=-5$, $\theta=\frac{\pi}{3}$; (3) $\rho=5$, $\theta=-\frac{\pi}{4}$;

and draw a figure in each case.

10. Transform the equations

$$x \cos \alpha + y \sin \alpha = a, \qquad x^2 + xy + y^2 = b^2,$$

from rectangular to polar co-ordinates.

11. Transform the equation $\rho^2 = a^2 \cos 2\theta$ from polar to rectangular co-ordinates.

12. A straight line joins the points (2, 3) and (−2, −3); find the co-ordinates of the points which divide the line into three equal parts.

13. If ABC is a triangle, and AB, AC are taken as axes of x and y, find the co-ordinates (i) of the bisection of BC, (ii) of the point where the perpendicular from A meets BC, and (iii) of the point where the line bisecting the angle BAC meets BC.

14. Find the co-ordinates of the same points, when AB is the axis of x, and a straight line drawn from A perpendicular to AB the axis of y.

15. The rectangular co-ordinates of a point S are h and k, and a straight line PS is drawn, making an angle θ with the axis of x; shew that the co-ordinates of P are

$$x = h + \rho \cos \theta, \qquad y = k + \rho \sin \theta,$$

where $SP = \rho$.

CHAPTER II.

General Equation of the first degree. The straight line.

23. *The general equation of the first degree between two variable quantities (x and y) represents a straight line.*

Every equation of the first degree is included in the form

$$Ax + By + C = 0 \text{..................} (A),$$

A, B, and C being quantities which do not involve x and y; they represent numbers, are always invariable for any particular equation, and are therefore called *constants*. Let now (x_1y_1), (x_2y_2), (x_3y_3) be three points in the line (whatever kind of line it may be) represented by equation (A), and let the abscissæ be in order of magnitude, and therefore, from (A), the ordinates also in the same order. Then, since the relation among the co-ordinates is true for every point of the line, it is true for these three, and we have

$$Ax_1 + By_1 + C = 0 \text{..................} (1),$$

$$Ax_2 + By_2 + C = 0 \text{..................} (2),$$

$$Ax_3 + By_3 + C = 0 \text{..................} (3),$$

from (2) – (1) $A(x_2 - x_1) + B(y_2 - y_1) = 0$,

from (3) – (1) $A(x_3 - x_1) + B(y_3 - y_1) = 0$,

which evidently gives us

$$\frac{y_3 - y_1}{x_3 - x_1} = \frac{y_2 - y_1}{x_2 - x_1} \text{..................} (4).$$

Now, if P, Q, R be the points (x_1y_1), (x_2y_2), (x_3y_3), and PLM be drawn parallel to Ox, equation (4) gives us

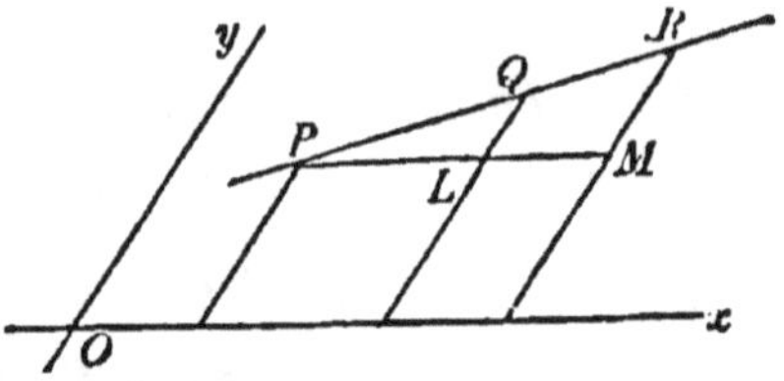

$$\frac{RM}{MP}=\frac{QL}{LP}\text{..............}(5);$$

hence the triangles PMR, PLQ are similar (Euc. VI. 6), and the angles RPM, QPL are equal, and therefore Q is on the straight line PR. In like manner it may be shewn that every other point in the *line* represented by the equation

$$Ax + By + C = 0$$

is in the same *straight line* with P and R. The line represented is therefore a straight line.

24. It has been already stated that A, B, C are fixed quantities for any particular line. They are, therefore, the quantities which distinguish one line from another; for the same symbols x and y are used, not only in the same line for different points, but also for points in different lines. A little care and practice are sufficient to prevent this apparent ambiguity from causing any confusion.

For instance,

$$Ax + By + C = 0\text{....................}(1),$$
$$ax + by + c = 0\text{....................}(2),$$

represent (except when one equation is formed from the other and therefore expresses no new relation between x and y) two different straight lines, since A, B, C are different from a, b, c; but x and y, though the same in both, have not the same meaning in both, for in the first case x and y represent the co-ordinates of any point in (1), and in the second the co-ordinates of any point in (2).

It is manifest that the position of the line does not

depend upon the absolute magnitude of A, B, C, since, if we multiply or divide the equation by any constant, it will still represent the same line. It is seen, indeed, by dividing the equation by one of the constants A, B, C, that there are in reality only two independent constants involved; for instance, if we divide by A, the equation becomes

$$x + \frac{B}{A}y + \frac{C}{A} = 0,$$

where $\frac{B}{A}$ and $\frac{C}{A}$ are the two constants that particularize the line; and we shall see hereafter, that two conditions, which determine these constants, are sufficient and necessary to fix the position of a straight line in a plane.

We shall hereafter, for the sake of brevity, often speak of "the line represented by the equation $Ax + By + C = 0$," as "the line $(Ax + By + C)$."

25. The converse of the preceding proposition is equally true, that all straight lines are represented by equations of the first degree; but, if we wish to reason about any particular straight line, with reference to two fixed axes, we must have some data by means of which we may construct it, such as the position (with reference to the axes) of one or more points through which it passes, the angle it makes with either of the axes, its distance from the origin, &c., &c.; and we shall find that, as we suppose these data to vary, the form of the equation $(Ax + By + C = 0)$ will vary too. We say *form*, because, as the line itself is supposed to remain the same, and only the means by which we determine its position to alter, it is plain that its equation must remain really the same also.

We shall now proceed to examine some of the most obvious of those conditions which fix the position of a line with reference to the axes, and find what form the equation will take in each particular case. We shall first deduce the forms from the general equation ($Ax + By + C = 0$), and afterwards verify our results by independent geometrical considerations.

26. (I) Let DT be the line, and let the lengths OD and OT (which are called its intercepts on the axes) be the data to determine its equation. It is plain that these are sufficient to do so, as there can be only *one* straight line passing through D and T.

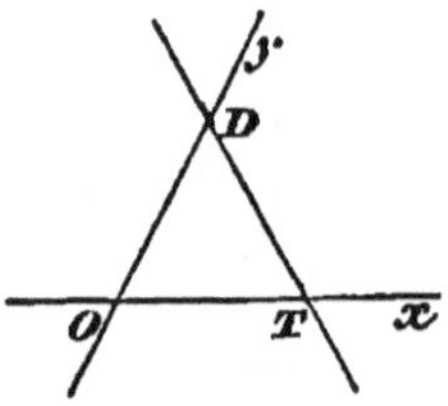

Let then $OD = b$, $OT = a$.

Now, in the equation

$$Ax + By + C = 0 \dots\dots\dots\dots\dots (A),$$

when $x = 0$, $y = -\frac{C}{B}$, which are the co-ordinates of the point D;

when $y = 0$, $x = -\frac{C}{A}$, ... T,

therefore $\quad -\frac{C}{B} = OD = b, \quad -\frac{C}{A} = OT = a;$

but equation (A) may be written in the form

$$\frac{x}{-\frac{C}{A}} + \frac{y}{-\frac{C}{B}} = 1 \dots\dots\dots\dots\dots (1),$$

which becomes $\quad \frac{x}{a} + \frac{y}{b} = 1,$

for these data; hence, if by dividing out we write any equation of the first degree in the form of (1), the quantities in the denominators will be the intercepts on the axes.

27. (II) Next let the data be the length of the perpendicular OE from the origin on DT, and the angle EOx which it makes with the axis of x. These being known, the point E is evidently known, and, as there can only be *one* straight line drawn through E at right angles to OE, these data are sufficient to determine the position of DT.

Let $OE=p$, angle $EOT=\alpha$, and $DOT=\omega$, the known angle between the axes.

Then, as before,

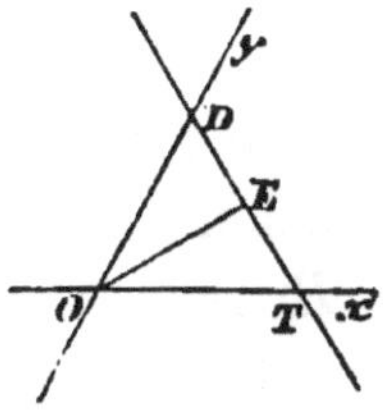

$$-\frac{C}{B}=OD, \text{ and } \therefore =\frac{p}{\cos(\omega-\alpha)},$$

$$-\frac{C}{A}=OT, \text{ and } \therefore =\frac{p}{\cos\alpha},$$

and therefore (A) becomes

$$\frac{x}{\frac{p}{\cos\alpha}}+\frac{y}{\frac{p}{\cos(\omega-\alpha)}}=1,$$

or $$x\cos\alpha+y\cos(\omega-\alpha)=p.$$

The coefficients of x and y in this equation are called the direction-cosines of the line. It must be carefully remembered that by α is meant the angle EOx which the perpendicular makes with the *positive* part of the axis of x, *i.e.* it is the angle through which Ox must be turned towards Oy, in order that Ox may coincide with OE, with x on the same side of O as E is; for instance, if the data to determine the position of the line were $OE=p$, angle $EOx=180°+\alpha$, E would lie in EO produced, at a distance $=p$ from O; but p will in this case, and always, be a positive quantity, since the positive direction of OE may be defined as that which marks the boundary of the angle α.

COR. 1. If $\omega = 90°$, or the axes be rectangular, the equation becomes

$$x \cos \alpha + y \sin \alpha = p,$$

a very useful form of the equation.

COR. 2. If the equation to any line be

$$Ax + By + C = 0 \quad \dots\dots\dots\dots (1),$$

and the equation to the same line be written in the form

$$x \cos \alpha + y \cos (\omega - \alpha) = p \quad \dots\dots\dots\dots (2),$$

or

$$x \cos \alpha + y \sin \alpha = p \quad \dots\dots\dots\dots (3),$$

according as the axes are oblique or rectangular, then, since equations (2) or (3) are really identical with (1), we see that A, B, and C are proportional to $\cos \alpha$, $\cos (\omega - \alpha)$, and $-p$ for oblique axes, and to $\cos \alpha$, $\sin \alpha$, and $-p$ for rectangular axes; and therefore, for the latter case,

$$\frac{A}{\cos \alpha} = \frac{B}{\sin \alpha} = -\frac{C}{p} \quad \dots\dots\dots\dots (4),$$

equations which are often useful. For example, if we wish to write equation (1) in the form (3), we have from (4)

$$\frac{\sin \alpha}{p} = -\frac{B}{C}, \quad \frac{\cos \alpha}{p} = -\frac{A}{C},$$

$$\frac{\cos^2 \alpha + \sin^2 \alpha}{p^2} = \frac{A^2 + B^2}{C^2},$$

or

$$p = \frac{C}{\sqrt{A^2 + B^2}},$$

since p is always a positive quantity. Therefore, from (4)

$$\cos \alpha = -\frac{A}{\sqrt{A^2 + B^2}}, \quad \sin \alpha = -\frac{B}{\sqrt{A^2 + B^2}}.$$

Hence equation (1) must be written, supposing C a positive quantity,

$$-\frac{A}{\sqrt{A^2+B^2}}x-\frac{B}{\sqrt{A^2+B^2}}y=\frac{C}{\sqrt{A^2+B^2}} \quad \text{......} \ (5).$$

Hence, any equation may be written in the form of (3), by so adjusting the signs, that the term not involving x or y may stand, as a positive quantity, as the right-hand member, and then dividing the whole equation by the square root of the sum of the squares of the coefficients of x and y.

For instance, the equation

$$3x-4y+7=0$$

will be written

$$-\frac{3}{5}x+\frac{4}{5}y=\frac{7}{5} \quad \text{....................} \ (6).$$

We say that p is always a positive quantity, because we have agreed above always to consider α as the angle which the perpendicular on the line, and not that perpendicular produced, makes with the positive part of the axis of x. If we were to remove this restriction, p might sometimes be a negative quantity, as we have explained (Art. 12) in the case of polar co-ordinates. For instance, equation (6) represents a line DT, where $\sin EOx=\frac{4}{5}$, $\cos EOx=-\frac{3}{5}$, and $OE=\frac{7}{5}$, but the equation may be written

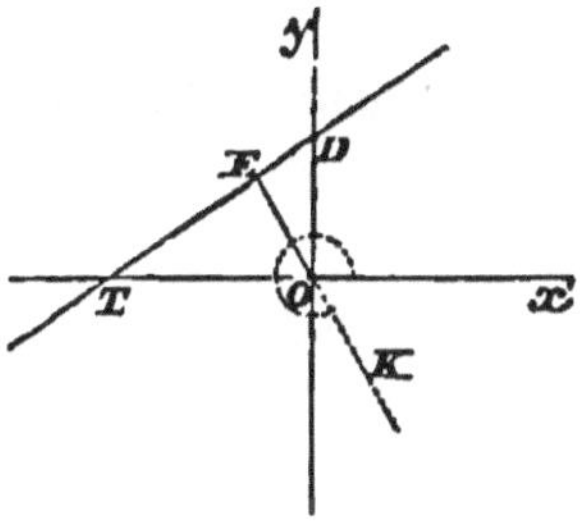

$$\frac{3}{5}x-\frac{4}{5}y=-\frac{7}{5},$$

where we must take the angle $KOx\,(=180°+EOx)$ whose sine $=-\frac{4}{5}$, and cosine $=\frac{3}{5}$, and then measure $OE=\frac{7}{5}$ in what is now the negative direction of the perpendicular. This will evidently give us the same line DT; but in the following pages, when we speak of the equation to a line written in the above form, we shall always suppose p to be positive.

28. (III) Again, let the data be the length of OD and the angle DTx. Now, although we can draw an infinite number of straight lines through D, there is only one that makes this particular angle with the axis of x; for, when we speak of the angle which a straight line makes with the axis of x, we always mean the angle which the part of the line above the axis makes with the axis produced in a positive direction, *i.e.* the angle DTx, not DTO. These data then determine the line. Let $OD = b$, $DTx = \alpha$, and $DOx = \omega$, as before.

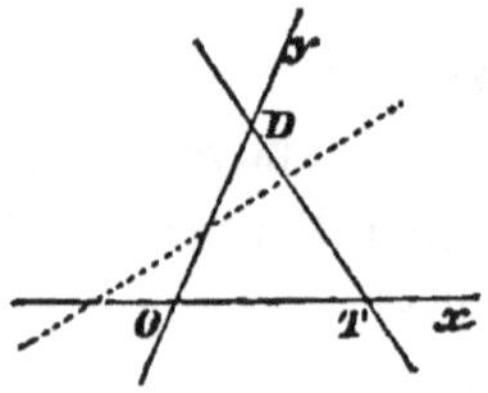

$$\text{Then } -\frac{C}{B} = b,$$

$$-\frac{C}{A} = OT = \frac{\sin(\alpha - \omega)}{\sin \alpha} b$$

$$= -\frac{\sin(\omega - \alpha)}{\sin \alpha} b,$$

and therefore
$$\frac{A}{B} = -\frac{\sin \alpha}{\sin(\omega - \alpha)};$$

but equation (A) may be written

$$y = -\frac{A}{B}x - \frac{C}{B},$$

and therefore the equation becomes for these data

$$y = \frac{\sin \alpha}{\sin(\omega - \alpha)} x + b,$$

which is often written

$$y = mx + b,$$

the constant m being a short way of writing the constant

$$\frac{\sin \alpha}{\sin(\omega - \alpha)}.$$

COR. 1. If $\omega = 90^\circ$, or the co-ordinates be rectangular,

$$m = \tan \alpha.$$

Also m will be positive only when α is less than ω. In the figure the dotted line has m positive; in DT it is negative.

COR. 2. Every equation of the first degree may, by dividing by the coefficient of y and transposing, be written in the form

$$y = mx + b,$$

where, for rectangular axes, m represents $\tan \alpha$, and for oblique axes, m represents $\frac{\sin \alpha}{\sin(\omega - \alpha)}$, and where, for all axes, b represents the intercept of the line on the axis of y, and will be positive or negative according as the line cuts that axis above or below the origin. The constant m may be called the *angular coefficient* of the line.

COR. 3. If the distance $OD = 0$, or the line pass through the origin, we shall have

$$-\frac{C}{B} = 0,$$

and the equation to the line will be

$$y = \frac{\sin \alpha}{\sin(\omega - \alpha)} x.$$

Hence the equation to any straight line passing through the origin is of the form

$$y = mx.$$

29. (IV) Next let the data be the point P' through which DT passes, and the angle DTx which it makes with Ox. Let the known co-ordinates of P' be x' and y', and let $\angle DTx = \alpha$, $DOx = \omega$, as before.

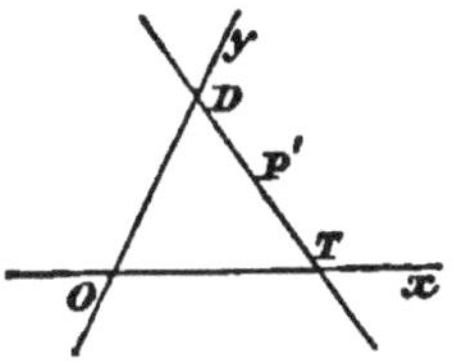

Then, since $(x' y')$ is a point on the line

$$Ax + By + C = 0,$$

therefore $$Ax' + By' + C = 0,$$

and $$A(x - x') + B(y - y') = 0,$$

or $$\frac{y - y'}{x - x'} = -\frac{A}{B};$$

but as before,

$$-\frac{C}{B} = OD, \text{ and } -\frac{C}{A} = OT;$$

and therefore $$\frac{A}{B} = \frac{OD}{OT} = -\frac{\sin \alpha}{\sin(\omega - \alpha)};$$

hence $$\frac{y - y'}{x - x'} = \frac{\sin \alpha}{\sin(\omega - \alpha)},$$

or $$\frac{x - x'}{\sin(\omega - \alpha)} = \frac{y - y'}{\sin \alpha}.$$

COR. 1. If $\omega = 90°$, or the co-ordinates be rectangular

$$\sin(\omega - \alpha) = \cos \alpha,$$

and the equation may be written

$$\frac{(x - x')}{\cos \alpha} = \frac{y - y'}{\sin \alpha},$$

or more commonly

$$y - y' = m(x - x'),$$

where $m = \tan \alpha$.

COR. 2. If the given point be the origin, and therefore $x' = 0$, $y' = 0$, this equation reduces to

$$y = mx,$$

as in III. Cor. 3.

30. (V) Next let the data be the co-ordinates of the two points P' and P'' through which the line passes. Let the co-ordinates of P' be x', y', and of P'', x'', y''; then, since P' and P'' are points on the line,

$$Ax + By + C = 0 \dots\dots\dots\dots (1),$$

therefore $$Ax' + By' + C = 0 \dots\dots\dots\dots (2),$$

and $$Ax'' + By'' + C = 0 \dots\dots\dots\dots (3);$$

from (1) – (2) $$A(x - x') + B(y - y') = 0,$$

from (3) – (2) $$A(x'' - x') + B(y'' - y') = 0,$$

whence we have for the required equation

$$\frac{y - y'}{x - x'} = \frac{y'' - y'}{x'' - x'}.$$

COR. If one of the points $(x''\ y'')$ be the origin, the equation becomes, since $x'' = 0$, $y'' = 0$,

$$\frac{y - y'}{x - x'} = \frac{y'}{x'},$$

or, after reduction,

$$\frac{y}{x} = \frac{y'}{x'},$$

which is therefore the equation to a line passing through the origin and $(x'\ y')$.

31. All these equations may be obtained independently from geometrical considerations, instead of being deduced from the general equation of the first degree; thus

(I) The intercepts $OD = b$, $OT = a$ being given. Let PM $(= y)$, OM $(= x)$ be the co-ordinates of any point P in the line; then the triangles DOT, PMT are always similar, whatever be the position of P, and we have

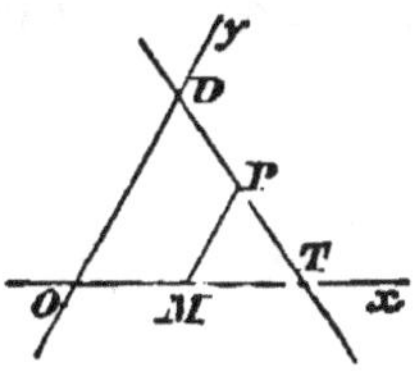

$$PM : MT = OD : OT,$$

or $$\frac{y}{a-x}=\frac{b}{a},$$

whence, as before, $$\frac{x}{a}+\frac{y}{b}=1.$$

This equation will be easily remembered, as each variable stands over the intercept on its own axis. It is the same, whether the co-ordinates are rectangular or oblique.

32. (II) Given the perpendicular (p) and the angle (α) it makes with the axis of x. Let P be any point in DT; OM, PM (x, y) its co-ordinates. Draw MS perpendicular to OE; then

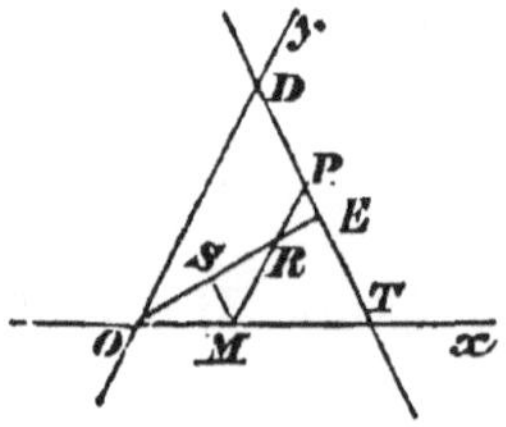

$$\cos\alpha=\frac{OS}{x}\;\ldots\ldots\ldots\ldots\ldots\ (1),$$

and $$\cos(\omega-\alpha)=\cos SRM$$

$$=\frac{SR}{RM},$$

and also $$=\frac{RE}{RP},$$

and therefore $$=\frac{SR+RE}{RM+RP},$$

$$=\frac{SE}{y}\;\ldots\ldots\ldots\ldots\ldots\ldots\ldots\ldots\ldots\ldots\ (2);$$

from (1) $OS=x\cos\alpha$; from (2) $SE=y\cos(\omega-\alpha)$;

therefore $$x\cos\alpha+y\cos(\omega-\alpha)=OS+SE$$
$$=p.$$

COR. If yOx be a right angle, we have

$$\cos\alpha=\frac{OS}{x},\ \sin\alpha=\cos SRM=\frac{SE}{y},$$

and we obtain the equation

$$x \cos a + y \sin a = p.$$

33. (III) The length $OD (= b)$ and the angle $DTx (= \alpha)$ being given.

Let $OM (= x)$ and $PM (= y)$ be the co-ordinates of any point P in the line; through D draw DQ parallel to Ox to meet MP produced in Q; then

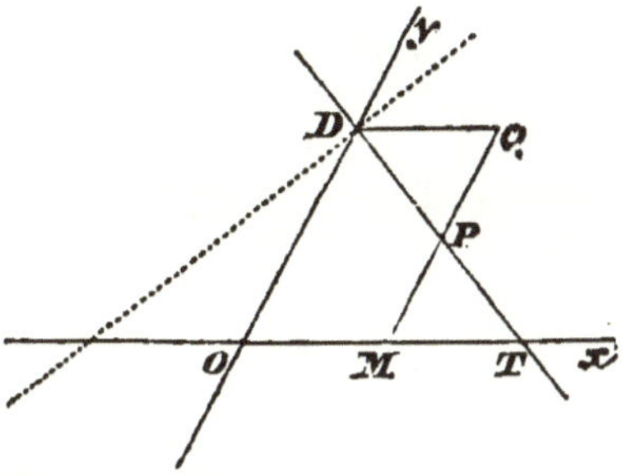

$$\frac{PQ}{DQ} = \frac{\sin PDQ}{\sin DPQ},$$

or $$\frac{QM - PM}{DQ} = \frac{\sin (180^\circ - \alpha)}{\sin (\alpha - \omega)},$$

or $$\frac{b - y}{x} = -\frac{\sin \alpha}{\sin (\omega - \alpha)},$$

whence $$y = \frac{\sin \alpha}{\sin (\omega - \alpha)} x + b,$$

or $$y = mx + b,$$

as before.

Cor. If yOx be a right angle, we have

$$\frac{PQ}{DQ} = \tan PDQ$$

$$= \tan (180^\circ - \alpha),$$

therefore $$\frac{b - y}{x} = -\tan \alpha,$$

or $$y = \tan \alpha . x + b.$$

It will be observed, that, in the figure we have chosen, m is a negative quantity, since $\sin (\omega - \alpha)$ and $\tan \alpha$ are both negative. If we take the dotted line, the geometrical construction will shew that m is positive in that case.

34. (IV) The co-ordinates (x', y') of the point P', and the angle DTx $(=\alpha)$ being given.

Let P be any point (xy) on the line, and draw $P'Q$ parallel to Ox, to meet the ordinate of P in Q;

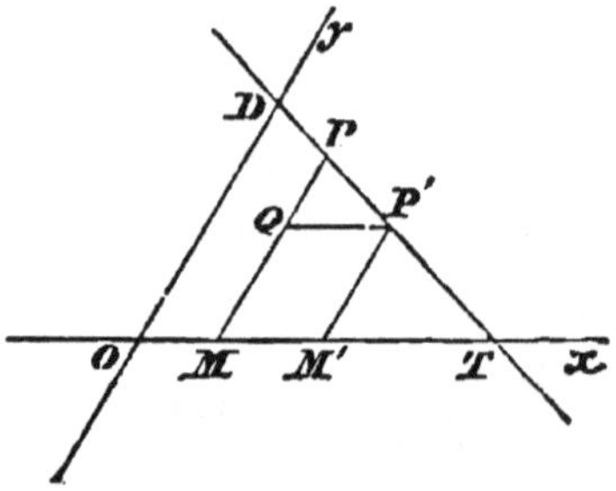

then $$\frac{PQ}{P'Q}=\frac{\sin PP'Q}{\sin QPP'},$$

or $$\frac{y-y'}{x'-x}=\frac{\sin(180^\circ-\alpha)}{\sin(\alpha-\omega)};$$

hence $$\frac{y-y'}{x-x'}=\frac{\sin\alpha}{\sin(\omega-\alpha)},$$

as before (Art. 29). This equation may be written

$$\frac{x-x'}{\sin(\omega-\alpha)}=\frac{y-y'}{\sin\alpha},$$

where each member of the equation is evidently $=\frac{P'P}{\sin\omega}$; hence we may write the equation thus,

$$\frac{x-x'}{c}=\frac{y-y'}{s}=l,$$

where $$s=\frac{\sin\alpha}{\sin\omega},\quad c=\frac{\sin(\omega-\alpha)}{\sin\omega},\quad l=P'P,$$

and where we must remember, that s, c are constants for any given line, but l is a variable quantity, and equals the length of the straight line between the points (xy) and $(x'y')$.

Cor. If the axes are rectangular,

$$\frac{x'-x}{\sin(\alpha-90^\circ)}=\frac{y-y'}{\sin(180^\circ-\alpha)},$$

and each member of the equation is equal to $P'P$; hence we may write the equation

$$\frac{x-x'}{c}=\frac{y-y'}{s}=l$$

where $$s=\sin\alpha,\quad c=\cos\alpha,\quad l=P'P.$$

The geometrical meaning of this equation should be carefully noted, as we shall frequently have occasion to use it hereafter. The formula will evidently make l positive or negative, according as the distance is measured from P' towards D or T; and, as $\sin\alpha$ and $\sin\omega$ are always positive, the sign of l is always the same as that of $y-y'$.

It is also useful to observe, that, when the co-ordinates are rectangular, this equation connects the rectangular and polar co-ordinates of any point (xy), if we consider $(x'y')$ the pole, and the initial line to be drawn through $(x'y')$ parallel to the axis of x; for we have

$$y=y'+sl, \quad x=x'+cl;$$

or, with the ordinary notation of polar co-ordinates,

$$y=y'+\rho\sin\theta, \quad x=x'+\rho\cos\theta.$$

35. (V) The co-ordinates of two points P' $(x'y')$ and P'' $(x''y'')$ being given.

Let P be any point (xy) on the line; and draw $P'Q''Q$ parallel to Ox, to meet the ordinates of P and P'' in Q and Q''; then

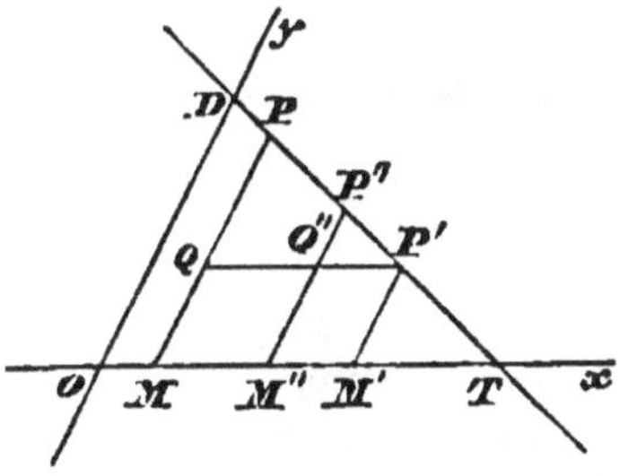

$$\frac{PQ}{QP'}=\frac{P''Q''}{Q''P'},$$

or
$$\frac{y-y'}{x'-x}=\frac{y''-y'}{x'-x''};$$

or
$$\frac{y-y'}{x-x'}=\frac{y''-y'}{x''-x'},$$

as before.

This equation may be written in the form

$$y-y'=m(x-x'),$$

where $$m=\frac{y''-y'}{x''-x'},$$

$$=\frac{\sin\alpha}{\sin(\omega-\alpha)},\text{ if the axes be oblique,}$$

$$=\tan\alpha,\text{ if the axes be rectangular,}$$

which agrees with the result of (IV).

COR. Since $m=\dfrac{\sin\alpha}{\sin(\omega-\alpha)}$, we have

$$m(\sin\omega\cos\alpha-\cos\omega\sin\alpha)=\sin\alpha;$$

therefore $$m(\sin\omega-\cos\omega\tan\alpha)=\tan\alpha;$$

whence $$\tan\alpha=\frac{m\sin\omega}{1+m\cos\omega}.$$

36. We will now give a few numerical examples on the preceding articles.

Ex. 1. *To find the equation to a straight line which cuts off intercepts on the axes of x and y equal to 3 and -5 respectively.*

Writing $a=3$, $b=-5$ in the equation of (I),

$$\frac{x}{a}+\frac{y}{b}=1,$$

we have, for the required equation,

$$\frac{x}{3}-\frac{y}{5}=1.$$

Ex. 2. *The perpendicular from the origin on a straight line $=5$, and makes an angle of 30^0 with the axis of x; find the equation to the line, (i) when the axes are inclined at an angle of 60^0, and (ii) when they are rectangular.*

Writing $\omega=60^0$, $\alpha=30^0$, and $p=5$ in the equation of (II),

$$x\cos\alpha+y\cos(\omega-\alpha)=p,$$

we have

$$\sqrt{3}(x+y)=10,$$

for the equation when the axes are inclined at an angle of 60^0; and if the axes are rectangular,

$$x\cos\alpha+y\sin\alpha=p,$$

or $$\sqrt{3}x+y=10,$$
is the equation required.

Ex. 3. *To find the equation to a straight line which makes an angle of 135° with the axis of x, and cuts off an intercept $=-3$ on the axis of y, (i) if the co-ordinates are rectangular, and (ii) if they are inclined at an angle of 45°.*

(i) Writing $b=-3$, and $m=\tan 135°=-1$ in the equation of (III),
$$y=mx+b,$$
we have, for the required equation,
$$y=-x-3.$$

(ii) Writing $\alpha=135°$, $\omega-\alpha=45°-135°=-90°$ in the equation of (III),
$$y=\frac{\sin\alpha}{\sin(\omega-\alpha)}x+b,$$
we have, for the required equation,
$$y=\frac{\frac{1}{\sqrt{2}}}{-1}x-3,$$
or
$$y=-\frac{1}{\sqrt{2}}x-3.$$

Ex. 4. *To find the equation to a straight line which passes through the points whose co-ordinates are*
$$x=4,\quad y=-2,\quad \textit{and}\ x=-3,\quad y=-5.$$

Writing 4 and -2 for x' and y', and -3 and -5 for x'' and y'' in the equation of (V)
$$\frac{y-y'}{x-x'}=\frac{y''-y'}{x''-x'},$$
we have, for the required equation,
$$\frac{y+2}{x-4}=\frac{-5+2}{-3-4},$$
or
$$7y-3x+26=0.$$

Ex. 5. *To find the equation to a straight line which passes through the origin and a point whose co-ordinates are*
$$x=3,\ y=-2.$$

Writing 3 and -2 for x' and y' in the equation of Art. 30, Cor.
$$\frac{y}{x}=\frac{y'}{x'},$$
we have
$$3y+2x=0,$$
for the equation required.

37. We have hitherto considered those points in the line only, whose co-ordinates are positive; but it may be seen, in every case, by noticing the direction in which the co-ordinates are drawn, that the equation is satisfied by every point in the line, if produced indefinitely both ways. We will take the form of equation given in (III), with rectangular co-ordinates, and suppose the equation to DT to be

$$y = mx + b,$$

where b, as we have shewn above, is a positive quantity and $= OD$, and m is, for this line, a negative quantity, and $= \tan \alpha$.

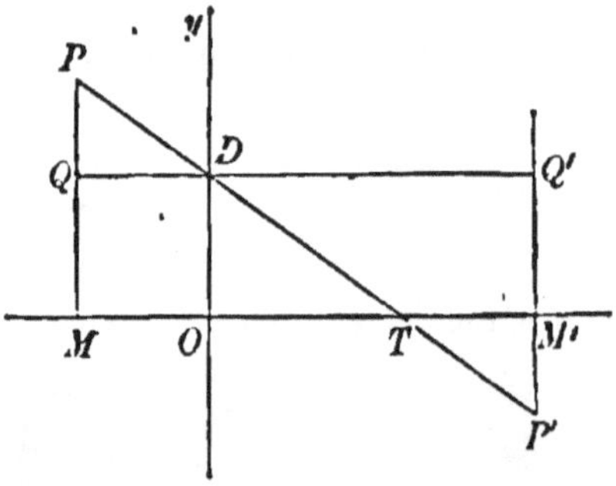

Hence, for the part between D and T, y is equal to the *difference* of the absolute lengths represented by b and mx, as was shewn in Art. 33; but, if we take P on the other side of the axis of y, so that x is negative, the quantity mx becomes positive, and y is now equal to the *sum* of the absolute lengths represented by mx and b. We may easily see from geometrical considerations that this is the case; for PQ is the quantity which now represents the absolute length of the quantity mx or $\tan \alpha \, . \, DQ$ (the signs being disregarded), and $QM = b$; therefore PM, or y, which $=$ the sum of PQ and QM, $=$ the sum of mx and b.

Again, if we take P' on the other side of the axis of x, the ordinate will, as in the portion DT, represent the difference of the absolute lengths of b and mx, since x is still positive and m is negative; and here $M'Q' = b$, and $P'Q' =$ the absolute length of mx or of $\tan \alpha \, . \, DQ'$, and we see, geometrically, that the ordinate $P'M'$ is the difference of these lines: also it is drawn upwards or in the negative direction, as it evidently should be, since of the two quantities b and mx, whose difference it is, the negative quantity, mx or $P'Q'$, is the greater.

38. In order to trace the straight line represented by any equation of the first degree, it is only necessary to determine two points, through which the line passes; the line joining these two points, and produced indefinitely both ways, will be the locus of the equation. The following examples will shew how this may be done most conveniently, (I) when the line *does not* pass through the origin, and (II) when the line *does* pass through the origin.

Ex. 1. *To trace the lines*

$$\text{(i)}\quad 5y+3x+15=0,$$
$$\text{(ii)}\quad 2y-5x+10=0.$$

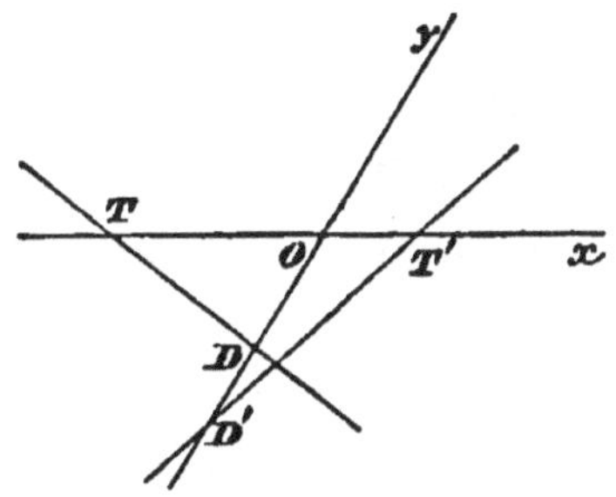

In (i), when $x=0$, $y=-3$, or the line cuts the axis of y at a distance OD $(=3)$ below the origin; when $y=0$, $x=-5$, or the line cuts the axis of x at a distance OT $(=5)$ to the left of the origin. Hence the line DT, produced indefinitely, is the locus of equation (i).

In (ii), when $x=0$, $y=-5$, and when

$$y=0,\ x=2;$$

hence, if we take $OD'=5$, and $OT'=2$, $D'T'$ produced indefinitely will be the locus of equation (ii).

Ex. 2. *To trace the lines*

$$\text{(i)}\quad 3x+4y=0,$$
$$\text{(ii)}\quad 2x-3y=0.$$

These lines pass through the origin, since the equations are satisfied by the values $x=0$, $y=0$. To find another point, we may take any value of one variable, and find the corresponding value of the other. Thus in (i) take $OM=-4$, then $PM=3$; hence P is a point in the line, and PO produced indefinitely is the locus of equation (i).

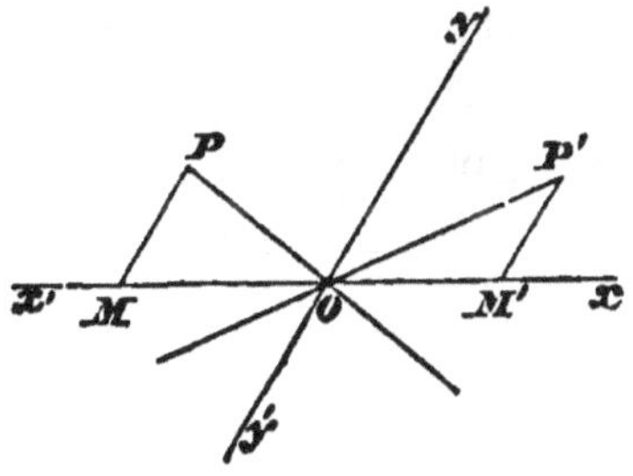

Again, in (ii) take $OM'=3$, then

$$P'M'=2;$$

hence P' is a point in the line, and $P'O$ produced indefinitely is the locus required.

39. *The equations to two straight lines being given, to find the co-ordinates of their point of intersection.*

Since the co-ordinates of every point on a line must satisfy its equation, the co-ordinates of the point through which both the lines pass will satisfy both equations, or they will be the values obtained by solving the equations as simultaneous.

Ex. *To find the co-ordinates of the point where the two lines,*

$$4x+3y-11=0,\ 3x+4y-10=0,$$

intersect.

The values of x and y, obtained by combining these two equations, are $x=2$, $y=1$, which are, therefore, the co-ordinates of the point of intersection of the lines.

40. We remarked (Art. 24) that the variables x and y had not the same meaning in two different equations. If, however, we combine these equations in any manner whatsoever, we tacitly introduce the condition, that the values of the variables *are* the same in each. All equations so obtained are therefore satisfied by the co-ordinates of the points which the loci have in common. It does not follow, that every point in the locus of an equation so obtained is one of the common points of the original loci. We may combine in such a manner as absolutely to determine these co-ordinates, or we may obtain some relation between them, expressed by an equation, and, as this equation must represent *some* line, straight or curved, and is satisfied by the co-ordinates of the point or points of intersection, it represents a line passing through those points.

We will take two simple examples of this.

(i) The values $x=2$, $y=1$, which satisfy the equations Art. 39. Ex.

$$4x+3y-11=0,\ 3x+4y-10=0\ldots\ldots\ldots(1),$$

will evidently satisfy the equation

$$4x + 3y - 11 + 3x + 4y - 10 = 0,$$

or

$$x + y - 3 = 0,$$

which is therefore the equation to a straight line, passing through the intersection of the two lines represented by equations (1).

(ii) Let the equations to two straight lines referred to rectangular axes be

$$y = mx + b\ldots\ldots\ldots(1), \qquad y = -\frac{1}{m}x + b'\ldots\ldots\ldots(2).$$

The geometrical meaning of these equations is (Art. 33) that the lines cut off intercepts on the axis of y equal to b and b', and that they make angles θ, θ' with the axis of x, so related that

$$\tan\theta \tan\theta' = -1\ldots\ldots\ldots\ldots\ldots\ldots(3).$$

Now, if θ and θ' be quantities whose value is absolutely known, we may treat (1) and (2) as simultaneous equations, and obtain definite values for x and y, the co-ordinates of the intersection of (1) and (2), in terms of the constants θ, θ', b, b'. But, if all we know about θ and θ' is that they are connected by equation (3), it is evident that for every value of θ there will be a corresponding value of θ', and that each pair will produce a new point of intersection. Suppose then that we combine (1) and (2), so as to eliminate this variable quantity m, (which $= \tan\theta$, by Art. 33,) instead of one of the variables x or y. We obtain the equation

$$(y - b)(y - b') + x^2 = 0\ldots\ldots\ldots\ldots\ldots\ldots\ldots(4).$$

Equation (4) is a relation between the co-ordinates of the intersection of (1) and (2); and, as it represents a line of *some* kind, that line must pass through the intersection. But, whatever value θ and θ' have in (1) and (2), we shall, by the same means, obtain equation (4), provided only that the relation (3) exist between them. Hence, (4) passes through *every* point of intersection which can be generated by the change of θ, and is called the *locus of the intersection* of (1) and (2).

41. We may here remark that *the student cannot be too careful in considering the geometrical meaning of every algebraic step he takes.* The very facility with which algebraic expressions are combined and manipulated, is often the most serious drawback to his progress.

42. In finding the co-ordinates of the points where two loci intersect, we shall sometimes fall upon impossible, and sometimes upon infinite values of the variables. We gather from the former, that the loci do not meet there; and the latter will often give us some information about the geometrical position of the loci. For instance, from the equations

$$y = mx + b \ldots\ldots\ldots\ldots(1), \qquad y = m'x + b' \ldots\ldots\ldots\ldots(2),$$

we obtain, for the abscissa of the intersection,

$$x = \frac{b' - b}{m - m'}.$$

If we suppose b and b' to remain the same, while m' approaches indefinitely near to m in value, the value of x becomes indefinitely great, and we infer from this that the lines become parallel, which is evidently true by Art. 28. It is in accordance then with the algebraic result, to say that *parallel straight lines meet at infinity;* but it must be clearly understood what this means.

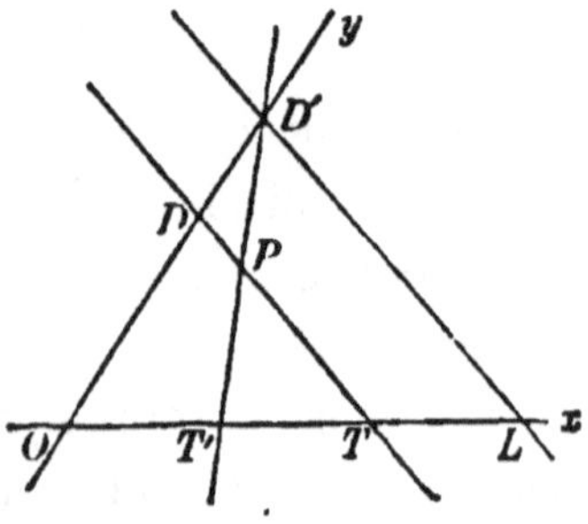

Suppose DT and $D'T'$ to be the two lines, and draw $D'L$ parallel to DT. Then, if we suppose OD' ($=b'$) to remain the same, and the point of intersection P to move along DT to an infinite distance, the line $D'T'$ tends to coincide with $D'L$ as its limiting position. The equation of the second degree will furnish similar examples, where it is very important to bear in mind this explanation.

43. We will now examine the meaning of the equation

$$Ax + By + C + k(ax + by + c) = 0 \text{..........} (1),$$

where k is any constant quantity, positive or negative, and where the equations

$$Ax + By + C = 0 \text{..................} (2),$$

$$ax + by + c = 0 \text{..................} (3),$$

represent two straight lines.

(i) Equation (1) will represent *some* straight line, since it is an equation of the first degree.

(ii) It will represent a straight line passing through the point where (2) and (3) intersect. For the co-ordinates of the point of intersection of (2) and (3) satisfy both these equations, and therefore evidently satisfy (1); hence the point is on the line represented by (1).

(iii) Equation (1) may be made to represent *any* straight line passing through the intersection of (2) and (3). For, since k is an arbitrary quantity, undetermined by the nature of the problem, equation (1) will represent a *system* of lines fulfilling one condition only, viz. that of passing through the intersection of (2) and (3). *Any* individual of the system, particularized by the fulfilment of a second condition, can have the appropriate equation assigned to it, by giving the proper value to k. For example, if any line of the system passed through a point $(x'y')$, we should obtain the value of k for that line from the equation

$$(Ax' + By' + C) + k(ax' + by' + c) = 0,$$

and, substituting this value in (1), we should have the equation required.

Cor. Similarly, if the equations to any two loci be

$$S=0,\ S'=0,$$

where S and S' stand for expressions involving x and y, the equation

$$S+kS'=0$$

will represent a locus passing through the points of intersection of the two first-mentioned loci.

Hence, if any equation involve an arbitrary constant k in the first degree only, so that the equation may be written in the form $S+kS'=0$, that equation will represent a series of loci which pass through the points of intersection of

$$S=0,\ S'=0.$$

Ex. 1. The equation

$$4x+3y-11+k(3x+4y-10)=0,$$

or
$$(4+3k)x+(3+4k)y-11-10k=0 \qquad \text{(1)},$$

will, according as we give different values to k, represent a series of straight lines passing through the point ($x=2$, $y=1$), which is the point of intersection of the two lines represented by

$$3x+4y-10=0 \qquad \text{(2)},$$

$$4x+3y-11=0 \qquad \text{(3)}.$$

Let it be required to find the equation to a straight line passing through the intersection of (2) and (3) and the point (1, 2). Then, putting

$$x=1,\ y=2$$

in (1), we get a value of k which will give us the equation required,

$$x+y-3=0.$$

Or, to find the equation to a straight line passing through the origin and the intersection of (2) and (3); putting

$$x=0,\ y=0$$

in (1), we get the equation

$$x-2y=0.$$

Ex. 2. The equation $y=mx+b$ may be written

$$(y-b)-mx=0,$$

and, if m is indeterminate, this will represent a series of straight lines passing through the intersection of the lines

$$y-b=0,\ x=0,$$

or cutting off an intercept $=b$ from the positive part of the axis of y, as we have already seen (Art. 28).

Ex. 3. The equation

$$y - y' - m(x - x') = 0$$

represents, if m is indeterminate, a system of straight lines passing through the intersection of the lines

$$y - y' = 0, \qquad x - x' = 0,$$

as we saw in Art. 29.

Ex. 4. *The three bisectors of the sides of a triangle, drawn from the opposite angles, meet in a point.*

Let ABC be the triangle, CD, BE, AF the bisectors; then, taking CA, CB as axes of x and y, the equations to these three straight lines (Euc. vi. 2, Arts. 26, 30, Cor.) are respectively,

$$\frac{y}{x} = \frac{a}{b}, \quad \frac{2x}{b} + \frac{y}{a} = 1, \quad \frac{x}{b} + \frac{2y}{a} = 1.$$

The equation to a straight line passing through the intersection of BE and AF must be of the form

$$\frac{2x}{b} + \frac{y}{a} - 1 + k\left(\frac{x}{b} + \frac{2y}{a} - 1\right) = 0,$$

or

$$\frac{x}{b}(2+k) + \frac{y}{a}(1+2k) - (k+1) = 0.$$

If we put $k = -1$, this becomes the equation to CD; hence CD passes through the intersection of BE and AF.

Ex. 5. *To find the condition that the line*

$$lx + my + n = 0 \quad\text{.............................}(1)$$

should pass through the intersection of the lines

$$Ax + By + C = 0 \quad\text{.............................}(2),$$

$$ax + by + c = 0 \quad\text{.............................}(3),$$

or that these three lines should meet in a point.

Equation (1) must, as in Ex. 1, be of the form

$$(A + ka)x + (B + kb)y + C + kc = 0;$$

hence (Arts. 24, 25),

$$\frac{A + ka}{l} = \frac{B + kb}{m} = \frac{C + kc}{n};$$

therefore

$$k = \frac{nB - mC}{mc - nb} = \frac{lC - nA}{na - lc},$$

the required condition, which becomes after reduction

$$l(Bc - bC) + m(Ca - cA) + n(Ab - aB) = 0.$$

44. *To find the general equation to the straight line, referred to polar co-ordinates.*

Writing $x = \rho \cos\theta$, $y = \rho \sin\theta$ (Art. 13), in the equation

$$Ax + By + C = 0,$$

we have

$$A\rho\cos\theta + B\rho\sin\theta + C = 0$$

for the general polar equation.

45. It is easily seen, that, as in the case of rectangular and oblique co-ordinates, the *form* of the general polar equation to the straight line will vary according to the data which determine the position of any particular line. The following data present the equation under its most convenient form.

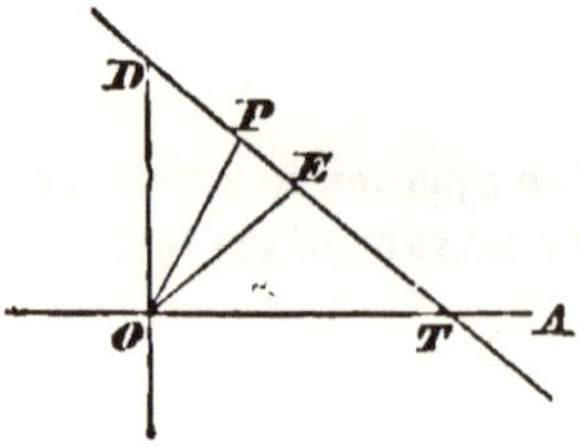

Let DT be the line, and let the perpendicular OE $(=p)$ from the pole, and the angle EOT $(=\alpha)$ which it makes with the initial line, be the data to determine its position. Now if

$$A\rho\cos\theta + B\rho\sin\theta + C = 0 \;\ldots\ldots\ldots\ldots (1),$$

be the equation to DT, we have, since E $(p\alpha)$ is a point on the line,

$$Ap\cos\alpha + Bp\sin\alpha + C = 0 \;\ldots\ldots\ldots\ldots (2).$$

From (1) and (2)

$$\frac{\rho\sin\theta - p\sin\alpha}{\rho\cos\theta - p\cos\alpha} = -\frac{A}{B} \;\ldots\ldots\ldots\ldots (3).$$

But from (1), when $\theta = 0$, $\rho = -\dfrac{C}{A}$, the point T,

and when $\theta = 90^\circ$, $\rho = -\dfrac{C}{B}$, ... D;

hence $$-\frac{C}{B} = OD,\ -\frac{C}{A} = OT,$$

and $$\frac{A}{B} = \frac{OD}{OT} = \cot \alpha;$$

and therefore, substituting in (3),

$$\rho \sin\theta \sin\alpha - p \sin^2\alpha = -\rho\cos\theta\cos\alpha + p\cos^2\alpha;$$

therefore $$\rho\cos(\theta - \alpha) = p,$$

or $$\rho = p \sec(\theta - \alpha),$$

the polar equation for the above data.

This equation may be deduced at once from the equation to the straight line in the form

$$x\cos\alpha + y\sin\alpha = p,$$

by writing $\rho\cos\theta$ for x, and $\rho\sin\theta$ for y, the data being exactly the same in the two cases.

The same equation may be obtained very simply from geometrical considerations; for, if P be any point $(\rho\theta)$ in the line, we have

$$OP = OE \,.\, \sec POE,$$

or $$\rho = p\sec(\theta - \alpha).$$

If $\alpha = 0$, or OE be taken for initial line, the equation is

$$\rho = p\sec\theta.$$

46. We will collect here, for the sake of reference, the different forms of the equation to the straight line, which we have investigated.

$$\frac{x}{a} + \frac{y}{b} = 1. \quad \text{Arts. 26 and 31.}$$

$$\left.\begin{aligned} x\cos\alpha + y\cos(\omega - \alpha) &= p \\ x\cos\alpha + y\sin\alpha &= p \end{aligned}\right\} \quad \text{Arts. 27 and 32.}$$

$$\left.\begin{array}{l} y = mx + b \\ y = mx \end{array}\right\} \quad \text{Arts. 28 and 33.}$$

$$\left.\begin{array}{c} \dfrac{x - x'}{c} = \dfrac{y - y'}{s} = l \\ (y - y') = m(x - x') \end{array}\right\} \quad \text{Arts. 29 and 34.}$$

$$\left.\begin{array}{c} \dfrac{y - y'}{x - x'} = \dfrac{y'' - y'}{x'' - x'} \\ \dfrac{y}{x} = \dfrac{y'}{x'} \end{array}\right\} \quad \text{Arts. 30 and 35.}$$

$$\left.\begin{array}{r} A\rho \cos\theta + B\rho \sin\theta + C = 0 \\ \rho = p \sec(\theta - \alpha) \end{array}\right\} \quad \text{Art 45.}$$

$$\theta = \text{constant} \qquad \text{Art. 19, v.}$$

EXAMPLES II.

1. DRAW the lines whose equations are

(1) $y = 5x + 2$, (2) $y - 7 = 5x + 3$, (3) $7y - 3x = 0$,

(4) $6 - x = 2y$, (5) $\dfrac{x}{3} + \dfrac{y}{11} = 2$, (6) $2x + 3 = 0$.

2. Find the equation to the straight line which passes through the points (2, 5) and (0, – 7).

3. The co-ordinates of the angular points of a triangle being given, find the equations to the three straight lines, each of which bisects two of the sides.

4. Two straight lines make each of them an angle of 45° with the axis of x, and their intercepts on the axis of y are 6 and 8; find the equation to the straight line which is equidistant from the two, the axes being rectangular.

5. Find the equation to a straight line on which the perpendicular from the origin = 6, and makes (1) an angle of 45°, and

(2) an angle of 225° with the axis of x, the axes being rectangular.

6. Determine the point of intersection of the two lines $(3y - x = 0)$ and $(2x + y = 1)$.

7. Find the equation to the straight line which passes through the point of intersection of the straight lines

$$x - 2y - a = 0, \quad x + 3y - 2a = 0,$$

and is parallel to the line $3x + 4y = 0$.

8. Find the equation to a straight line which is equidistant from the two lines represented by the equation $y = mx + c \pm c'$.

9. Find the equation to the straight line that joins the points of intersection of the two pairs of lines

$$\left.\begin{aligned} 2x + 3y - 4a &= 0, \\ 2x + y - a &= 0, \end{aligned}\right\} \text{ and } \left\{\begin{aligned} x + 6y - 7a &= 0, \\ 3x - 2y + 2a &= 0. \end{aligned}\right.$$

10. Find the length of the perpendicular from the origin on the line $a(x - a) + b(y - b) = 0$, and the portion intercepted by the axes, which are rectangular.

11. The rectangular co-ordinates of two points are 3, 5 and 4, 4 respectively; find the equation to a straight line which bisects the distance between them, and makes an angle of 45° with the axis of x.

12. Find the equation to a straight line which passes through a given point (ab), and makes equal angles with the axes.

13. Find the equations to the diagonals of the parallelogram formed by the four lines $x = a$, $x = a'$, $y = b$, $y = b'$.

14. A straight line, inclined to the axis of x at an angle of 150°, cuts the positive axes of rectangular co-ordinates in A and B; find the equation to a straight line bisecting AB and passing through the origin.

15. Find the equations to the four sides of a square, the co-ordinates of two of its opposite angular points being (2, 3) and (3, 4), the co-ordinates being rectangular.

16. Find the distance of the origin of co-ordinates from the line $\frac{x}{2}+\frac{y}{3}=1$, the axes being rectangular.

17. Find the equation to a straight line which passes through the intersection of the lines $x=a$, $x+y+a=0$, and through the origin.

18. The axes of co-ordinates being inclined to each other at an angle of 60°, find the equation to a straight line parallel to the line $(x+y=3a)$, and at a distance from it equal to $\frac{1}{2}a\sqrt{3}$.

19. Shew that the lines $y=2x+3$, $y=3x+4$, $y=4x+5$, all pass through one point.

20. Find the value of m, in order that the line $(y=mx+3)$ may pass through the intersection of the lines $(y=x+1)$ and $(y=2x+2)$.

21. A straight line cuts off intercepts on the axes, the sum of the reciprocals of which is a constant quantity; shew that all straight lines which fulfil this condition pass through a fixed point.

22. A straight line slides along axes of x and y, and the difference of the intercepts is always proportional to the area it encloses; shew that the line always passes through a fixed point.

23. If the distance of a point from the origin = twice its distance from the axis of x, shew that it always lies in one of two straight lines that pass through the origin; axes rectangular.

24. Find the cosine of the angle which the line $(Ax+By+C)$ makes with the axis of x, the axes being inclined at an angle of 45°.

25. If a straight line cuts the (rectangular) axes of x and y at equal distances from the origin, and a straight line be drawn from the origin, dividing it in the ratio $m:n$, find the tangent of the angle which this latter line makes with the axis of x.

26. An equilateral triangle, whose side $= a$, has its vertex at the origin, and its sides equally inclined to the positive directions of rectangular axes; find the co-ordinates of the angles, and thence of the point bisecting the base.

27. Shew that the three straight lines which bisect the angles of an equilateral triangle meet in a point, taking two of the sides as axes.

28. The base of a triangle and the straight line joining the bisection of the base with the vertex being axes, form the equations to the straight lines which join the bisection of the other sides with the opposite angles, and find their co-ordinates of intersection.

29. Find the polar co-ordinates of the point of intersection of the lines whose equations are

$$\rho = 2a \sec\left(\theta - \frac{\pi}{2}\right) \text{ and } \rho = a \sec\left(\theta - \frac{\pi}{6}\right),$$

and the angle between them.

30. Trace the line whose polar equation is

$$\frac{1}{\rho} = 2a \cos\left(\theta + \frac{\pi}{6}\right).$$

31. Shew that the polar equation to a straight line, passing through the points $(\rho'\theta')$, $(\rho''\theta'')$, is

$$\rho\rho' \sin(\theta - \theta') + \rho'\rho'' \sin(\theta' - \theta'') + \rho''\rho \sin(\theta'' - \theta) = 0.$$

What is the geometrical interpretation of this equation?

CHAPTER III.

The straight line continued.—Angles.—Perpendiculars.—Equations representing straight lines.

47. *To find the angle between two straight lines whose equations are given, the axes being supposed rectangular.*

Let $\tan TPT' = t$, and let the equations to DT, $D'T'$ be

$$y = mx + b, \ y = m'x + b',$$

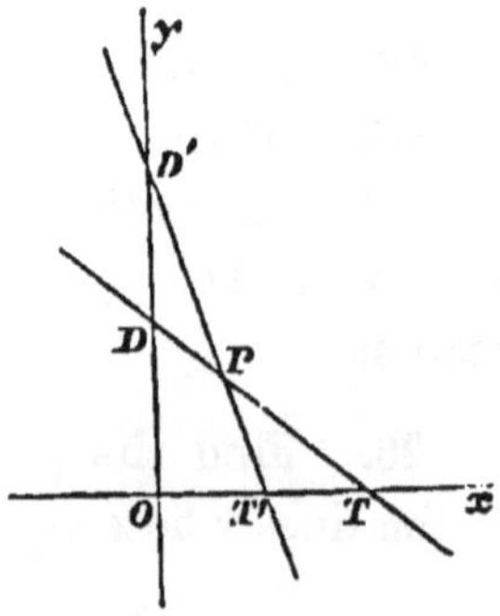

since any equation may be written in this form; then (Art. 33)

$$m = \tan DTx, \ m' = \tan D'T'x,$$

and $\tan TPT' = \tan (DTx - D'T'x)$,

or $$t = \frac{m - m'}{1 + mm'},$$

which determines the angle TPT'.

Cor. 1. Hence we see that, if $m = m'$, $\tan TPT' = 0$, or the lines are parallel. Also, if $m' = -\frac{1}{m}$, $\tan TPT' = \infty$, or the angle TPT' is a right angle.

The condition then that the two lines

$$y = mx + b, \ y = m'x + b'$$

should be parallel, is

$$m = m',$$

and that they should be perpendicular to each other, is

$$m' = -\frac{1}{m}.$$

COR. 2. If the two lines are

$$Ax + By + C = 0, \quad ax + by + c = 0,$$

the conditions of parallelism and perpendicularity are respectively

$$\frac{A}{B} = \frac{a}{b},$$

$$\frac{A}{B} = -\frac{b}{a} \text{ or } Aa + Bb = 0,$$

hence *two straight lines are perpendicular, if the coefficients of x and y are interchanged, and the sign of one of them changed.*

Ex. *To find the angle between the lines*

$$2y + x + 1 = 0, \quad 3y - x - 1 = 0,$$

the axes being rectangular.

Here $m = -\frac{1}{2}$, $m' = \frac{1}{3}$, and, if α be the angle between the lines,

$$\tan\alpha = \frac{-\frac{1}{2} - \frac{1}{3}}{1 - \frac{1}{2}\cdot\frac{1}{3}} = -1,$$

or $\alpha = 135°$.

48. The condition of parallelism, deduced in the last article, is evident independently, and we have, in fact, assumed it in previous articles; for m and m' are the tangents of the angles which the lines make with the axis of x, and those angles are equal, if the lines be parallel. The condition of perpendicularity may be proved independently; for, if we consider TPT' a right angle, we have

$$m' = \tan D'T'x = -\cot DTx$$
$$= -\frac{1}{m}.$$

*49. *To find the angle between two straight lines, the axes being oblique.*

Let the lines be as before

$$y = mx + b, \quad y = m'x + b';$$

making angles α and α' with the axis of x; and let the angle between the axes be ω. Then (Art. 35, Cor.)

$$\tan\alpha = \frac{m\sin\omega}{1+m\cos\omega}, \qquad \tan\alpha' = \frac{m'\sin\omega}{1+m'\cos\omega};$$

hence, if β be the angle between the lines,

$$\tan\beta = \tan(\alpha-\alpha')$$

$$= \frac{\dfrac{m\sin\omega}{1+m\cos\omega} - \dfrac{m'\sin\omega}{1+m'\cos\omega}}{1+\dfrac{mm'\sin^2\omega}{(1+m\cos\omega)\,(1+m'\cos\omega)}}$$

$$= \frac{(m-m')\sin\omega}{1+(m+m')\cos\omega+mm'}.$$

Hence the condition of parallelism is as before $m = m'$; and the condition of perpendicularity is

$$1+(m+m')\cos\omega+mm' = 0.$$

The condition of perpendicularity is much more complicated, when the axes are oblique, than when they are rectangular, as, indeed, are all formulæ in which angles are involved. For this reason we shall choose rectangular axes in the solution of questions concerning angles, and shall not extend the remaining articles of this chapter to the case of oblique co-ordinates.

50. *To find the equation to a straight line, which shall make a given angle with a given straight line* ($y = mx + b$), *the axes being rectangular.*

Let α be the given angle, and let $\tan\alpha = t$, and let the equation to the given line (DT) be

$$y = mx + b \ldots\ldots\ldots\ldots (1),$$

and the equation of the required line

$$y = m'x + b' \ldots\ldots\ldots\ldots (2),$$

where m' is to be found by the conditions of the problem.

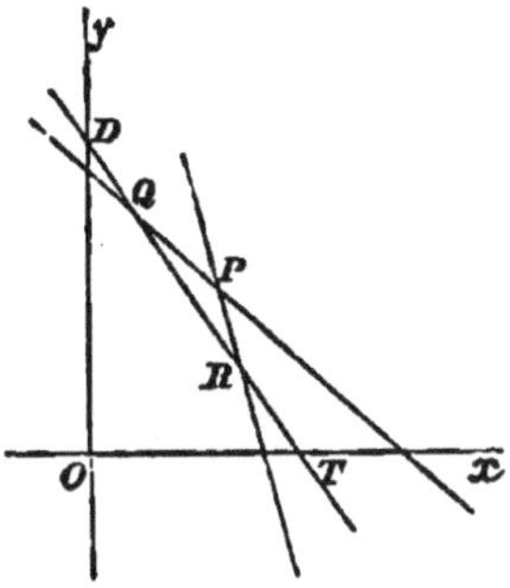

Then, since the required line may lie either as PQ or PR, we shall have, by Art. 47,

$$t = \frac{m' - m}{1 + mm'}, \quad \text{or} = \frac{m - m'}{1 + mm'}.$$

Hence

$$m' = \frac{m \pm t}{1 \mp mt},$$

and the required equation is

$$y = \frac{m \pm t}{1 \mp mt} x + b' \ldots\ldots\ldots\ldots\ldots\ldots\ldots (3),$$

where b' still remains undetermined, as an infinite number of straight lines may be drawn fulfilling this condition.

If we add another condition, that the line should pass through a point $P(x'y')$, the equation will be (Art. 29, Cor. 1)

$$y - y' = m'(x - x')$$

$$= \frac{m \pm t}{1 \mp mt}(x - x') \ldots\ldots\ldots\ldots\ldots\ldots\ldots (4),$$

and it may be seen from the figure, that there are, generally, two straight lines fulfilling these conditions.

COR. 1. If $t=0$, or the problem be *to find the equation to a straight line which passes through a given point, and is parallel to a given straight line,* the equation is

$$y-y'=m(x-x').$$

COR. 2. If $t=\infty$, or the problem be *to find the equation to a straight line which passes through a given point, and is perpendicular to a given line,* the equation is

$$y-y'=-\frac{1}{m}(x-x'),$$

for, if t becomes $=\infty$, we have

$$\frac{m\pm t}{1\mp mt}=\frac{\frac{m}{t}\pm 1}{\frac{1}{t}\mp m}=-\frac{1}{m}.$$

EX. *To find the equation to the straight lines which pass through the point* (1, 2), *and make an angle of* 45° *with the line*

$$3x+4y+7=0.$$

Here, $m=-\frac{3}{4}$, $t=1$; hence, according as we take upper or lower signs in equation (4), we have

$$7y-x-13=0, \quad y+7x-9=0,$$

for the required equations.

51. The equation obtained in Cor. 1, is the same as that of Art. 34, where its geometrical meaning is explained.

The geometrical meaning of the equation of Cor. 2, may be seen thus:

Let DT be the given line ($y=mx+b$), P' ($x'y'$) the given

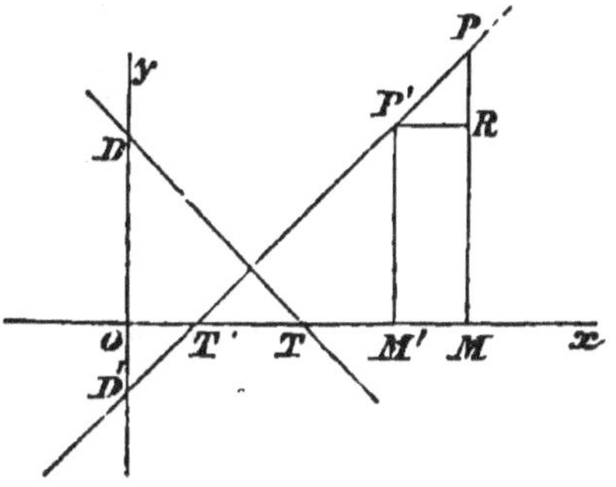

point, through which the line $D'T'$ is drawn perpendicular to DT, and let P be any other point (xy) on $D'T'$; then, if $P'R$ be drawn parallel to Ox, to meet the ordinate PM in R, we have

$$\frac{PR}{P'R} = \tan PT'x = -\cot DTx,$$

or, since $\tan DTx = m$,

$$\frac{y-y'}{x-x'} = -\frac{1}{m}.$$

52. *To find the length of a straight line drawn from a given point $(x'y')$ to meet the line $(Ax + By + C = 0)$.*

The equation to a straight line through $(x'y')$ is (Art. 34)

$$\frac{x-x'}{c} = \frac{y-y'}{s} = l \text{..................} (1),$$

where l is the distance of any point (xy) from the fixed point $(x'y')$. If we substitute for x and y from (1) in the equation to the given line, the resulting equation will refer to the point where the two lines intersect, and we shall have for the distance (l) from $(x'y')$ to that point, the equation

$$A(cl + x') + B(sl + y') + C = 0,$$

or
$$(Ac + Bs)\,l + Ax' + By' + C = 0,$$

$$\text{therefore } l = -\frac{Ax' + By' + C}{Ac + Bs} \text{................} (2).$$

COR. If the axes are rectangular, and the line (1) is perpendicular to $(Ax + By + C = 0)$, we have (Art. 47)

$$\frac{s}{c} = \frac{B}{A} \text{.........................} (3),$$

since the two lines make with the axis of x angles whose tangents are, respectively, equal to

$$\frac{s}{c} \text{ and } -\frac{A}{B}.$$

But, since s and c are the sine and cosine of the angle which the line (1) makes with the axis of x,

$$s^2 + c^2 = 1,$$

from which and from (3) we have

$$c = \pm \frac{A}{\sqrt{A^2 + B^2}}, \quad s = \pm \frac{B}{\sqrt{A^2 + B^2}};$$

hence from (2) we have for the perpendicular,

$$l = \pm \frac{Ax' + By' + C}{\sqrt{A^2 + B^2}} \quad \text{.................} \quad (4),$$

according as we take c and s with the lower or upper signs.

As we are now considering the *magnitude* only of the perpendicular, the algebraic sign with which it is affected is immaterial. We shall see hereafter, that it is sometimes necessary to select the appropriate sign for l; but for this purpose we shall use a geometrical construction, and the equation to the line in a less general form.

COR. 1. Since $Ax' + By' + C$ is the only part in (4) that varies with the position of $(x'y')$, it follows that the $Ax + By + C$ of any point (xy) varies as the distance of the point from the line $Ax + By + C = 0$.

Similarly from equation (2) the lengths of *parallel* straight lines drawn from a point (xy) to meet the line $(Ax + By + C = 0)$, vary as $Ax + By + C$, since $Ac + Bs$ does not vary in that case.

COR. 2. If the point $(x'y')$ be the origin, $x'=0$, $y'=0$, and we have, for the distance of the line from the origin,

$$l=\frac{C}{\sqrt{(A^2+B^2)}}.$$

Ex. *To find the length of the perpendicular from the point* $(x=3,\ y=5)$ *on the line* $(3y-7x+9=0)$.

$$l=\frac{3\times5-7\times3+9}{\sqrt{7^2+3^2}}=\frac{3}{\sqrt{58}}.$$

53[1]. *To find the area of a triangle in terms of the co-ordinates of the angular points, the axes being rectangular.*

Let the three points be (xy), $(x'y')$, $(x''y'')$. The equation to the straight line joining $(x'y')$ and $(x''y'')$ is (Art. 35)

$$(y-y')(x''-x')-(x-x')(y''-y')=0,$$

and the perpendicular upon this from the point (xy) is (Art. 52)

$$\pm\frac{(y-y')(x''-x')-(x-x')(y''-y')}{\{(x''-x')^2+(y''-y')^2\}^{\frac{1}{2}}};$$

and the distance between $(x'y')$ and $(x''y'')$ is the denominator of this fraction. But this perpendicular multiplied by this distance is double of the area of the triangle. Hence

$$\pm\frac{1}{2}\{(y-y')(x''-x')-(x-x')(y''-y')\}$$

is the area of the triangle, the upper or lower sign being used, according as the expression is positive or negative.

If the three points (xy), $(x'y')$, $(x''y'')$ are in the same straight line, the area becomes $=0$, which gives, as it ought, the equation of Art. 35. This gives an easy way of remembering the formula.

For the case of oblique axes see Chap. IV. Ex. 8.

[1] From Salmon's *Conic Sections.*

54. *To find the length of the perpendicular from a point $(x'y')$ on the line $(x\cos\alpha + y\sin\alpha = p)$, the axes being rectangular.*

Let DT be the line

$$x\cos\alpha + y\sin\alpha = p,$$

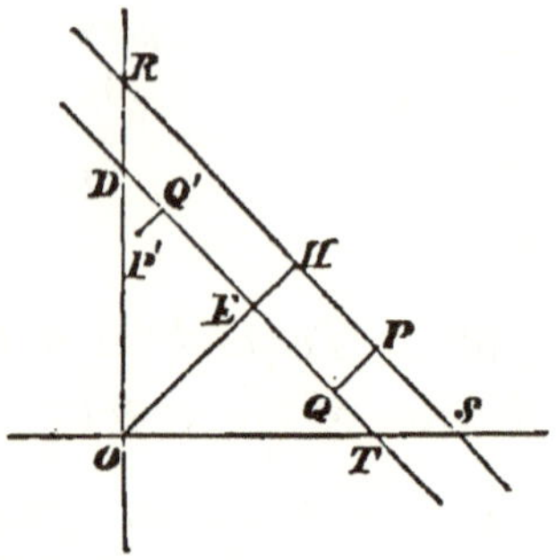

and P the point $(x'y')$. Draw a line RS through P parallel to DT, and draw PQ, OEH perpendicular to the lines DT and RS; then the equation to RS must be

$$x\cos\alpha + y\sin\alpha = OH$$

$$= p' \text{ suppose};$$

but since $(x'y')$ is a point in RS, we have

$$x'\cos\alpha + y'\sin\alpha = p'.$$

Now $PQ = HE = p' - p$,

therefore $$PQ = x'\cos\alpha + y'\sin\alpha - p,$$

and is known.

If the point $(x'y')$ be on the other side of the line, as P', the length of the perpendicular $P'Q'$ will evidently be $= p - p'$, or $= p - x'\cos\alpha - y'\sin\alpha$. Hence, if the equation to a line be $x\cos\alpha + y\sin\alpha = p$, where p is a positive quantity, the length of the perpendicular from $(x'y')$ is

$$\pm(x'\cos\alpha + y'\sin\alpha - p),$$

the lower sign being used, when $(x'y')$ is on the origin side of the line.

55. Thus we see that the $x\cos\alpha + y\sin\alpha - p$ of any point (xy) is negative or positive, according as (xy) is or is not on the origin side of the line, and vanishes when (xy) is *on* the line. Hence when we have once adopted a certain name for a line, such as $x\cos\alpha + y\sin\alpha - p$, we may say

that the line has a positive and a negative side; the positive side being that, for any point on which the expression called the name of the line is positive, so that the perpendicular on that side is written with the same sign as the adopted name. The origin side is negative in Art. 54; if we called the line $p - x\cos\alpha - y\sin\alpha$, the origin side would be positive. The sign of the origin side, then, is the sign of the expression which we have called the name of the line, when x and y each $= 0$.

COR. 1. Similarly, as in Art. 52, Cor. 1, we see that the $Ax + By + C$ of any point (xy) changes sign, as the point crosses the line $Ax + By + C = 0$; so that the origin side is the positive or negative side of that line, according as C is positive or negative.

COR. 2. It is easy to see that the results of Arts. 52 and 54 differ in appearance only; for, if the equation

$$Ax + By + C = 0$$

is written in the form $x\cos\alpha + y\sin\alpha - p = 0$, it becomes (Art. 27, Cor. 2)

$$-\frac{A}{\surd A^2 + B^2}x - \frac{B}{\surd A^2 + B^2}y - \frac{C}{\surd A^2 + B^2} = 0;$$

and the perpendicular on this line, obtained by Art. 54, coincides with the result of Art. 52.

Ex. If the equation to a line be

$$3y - 4x + 3 = 0,$$

the expression $3y - 4x + 3$ is positive on the origin side, and negative on the other, and the expressions

$$\frac{3y' - 4x' + 3}{5}, \quad \frac{-3y' + 4x' - 3}{5},$$

represent perpendiculars from points $(x'y')$ on the origin side and the other side, respectively.

56. The reasoning of the preceding article supposes that the equation can be written in the form

$$x\cos\alpha + y\sin\alpha - p = 0,$$

where p is a positive quantity, and α is defined by Art. 27, Cor. 2; but, when the line passes through the origin, this definition of α evidently fails, and the equation $Ax + By = 0$, written in the form required, is either

$$\frac{A}{\sqrt{A^2+B^2}}x + \frac{B}{\sqrt{A^2+B^2}}y = 0,$$

or

$$-\frac{A}{\sqrt{A^2+B^2}}x - \frac{B}{\sqrt{A^2+B^2}}y = 0.$$

The length of the perpendicular from $(x'y')$ is still

$$\pm(x'\cos\alpha + y'\sin\alpha), \text{ or } \pm\frac{Ax' + By'}{\sqrt{A^2+B^2}};$$

and we require to know, on which side of the line the expression adopted as its name is positive. If the equation be written in the form $y - mx = 0$, it is evident that, for any point *above* the line, $y - mx$ is positive; for y will then have a larger positive or a smaller negative value, than it has *on* the line for the same abscissa. Similarly, for any point *below* the line, $y - mx$ is negative, where we consider the positive part of the axis of y as above the line. Hence we obtain the following rule. Write the coefficient of y positive; then the positive part of the axis of y is on the positive side of the line.

Ex. If the equation to the line be

$$3y - 4x = 0,$$

the expression $3y - 4x$ is positive for points above the line, and negative for points below it; and the expressions

$$\frac{3y' - 4x'}{5}, \quad \frac{-3y' + 4x'}{5},$$

represent perpendiculars from points $(x'y')$ above and below the line, respectively.

*57. It has already been shewn (Art. 43) that the equation

$$Ax + By + C + k(ax + by + c) = 0 \text{.........(1)}$$

is the equation to a straight line passing through the intersection of the lines

$$Ax + By + C = 0, \quad ax + by + c = 0 \text{.........(2)}.$$

Now equation (1) admits of a very simple geometrical interpretation, if the lines (2) be written in the form

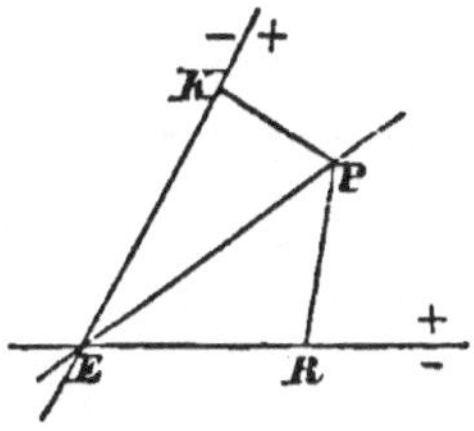

$$x \cos \alpha + y \sin \alpha - p = 0 \text{......(3)},$$
$$x \cos \beta + y \sin \beta - q = 0 \text{......(4)};$$

for let EK and ER be the lines (3) and (4), and suppose the origin somewhere in the angle opposite to KER, so that the positive and negative sides of the lines are as in the figure. Then, if we take any point, as P, $(x'y')$, the perpendiculars PK and PR will be represented (Art. 54) by the expressions

$$x' \cos \alpha + y' \sin \alpha - p, \quad x' \cos \beta + y' \sin \beta - q,$$

and the equation

$$x' \cos \alpha + y' \sin \alpha - p - k(x' \cos \beta + y' \sin \beta - q) = 0 \text{......(5)}$$

asserts that

$$\frac{PK}{PR} = \frac{k}{1},$$

and hence the locus of the point P, or of the equation

$$x \cos \alpha + y \sin \alpha - p - k(x \cos \beta + y \sin \beta - q) = 0 \text{......(6)}$$

is a straight line passing through the intersection of (3) and (4), and such that, if perpendiculars be dropped from any point in it upon the lines (3) and (4), those perpendiculars will be to one another in the ratio of k to 1. Hence also the sines of the angles which (6) makes with (3) and (4) are in the ratio of k to 1.

*58. It will now be seen, why we have (Art. 54) considered so carefully the *signs* of the perpendiculars, or rather, investigated a method by which we may always be able to write the expressions for them, *so that those expressions shall represent positive quantities;* for if the lines in question be DT and $D'T'$, it is evident that the reasoning of Article 54 will give us, for the point $P(x'y')$, if its perpendiculars on DT and $D'T'$ are as k to 1, the equation

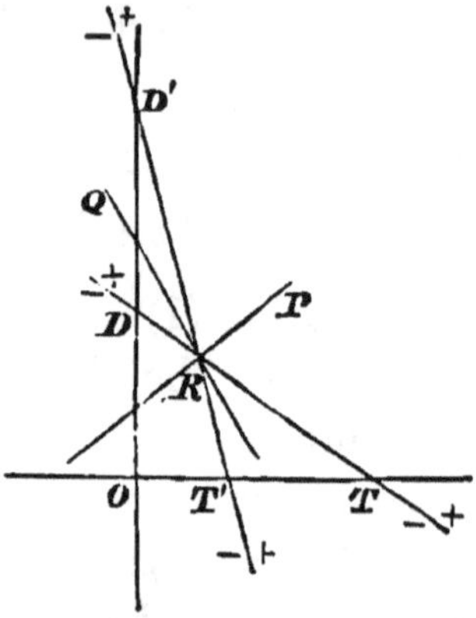

$$\frac{x'\cos\alpha + y'\sin\alpha - p}{x'\cos\beta + y'\sin\beta - q} = \frac{k}{1},$$

or $\quad x\cos\alpha + y\sin\alpha - p - k(x\cos\beta + y\sin\beta - q) = 0 \ldots (7)$,
as the equation to PR; whereas the point Q would, under the same circumstances, give us the equation

$$\frac{x'\cos\alpha + y'\sin\alpha - p}{-(x'\cos\beta + y'\sin\beta - q)} = \frac{k}{1},$$

or $\quad x\cos\alpha + y\sin\alpha - p + k(x\cos\beta + y\sin\beta - q) = 0 \ldots (8)$,
as the equation to QR.

COR. If $k = 1$, or the perpendiculars are equal, equations (7) and (8) will represent the straight lines which *bisect* the angles $D'RT$ and $D'RD$. These lines are easily seen to be at right angles, by the condition of Art. 47, Cor.

*59. In order to select the proper equation for any particular line passing through R, where the ratio of the perpendiculars or sines is given, the following rule, which may be readily deduced from Art. 58, is universally true.

If any point in the required line is on the origin side of *both* or *neither* of the given lines, the equation is of the form

(7); if it is on the origin side of *one only*, the equation is of the form (8).

The equations to the lines, if given in any form, may (Art. 27, Cor. 2) be reduced to the form here used, and the above test applied. Arts. 54—59 are equally true for oblique axes, if (Art. 32) we use the equation $x\cos\alpha + y\cos(\omega-\alpha) = p$; but in that case the reduction of any equation to the required form is more complicated.

*60. Equations (6) and (7) (Arts. 57, 58) usually present some difficulty to the student, because we use the same symbols, x and y, to represent the co-ordinates of the point P, and the co-ordinates of any point on the given lines. In these articles, we have endeavoured to avoid this difficulty, by first calling the co-ordinates of P x' and y', considering P as one fixed point. When we have obtained relations (5) between these co-ordinates, by means of the conditions of the problem, we may evidently (Art. 21) write x and y for x' and y', in the equation to the *locus* of P, without any fear of confusion. We shall hereafter frequently speak of the perpendicular from the point (xy) on the line $x\cos\alpha + y\sin\alpha - p = 0$, bearing in mind this explanation.

*Ex. *To find the equations (axes rectangular) to the straight lines which bisect the supplementary angles between the two lines,*

$$y - \sqrt{3}x - 5 = 0, \qquad \sqrt{3}y - x + 6\sqrt{3} = 0.$$

Let DR, $D'R$ be the lines, which will evidently lie as in the figure; then, comparing their equations with the equation $x\cos\alpha + y\sin\alpha = p$, and writing them in that form, we have (Art. 27, Cor. 2)

for DR, $\qquad -\frac{\sqrt{3}}{2}x + \frac{1}{2}y = \frac{5}{2}$,

for $D'R$ $\qquad \frac{1}{2}x - \frac{\sqrt{3}}{2}y = 3\sqrt{3}$.

Hence we have for the equation to the bisector PR, any point of which is on the origin side of *both* or *neither* of the given lines,

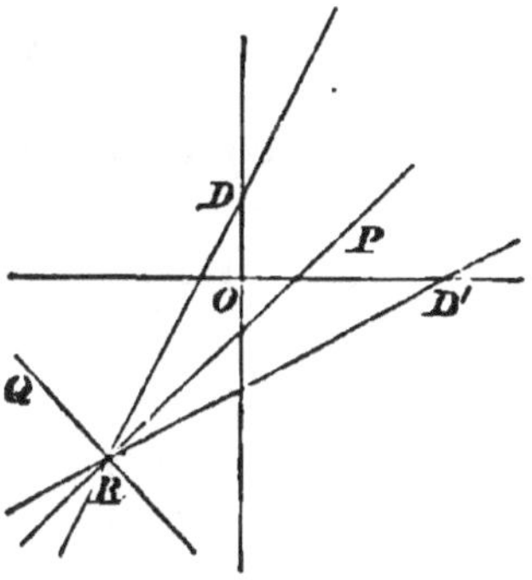

$$\tfrac{1}{2}x - \frac{\sqrt{3}}{2}y - 3\sqrt{3} = -\frac{\sqrt{3}}{2}x + \tfrac{1}{2}y - \tfrac{5}{2},$$

or

$$2y - 2x = 11\sqrt{3} - 23.$$

For the bisector RQ, any point of which is on the origin side of *one only*,

$$\tfrac{1}{2}x - \frac{\sqrt{3}}{2}y - 3\sqrt{3} = \frac{\sqrt{3}}{2}x - \tfrac{1}{2}y + \tfrac{5}{2},$$

or

$$2x + 2y = -23 - 11\sqrt{3}.$$

61. *Equations representing straight lines.*

We shall now notice a few particular cases, where equations of a degree higher than the first may be interpreted by means of the preceding articles.

If we have the equations to two straight lines

$$Ax + By + C = 0, \qquad ax + by + c = 0,$$

it is evident that the equation of the second degree

$$(Ax + By + C)(ax + by + c) = 0,$$

will represent both the lines, for the co-ordinates of any point in either of the lines, substituted in the equation, will make one of the factors vanish, and the equation will be satisfied. Similarly, if we multiply together the equations to n straight lines, we shall obtain an equation of the nth degree, which represents them all. Conversely, if any equation of the nth degree can be separated into n factors of the first degree, it represents n straight lines.

62. *To find the condition that an equation of the second degree should represent two straight lines.*

Let the equation of the second degree, in its most general form,

$$Ax^2 + 2Hxy + By^2 + 2Gx + 2Fy + C = 0 \ldots\ldots\ldots (1),$$

be written as a quadratic in y,

$$By^2 + 2(Hx + F)y + Ax^2 + 2Gx + C = 0.$$

Solving this equation, we have

$$y = \frac{-(Hx+F) \pm \sqrt{(Hx+F)^2 - B(Ax^2+2Gx+C)}}{B}.$$

The expression under the root becomes

$$(H^2 - AB)\,x^2 + 2(HF - BG)\,x + F^2 - BC,$$

and, in order that equation (1) may break up into two equations of the first degree, it is sufficient and necessary that this expression should be a *perfect square* for all values of x; hence the condition is

$$(HF - BG)^2 = (H^2 - AB)(F^2 - BC),$$

or, expanding and dividing by B,

$$ABC + 2FGH - AF^2 - BG^2 - CH^2 = 0 \ldots\ldots\ldots\ldots(2).$$

If the quantity under the root do not contain x at all, the equation will represent two *parallel* straight lines; for, in this case, the coefficient of x in the right-hand member of the equation, which is (Art. 28, Cor. 2) the *angular coefficient* of the lines, is not affected by the alteration of the sign of the radical. These lines are imaginary, or the locus impossible, when the quantity under the root is negative.

Ex. 1. The equation

$$y^2 - 3xy + 2x^2 + 3y - 4x + 2 = 0,$$

may be written

$$y^2 - 3(x-1)\,y + 2(x-1)^2 = 0.$$

Solving for y, we have the equations

$$y - 2x + 2 = 0, \qquad y - x + 1 = 0,$$

which are the equations required, since the original equation is obtained by multiplying these two equations together.

Ex. 2. The equation

$$3x^2 + 2xy + y^2 + 6x + 2y + 3 = 0,$$

when solved becomes

$$y + x + 1 \pm \sqrt{-2}\,(x+1) = 0,$$

which represents two imaginary straight lines, which may be said to intersect in the only point for which the equation is satisfied, viz.

$$y+x+1=0, \quad x+1=0, \quad \text{or } x=-1, \quad y=0.$$

Ex. 3. The equation

$$4x^2+8xy+4y^2+3x+3y=0$$

represents the two parallel lines,

$$y+x=0, \quad 4y+4x+3=0.$$

63 (i). The equation

$$Ax^2+2Hxy+By^2=0 \quad\dots\dots\dots\dots(1),$$

may be written

$$A+2H\left(\frac{y}{x}\right)+B\left(\frac{y}{x}\right)^2=0 \quad\dots\dots\dots\dots(2).$$

Solving this as a quadratic in $\frac{y}{x}$, we obtain

$$\frac{y}{x}=\frac{-H\pm\sqrt{H^2-AB}}{B};$$

and, if we call these values μ and μ', equation (1) is the same as

$$(y-\mu x)(y-\mu' x)=0,$$

which represents two straight lines, $y=\mu x$ and $y=\mu' x$, through the origin, possible and different, possible and coincident, or imaginary, according as $H^2-AB>=<0$.

If m is written for $\frac{y}{x}$ in (2), μ and μ' are the roots of the equation

$$Bm^2+2Hm+A=0.$$

Similarly every homogeneous equation of the nth degree may be written in the form

$$A_0\left(\frac{y}{x}\right)^n+A_1\left(\frac{y}{x}\right)^{n-1}+\dots\dots A_{n-1}\left(\frac{y}{x}\right)+A_n=0,$$

and will have n roots, possible or impossible,

$$\frac{y}{x} = \mu, \quad \frac{y}{x} = \mu', \&c.,$$

which will give n straight lines, possible or imaginary, passing through the origin.

63 (ii). Suppose we have two equations

$$Ax^2 + 2Hxy + By^2 + 2Gx + 2Fy + C = 0 \ldots\ldots\ldots\ldots\ldots(1),$$

$$ax + by = 1 \ldots\ldots\ldots\ldots\ldots(2);$$

then the equation

$$Ax^2 + 2Hxy + By^2 + 2(Gx + Fy)(ax + by) + C(ax + by)^2 = 0 \,.\, (3)$$

is homogeneous and of the second degree, and therefore represents two straight lines passing through the origin. But the co-ordinates of any point common to (1) and (2) will evidently satisfy (3); hence (3) passes through the intersections of (1) and (2). Hence (3) represents two straight lines joining the origin to the points of intersection of the loci represented by (1) and (2). The equation to *any* straight line may be written in the form (2); and the meaning of (1) will be fully explained hereafter.

64. The equation

$$Ax^2 + Bx + C = 0$$

will give two solutions of the form $x =$ constant, and will therefore represent two straight lines, possible or imaginary, parallel to the axis of y. In like manner the equation

$$Ay^2 + By + C = 0$$

represents two straight lines parallel to the axis of x.

Similarly any equation of a higher degree, which involves only one of the variables, will represent a series of straight lines, possible or imaginary, parallel to one of the axes.

65. If an equation of the second order can be written in the form

$$(Ax + By + C)^2 + (ax + by + c)^2 = 0,$$

it can be satisfied by those values only of x and y, which make

$$Ax + By + C = 0, \text{ and } ax + by + c = 0;$$

for otherwise we should have the sum of two positive quantities equal to zero. Hence the locus is a point, or may be considered as two imaginary lines

$$Ax + By + C \pm \sqrt{-1}\,(ax + by + c) = 0,$$

which intersect in a real point.

66. If an equation of the second order can be written in the form

$$(Ax + By + C)^2 + (ax + by + c)^2 = -P,$$

where P is positive, it can be satisfied by no real values of x and y, since the sum of the two quantities on the left-hand side can in no case be negative; the locus is therefore entirely imaginary.

67. *To find the angle between the straight lines represented by the equation*

$$Ax^2 + 2Hxy + By^2 = 0 \quad\text{..................(1),}$$

the axes being rectangular.

If the lines are $y = \mu x$ and $y = \mu' x$, and θ the angle between them, then (Art. 47)

$$\tan\theta = \frac{\mu - \mu'}{1 + \mu\mu'};$$

also (Art. 63) μ and μ' are the roots of the equation

$$Bm^2 + 2Hm + A = 0 \quad\text{..................(2);}$$

hence $$-(\mu + \mu') = \frac{2H}{B}, \quad \mu\mu' = \frac{A}{B} \quad\text{..................(3);}$$

therefore $$(\mu - \mu')^2 = (\mu + \mu')^2 - 4\mu\mu'$$
$$= \frac{4(H^2 - AB)}{B^2},$$

and $$1 + \mu\mu' = \frac{A + B}{B};$$

therefore $$\tan\theta = \frac{2(H^2 - AB)^{\frac{1}{2}}}{A + B}.$$

The lines are at right angles, if $A + B = 0$.

*COR. 1. If the axes are inclined at an angle ω, we have (Art. 49)

$$\tan\theta = \frac{(\mu - \mu')\sin\omega}{1 + (\mu + \mu')\cos\omega + \mu\mu'}$$
$$= \frac{2(H^2 - AB)^{\frac{1}{2}}\sin\omega}{A + B - 2H\cos\omega}.$$

The lines are at right angles, if

$$A + B - 2H\cos\omega = 0.$$

COR. 2. With axes of any inclination, if the equation representing two straight lines were

$$Ax^2 + 2Hxy + By^2 + 2Gx + 2Fy + C = 0 \ldots\ldots\ldots(4),$$

it could be written in the form

$$(y - \mu x - \beta)(y - \mu' x - \beta') = 0 \ldots\ldots\ldots\ldots(5),$$

or $$\mu\mu' x^2 - (\mu + \mu')xy + y^2 + Px + Qy + R = 0 \ldots\ldots(6),$$

where P, Q, and R are constants. Hence, if (4) and (6) are in reality the same equation,

$$\frac{\mu\mu'}{A} = -\frac{\mu + \mu'}{2H} = \frac{1}{B},$$

which equations for μ and μ' are the same as (3); hence μ and μ' are as before the roots of

$$Bm^2 + 2Hm + A = 0,$$

and the straight lines represented by (4) are *parallel* to those represented with the same axes by (1), and contain the same angle.

Cor. 3. The straight lines represented by (1) are coincident, and those represented by (4) are parallel, if

$$H^2 - AB = 0,$$

in which case

$$Ax^2 + 2Hxy + By^2$$

is a perfect square.

*68. *To find the equation to the straight lines which bisect the angles between the straight lines*

$$Ax^2 + 2Hxy + By^2 = 0 \text{.....................} (1),$$

the axes being rectangular.

Let the lines (1) be $y - \mu x = 0$, $y - \mu' x = 0$, where μ and μ' are the roots of the equation $Bm^2 + 2Hm + A = 0$. Then the equations to the bisectors are (Art. 58, Cor.)

$$\frac{y - \mu x}{\sqrt{1 + \mu^2}} - \frac{y - \mu' x}{\sqrt{1 + \mu'^2}} = 0, \quad \frac{y - \mu x}{\sqrt{1 + \mu^2}} + \frac{y - \mu' x}{\sqrt{1 + \mu'^2}} = 0,$$

or, when expressed as one locus,

$$\frac{(y - \mu x)^2}{1 + \mu^2} - \frac{(y - \mu' x)^2}{1 + \mu'^2} = 0,$$

or, simplifying and dividing by $\mu' - \mu$,

$$y^2 (\mu + \mu') + 2xy (1 - \mu\mu') - x^2 (\mu + \mu') = 0,$$

or, (Art. 67)

$$x^2 - \frac{A - B}{H} xy - y^2 = 0 \text{.................} (2).$$

The condition (Art. 63) that the lines (2) should be real is $(A - B)^2 + 4H^2 > 0$; hence the bisectors are always real, whether the original lines are real or imaginary. They are at right angles by Art. 67.

EXAMPLES III.

1. FIND the equation to the straight lines which pass through the point (1, 3), and make an angle of 30° with the line $(2y - x + 1 = 0)$; axes being rectangular.

2. Draw the lines represented by the equation

$$(2y - x + c)\ (3y + x - c) = 0,$$

and determine (1) where they intersect, and (2) at what angle; the axes being rectangular.

3. Find the equation to a straight line which passes through the point $(c, 0)$, and makes an angle of 45° with the line $(bx - ay = ab)$; axes being rectangular.

4. Find the equation to a straight line which is perpendicular to the line $(8y + 5x - 3 = 0)$, and cuts the axis of y at a distance $= 8$ from the origin; axes being rectangular.

5. Find the cosine of the angle between the lines

$$(y - 4x + 8 = 0) \text{ and } (y - 6x + 9 = 0);$$

axes being rectangular.

6. Find the angle between the lines $(4y + 3x + 5 = 0)$ and $(4x - 3y + 6 = 0)$; axes being rectangular.

*7. Find the equations to the straight lines which pass through the intersection of the lines $(y = 2x + 4)$, $(y = 3x + 6)$, and bisect the supplementary angles between them; axes being rectangular.

8. What is the geometrical signification of the equations

$$x^2 + y^2 = 0, \qquad xy = 0\,?$$

*9. Find the equations to the straight lines which bisect the angles between the lines $(12x + 5y = 8)$ and $(3x - 4y = 3)$; axes being rectangular.

10. Shew that the lines represented by the equation

$$x^2 - xy - 6y^2 + 2x - y + 1 = 0,$$

are inclined to one another at an angle of 45°; axes being rectangular.

11. The equation $2y^2 - 3xy - 2x^2 - 3y + 6x = 0$, represents two straight lines at right angles; axes being rectangular.

12. The equation $y^2 - 2xy \sec\theta + x^2 = 0$, represents two straight lines inclined to one another at an angle θ; axes being rectangular.

13. What is the inclination of the co-ordinate axes, when the lines represented by $y^2 - x^2 = 0$, are perpendicular to one another?

*14. The equations to two straight lines are

$$x + 3y - a = 0 \ldots\ldots(1), \qquad y - x + a = 0 \ldots\ldots(2);$$

find the equations to the straight lines which pass through the intersections of (1) and (2), so that the ratio of the sines of the inclination of each to (1) and (2) may be as $1 : \sqrt{5}$.

15. What must be the inclination of the axes in order that the lines $(xy - 3y - 2x + 6 = 0)$ may include an angle of 135°?

16. Find the equations to the two straight lines which pass through the origin, and divide into three equal parts the distance between the points in which the axes of co-ordinates are intersected by the line $(x + y = 1)$.

17. Find the distance of the point of intersection of the lines $(3x + 2y + 4 = 0)$, $(2x + 5y + 8 = 0)$, from the line $(y = 5x + 6)$; the axes being rectangular.

18. Find the perpendicular distance between the lines

$$Ax + By + C = 0, \quad Ax + By + C' = 0;$$

the axes being rectangular.

19. On which sides of the line $3x - 2y = 1$ do the points $(1, 2)$, $(3, -4)$ lie? Shew that they lie on the same side of the line $y = 3x$.

20. The perpendicular from a point $(x'y')$ on the line $3x + 3 = 4y$, is $\frac{1}{5}(4y' - 3x' - 3)$. On which side of the line does $(x'y')$ lie?

21. Shew that the equation $(3x + y)(2x - 3y + 8) = 4$, represents a locus which lies entirely in two of the angles formed by the lines $(3x + y = 0)$ and $(2x - 3y + 8 = 0)$.

22. Determine the loci represented by the polar equations

(i) $\theta - a = 0$. (ii) $\sin(\theta - a) = 0$.

(iii) $(\rho - a)(\rho^2 - a^2 - a\rho \tan\theta \sin\theta) = 0$.

23. Find the angle between the lines

$$\frac{c}{\rho} = \cos\theta - 2\sin\theta, \qquad \frac{d}{\rho} = \cos\theta + 3\sin\theta.$$

24. Interpret the equations:

(i) $\sin 2\theta = 0$, (ii) $\cos 3\theta = 0$, (iii) $\theta^3 + \alpha\theta^2 + \beta\theta + \gamma = 0$.

25. Find the area of the triangles whose angular points are

(i) $(0, b)$, $(a, 0)$, (a, b); (ii) $(a, 2a)$, $(2a, 3a)$, $(3a, -4a)$;

(iii) $(0, 0)$, (x, y), (x', y').

26. Find the area of the triangle contained by the three straight lines

$$3x + 4y = 12, \qquad 4x + 3y = 12, \qquad x + y = 3.$$

CHAPTER IV.

Transformation of Co-ordinates.

69. When the position of a point or the equation to a curve is given, with reference to any particular system of co-ordinates, it is frequently necessary to find that position or equation with regard to some other system. We have already shewn how to pass from rectangular to polar co-ordinates, and the converse, and we shall now shew how we may perform other transformations, such as altering the origin, passing from rectangular to oblique co-ordinates, and others of the same nature.

70. *To transfer the origin of co-ordinates to a point $(x'y')$ without altering the direction of the axes.*

Let $O'X$, $O'Y$ be the new axes, respectively parallel to Ox and Oy, the old ones; let the co-ordinates of any point P, referred to the old axes, be x, y, and, when referred to the new, X, Y;

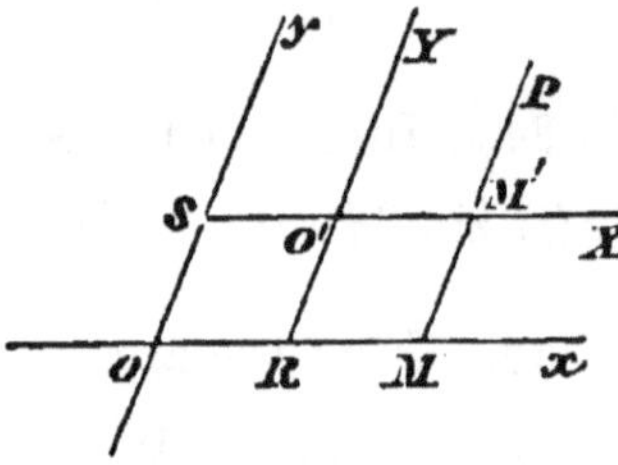

$$OR = x', \ O'R = y', \text{ and}$$

$$PM = PM' + O'R, \ OM = O'M' + OR,$$

$$\text{or } y = Y + y', \quad x = X + x'.$$

These formulæ are true for rectangular and oblique axes.

Hence, to find what the equation to any locus becomes, when the origin is transferred to a point $(x'y')$, the new axes remaining parallel to the old, we must write $x+x'$ for x, and $y+y'$ for y.

Ex. *To find what the equation*

$$y^2+4y-4x+8=0$$

becomes, when the origin is transferred to a point whose co-ordinates are

$$x=1, \qquad y=-2.$$

Writing $x+1$ for x, and $y-2$ for y, we obtain the equation

$$y^2=4x.$$

71. *To find what the co-ordinates of a point become, if the axes, being rectangular, are turned through a given angle* (α), *the origin remaining the same.*

Let Ox, Oy, OM, PM be the old axes and co-ordinates of any point P, OX, OY, OM', PM' the new ones; let angle $XOx=\alpha$, and let the old co-ordinates be x, y, the new X, Y; then, if $M'R$, $M'S$ be drawn parallel to Ox, Oy respectively, we have

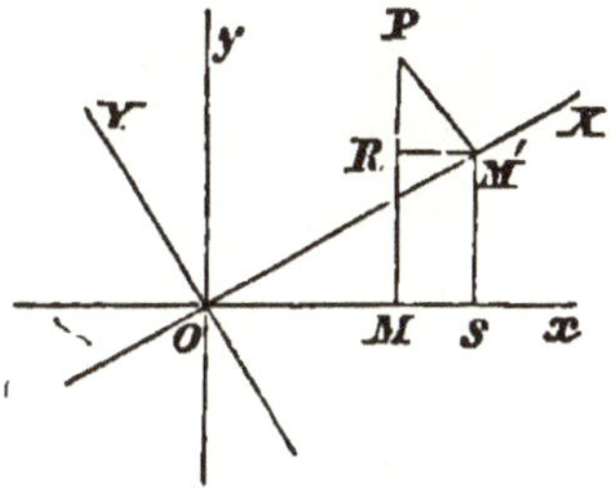

$$OM=OS-RM',$$

$$PM=PR+M'S,$$

or, since angle $RPM'=$ angle $XOx=\alpha$,

$$x=X\cos\alpha-Y\sin\alpha, \quad y=Y\cos\alpha+X\sin\alpha;$$

hence, to find what the equation to any locus becomes, when referred to the new axes, we must write

$$x\cos\alpha-y\sin\alpha \text{ for } x, \text{ and } y\cos\alpha+x\sin\alpha \text{ for } y.$$

It will be observed that $x^2+y^2=X^2+Y^2$, which ought to be the case, since each is equal to OP^2.

The student should remember this figure, and be able

readily to write the formulæ from it. They may be verified by observing that

$$\text{if } \alpha = 0 \text{ we must have } x = X,\ y = Y;$$

$$\text{if } \alpha = \frac{\pi}{2} \quad \text{,,} \quad \text{,,} \quad x = -Y,\ y = X.$$

Ex. *To find what the equation* $x^2 - y^2 = a^2$ *becomes, when the axes are moved through an angle of* 45°.

Here, $$\sin\alpha = \cos\alpha = \frac{1}{\sqrt{2}},$$

and we must write

$$\frac{1}{\sqrt{2}}(x-y) \text{ for } x, \text{ and } \frac{1}{\sqrt{2}}(x+y) \text{ for } y;$$

hence, the equation becomes

$$(x-y)^2 - (x+y)^2 = 2a^2,$$

or $$2xy + a^2 = 0.$$

72. *To find what the co-ordinates of a point become, when the axes are changed from one oblique system to another, the origin remaining the same.*

Using the same notation as in Art. 71 and a similar figure, we have

$$PM = M'S + PR$$

$$= OM'\frac{\sin(Xx)}{\sin(xy)} + PM'\frac{\sin(Yx)}{\sin(xy)};$$

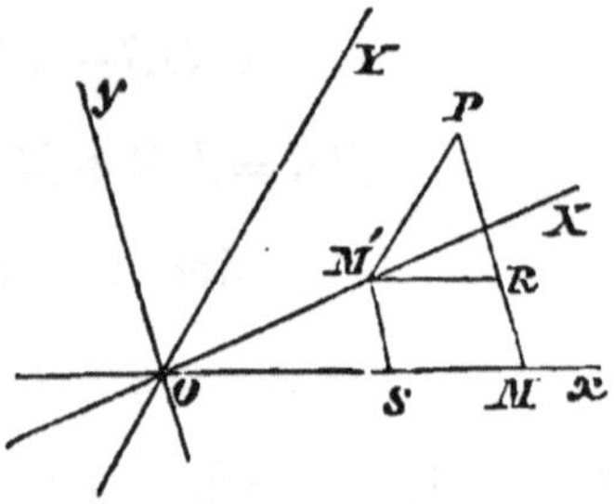

or $y\sin(yx) = X\sin(Xx) + Y\sin(Yx)$, where by sin (xy) we mean the sine of the angle which the axis of x makes with the axis of y, and by sin (Xx) the sine of the angle which the axis of X makes with the axis of x; and so with the rest.

Similarly, $OM = OS + M'R$

$$= OM'\frac{\sin(Xy)}{\sin(xy)} + PM'\frac{\sin(Yy)}{\sin(xy)},$$

or $$x\sin(xy) = X\sin(Xy) + Y\sin(Yy).$$

These formulæ include the particular cases, where we wish to pass from rectangular to oblique, or from oblique to rectangular axes. The signs are those appropriate to the figure, where (xy), (Xy), (Yy) are all measured on the same side of Oy, and (yx), (Xx), (Yx) on the same side of Ox. It must be remembered, that, in speaking of the angles made by the axes, we mean the angles made by their positive directions, and that (xy), (yx) are the same angle.

It is very rarely necessary to make a transformation such as the above; and for those which actually occur, it is generally easier to obtain from the figure the formula required, than to adapt those which we have obtained above. We will give a simple case which is often useful.

73. *To transform an equation from a rectangular to an oblique system, the axis of x remaining the same, and the new axis of y being inclined at an angle ω to the axis of x.*

Using the same notation as before, and a similar figure, we have

$$OM = OM' + M'M$$
$$= OM' + PM' \cos PM'M,$$

or $$x = X + Y\cos\omega;$$

$PM = PM' \sin PM'M$, or $y = Y \sin \omega$.

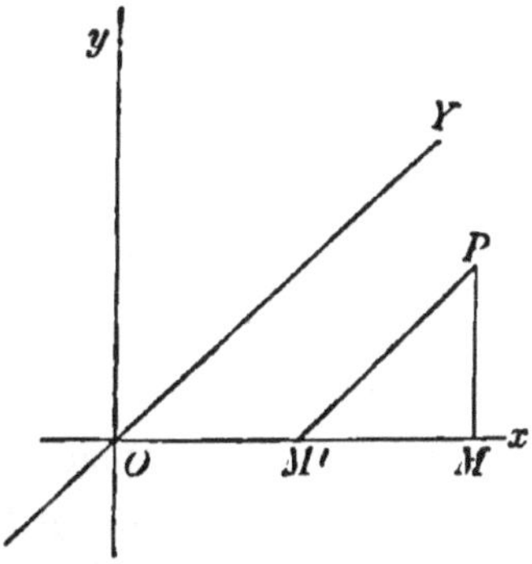

Similarly, if the transformation were from the axes Ox, OY to Ox, Oy, we should have

$$X = x - y\cot\omega, \quad Y = y \operatorname{cosec}\omega.$$

Ex. Let the equation to be transformed from rectangular to oblique axes be

$$Ax^2 + 2Hxy + By^2 + C = 0.$$

Writing $x + y\cos\omega$ for x, and $y\sin\omega$ for y, we have

$$\left.Ax^2 + 2A\cos\omega \atop + 2H\sin\omega\right| \left.\begin{matrix} xy + A\cos^2\omega \\ + 2H\sin\omega\cos\omega \\ + B\sin^2\omega \end{matrix}\right| y^2 + C = 0.$$

The following results of this transformation, which is a particular case of Art. 76 (ii), should be noted:

$$\frac{a+b-2h\cos\omega}{\sin^2\omega} = A + B, \qquad \frac{h^2 - ab}{\sin^2\omega} = H^2 - AB,$$

where a, $2h$, b denote the coefficients of x^2, xy, y^2 after the transformation.

74. If we wish to transfer the origin to a point $(x'y')$, and then change the direction of the axes, we have only to add x' and y' to the formulæ of Arts. 71—73. Thus, if the changes of Arts. 70, 71 have taken place,

$$x = x' + X\cos\alpha - Y\sin\alpha, \quad y = y' + Y\cos\alpha + X\sin\alpha.$$

75. In the case of polar co-ordinates, if we wish to turn the initial line through an angle α, we must write $\theta + \alpha$ for θ in the equation, since the new θ is less than the old θ by the angle α.

76 (i). The student must bear in mind that we make no change in the *locus*, by changing the origin or axes. The assemblage of points represented by the new equation is precisely the same as that represented by the old, but the manner in which the axes are placed with regard to them is changed, and therefore the equation, which expresses this *relative* position, is not the same as before.

We may notice also that the *degree* of an equation cannot be altered by transformation. For let Ax^py^q represent any term of an equation of the nth degree, where $p + q = n$: then, since (Art. 74) the old co-ordinates in terms of the new are expressions of the first degree only, we should have

$$Ax^py^q = A\,(ax' + by' + c)^p(dx' + ey' + f)^q,$$

and no term resulting from this multiplication can be of a

higher degree than $p+q$ or n. Hence, the equation cannot be *raised*. It, consequently, cannot be *depressed;* for, if that were possible, we might, by re-transforming it, *raise* it, which has been proved to be impossible.

*76 (ii). †Suppose we have an expression

$$Ax^2 + 2Hxy + By^2,$$

referred to axes inclined at an angle ω, and we transform it to other axes inclined at an angle ϕ, by making the substitutions of Art. 72; and suppose it to become

$$aX^2 + 2hXY + bY^2,$$

where x, y and X, Y are the co-ordinates of a point P with the old and new axes respectively. Then the same substitution will make

$$x^2 + 2xy\cos\omega + y^2 \text{ become } X^2 + 2XY\cos\phi + Y^2,$$

since each of these expressions is equal to OP^2.

Now, if we take the expression

$$Ax^2 + 2Hxy + By^2 + k(x^2 + 2xy\cos\omega + y^2)\ldots\ldots(1),$$

or $$(A+k)x^2 + 2(H + k\cos\omega)xy + (B+k)y^2.$$

where k is any constant; then, if this expression is a perfect square, it will plainly not be altered in this respect by the substitutions for x and y‡. But from above, after these substitutions (1) will become

$$aX^2 + 2hXY + bY^2 + k(X^2 + 2XY\cos\phi + Y^2)\ldots\ldots(2).$$

Hence, if k be chosen so as to make (1) a perfect square, then (2) will be a perfect square for the same value of k.

† From Salmon's *Conic Sections*.

‡ For example, $(Px+Qy)^2$ becomes after such a substitution
$\{P(aX+bY)+Q(cX+dY)\}^2$, or $\{(Pa+Qc)X+(Pb+Qd)Y\}^2$.

Now (1) will be a perfect square, if

$$(H + k\cos\omega)^2 - (A + k)(B + k) = 0,$$

or $$k^2 + \frac{A + B - 2H\cos\omega}{\sin^2\omega}k - \frac{H^2 - AB}{\sin^2\omega} = 0 \text{......(3)}.$$

Similarly (2) will be a perfect square, if

$$k^2 + \frac{a + b - 2h\cos\phi}{\sin^2\phi}k - \frac{h^2 - ab}{\sin^2\phi} = 0 \text{......(4)}.$$

But, since both become perfect squares for the *same* value of k, the two quadratics, (3) and (4), must be identical. Hence, by equating the coefficients of corresponding terms in (3) and (4), we prove the following proposition:

If, by any change in the direction of the axes, the expression

$$Ax^2 + 2Hxy + By^2$$

be transformed into

$$aX^2 + 2hXY + bY^2;$$

then, if ω and ϕ are the angles of inclination of the old and new axes, respectively, we shall have

$$\frac{A + B - 2H\cos\omega}{\sin^2\omega} = \frac{a + b - 2h\cos\phi}{\sin^2\phi},$$

and $$\frac{H^2 - AB}{\sin^2\omega} = \frac{h^2 - ab}{\sin^2\phi}.$$

If ω and ϕ are both right angles, these equations take the simpler forms

$$A + B = a + b, \quad H^2 - AB = h^2 - ab.$$

Compare Art. 73, Ex. and Art. 141 with this Article.

EXAMPLES IV.

1. TRANSFORM the origin to a point (ab) in the equation

$$x^2 + y^2 - 2ax - 2by + a^2 + b^2 - c^2 = 0.$$

2. Transform the equation

$$x^2 \cos^2 a - y^2 \sin^2 a = a^2$$

by turning the (rectangular) axes through an angle a.

3. Transform the equations $x + y = c$ and $x^2 - y^2 = 0$, by turning the (rectangular) axes through an angle of 45°.

4. If both systems be rectangular, and the equation to the old axis of y referred to the new axes is $x - y = 0$, the old and new axes are inclined to one another at an angle of 45°.

5. Transform the equation $y^2 + 4ay \cot a - 4ax = 0$, from a rectangular system to an oblique system inclined at an angle a, retaining the same origin and axis of x.

6. Transform the equation $2x^2 - 5xy + 2y^2 = 4$, from axes inclined at an angle of 60°, to the right lines which bisect the angles between the axes.

7. Transform the same equation to rectangular axes, retaining the old axis of x.

8. Shew by the transformation of Art. 73, that, when the axes are inclined at an angle ω, the expression of Art. 53 for the area of a triangle must be multiplied by $\sin \omega$.

9. If the origin and axis of x are taken as the pole and the initial line, and the angle between the axes is ω, shew that we may transform from Cartesian to polar co-ordinates by the formulæ

$$x = \rho \frac{\sin(\omega - \theta)}{\sin \omega}, \quad y = \rho \frac{\sin \theta}{\sin \omega}.$$

CHAPTER V.

Geometrical Applications.

77. WE shall now give a few examples of the application of the formulæ we have obtained to the solution of geometrical problems. In attempting to solve these problems algebraically, the student will find that much depends upon a judicious selection of the origin and axes, and the application of the proper equations and formulæ. He should in every case consider the problem well, before he attempts the solution, and form a *definite plan*, before he begins. He may very possibly not be able to carry out his original scheme, but his attempts to do so will probably suggest some method by which he may solve the problem; and he will, at any rate, avoid a practice very common to beginners, of working without any definite aim, and consequently introducing and combining equations and formulæ that only serve to embarrass him, without in any way aiding him in the solution.

78. *To shew that the perpendiculars drawn from the vertices on the opposite sides of a triangle meet in a point.*

Let ABC be the triangle, CD, BE, AF the perpendiculars, and assume Ax, Ay as rectangular axes; let the co-ordinates of C be $AD=x'$, $CD=y'$, and let $AB=x''$. Now the proposition is proved, if we can shew that the abscissa of the point where AF and BE intersect is $=x'$, for they will then evidently intersect in CD. In order to shew this, we must find their equations, which

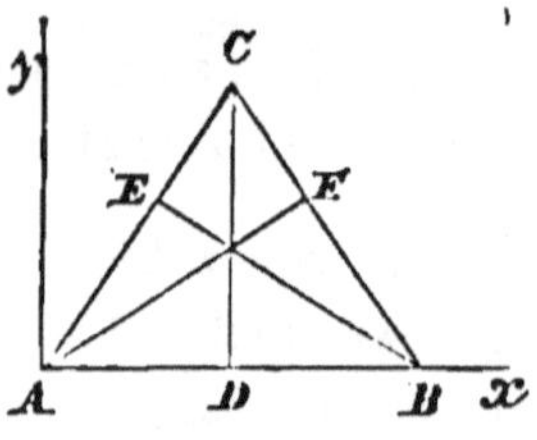

we shall obtain (Art. 50, Cor. 2) by observing that they each pass through a given point, and are perpendicular to a given line. We must then first find the equations to the lines AC, BC, to which they are perpendicular.

Since AC passes through the origin and the point C $(x'y')$, its equation is (Art. 30, Cor.)

$$y = \frac{y'}{x'}x \text{..............................} (1);$$

and since BE passes through B $(x''0)$, and is perpendicular to (1), its equation is (Art. 50, Cor. 2)

$$y = -\frac{x'}{y'}(x - x'') \text{........................} (2).$$

Also, since BC passes through the points B $(x''0)$ and C $(x'y')$, its equation is

$$y = \frac{y'}{x' - x''}(x - x'') \text{.....................} (3);$$

and since AF passes through the origin $(0, 0)$, and is perpendicular to (3), its equation is

$$y = -\frac{x' - x''}{y'}x \text{.........................} (4).$$

At the point where (2) and (4) intersect, their ordinates must be identical; hence, equating their values, we must have, at that point,

$$\frac{x'}{y'}(x - x'') = \frac{x' - x''}{y'}x,$$

whence, at the point of intersection, $x = x'$, which proves the proposition.

The student may exercise himself by solving this problem with DC as axis of y instead of Ay: then, assuming DC, DA, DB to be known, he may express the equations to AC and CB in terms of the portions of the axes they cut off (Art. 26). It will then remain to prove that AF and BE intersect in a point whose abscissa $= 0$.

79. *To shew that the three perpendiculars through the middle points of the sides of a triangle meet in a point.*

Using the same axes and notation as in the last problem, we must find the co-ordinates of the three middle points M, M', M''; we can then find the equations to the two perpendiculars MP, $M''P$, from the condition that they each pass through a given point, and are perpendicular to a given line; if we then shew that the abscissa of their point of intersection $= AM'$, we shall have proved that they intersect in the perpendicular from M'.

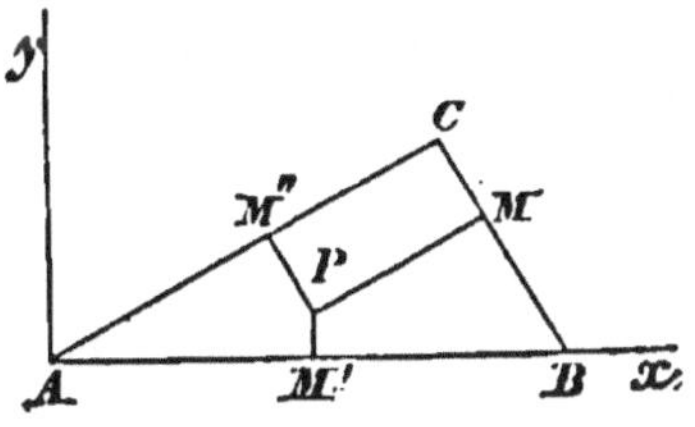

The co-ordinates of M'' are evidently $\frac{x'}{2}$, $\frac{y'}{2}$; also (Art. 10) the co-ordinates of M, the point of bisection of the line contained between the points B $(x''0)$ and C $(x'y')$, are

$$x=\frac{x'+x''}{2},\quad y=\frac{y'}{2}.$$

As before, the equation to AC is $y=\frac{y'}{x'}x$ and the equation to $M''P$ which passes through M'' $\left(\frac{x'}{2}\,\frac{y'}{2}\right)$, and is perpendicular to AC, is

$$y-\frac{y'}{2}=-\frac{x'}{y'}\left(x-\frac{x'}{2}\right)\ldots\ldots\ldots\ldots\ldots\ldots(1).$$

Also, as before, the equation to BC is

$$y=\frac{y'}{x'-x''}(x-x'')\ldots\ldots\ldots\ldots\ldots\ldots\ldots\ldots(2);$$

and the equation to MP, which passes through M $\left(\frac{x'+x''}{2}\,\frac{y'}{2}\right)$, and is perpendicular to (2), is

$$y-\frac{y'}{2}=-\frac{x'-x''}{y'}\left(x-\frac{x'+x''}{2}\right)\ldots\ldots\ldots\ldots(3).$$

At the point where (1) and (3) intersect, their ordinates must be identical; hence, equating their values, we have

$$\frac{x'}{y'}\left(x-\frac{x'}{2}\right)=\frac{x'-x''}{y'}\left(x-\frac{x'+x''}{2}\right);$$

which gives $x=\frac{x''}{2}$, as the abscissa of the point of intersection; but this abscissa belongs to some point in the perpendicular from M', which proves the proposition.

80. *In the figure of* Euc. I. 47, *if KH and FG be produced to meet in M, and MA produced to meet BC in T, shew that MT is perpendicular to BC.*

Take AB, AC produced indefinitely, as axes of x and y, and denote the sides of the triangle opposite to A, B, C by a, b, c; then the co-ordinates of M are $x=-b$, $y=-c$, and the equation to MA, which passes through the origin and M, is (Art. 30, Cor.)

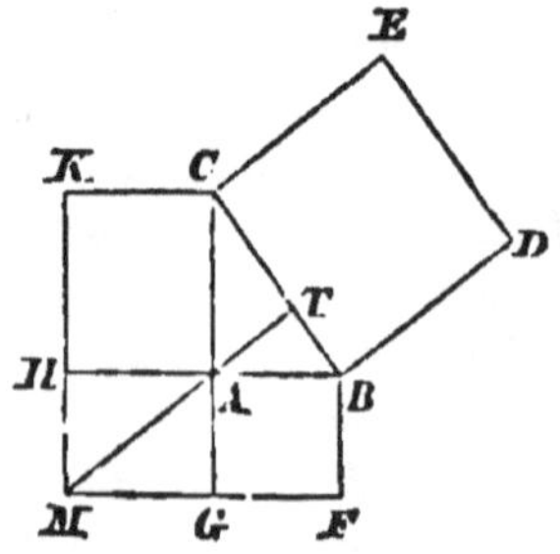

$$y=\frac{c}{b}x;$$

also the line BC cuts off from the axes of x and y intercepts $=c$ and b respectively; hence (Art. 26) its equation is

$$\frac{x}{c}+\frac{y}{b}=1, \quad \text{or} \quad y=-\frac{b}{c}x+b,$$

which, by Art. 47, represents a line perpendicular to MA.

81. *A straight line is drawn parallel to the base of a triangle, and its extremities joined transversely to the base; find the locus of the intersection of the joining lines.*

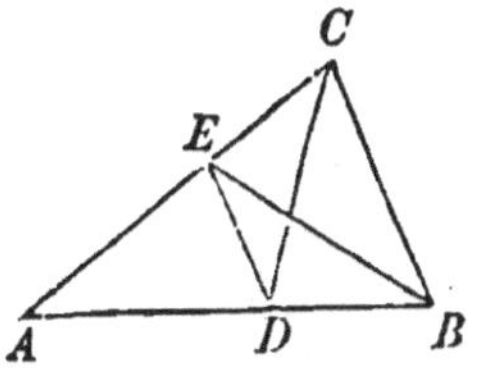

Let ABC be the triangle, CB the base, and ED parallel to it. Take AB, AC as axes, and let $AB=b$, $AC=c$, $AE=e$, $AD=d$; then the equations to EB and CD are

$$\frac{x}{b}+\frac{y}{e}=1\text{......(1)}, \qquad \frac{x}{d}+\frac{y}{c}=1\text{......(2)},$$

also (Euc. VI. 2), $$\frac{b}{c}=\frac{d}{e}\text{....................(3)},$$

$$\text{from (1) } e=\frac{by}{b-x}; \text{ from (2) } d=\frac{cx}{c-y};$$

$\therefore$ from (3) $$\frac{b}{c}=\frac{cx\,(b-x)}{by\,(c-y)},$$

whence $$c^2x^2-b^2y^2-bc\,(cx-by)=0,$$

or $$(cx-by)\,(cx+by-bc)=0\text{..............(4)}.$$

Now (4) represents two straight lines,

$$\frac{x}{b}-\frac{y}{c}=0, \qquad \frac{x}{b}+\frac{y}{c}=1.$$

The former of these is a straight line passing through the origin and the bisection of BC, for the co-ordinates of that point, $x=\frac{b}{2}$, $y=\frac{c}{2}$, satisfy the equation; and it is evidently the locus required. The latter represents the base BC, which is *not* the locus of the intersections of EB and CD.

82. The result of Art. 81 should be noticed. As is frequently the case in the solution of algebraical questions, we have arrived at two results, one which we were seeking, and another which does not satisfy the geometrical conditions of the problem. As shewn in Art. 40, the values of x and y which satisfy equations (1) and (2) with the condition (3), will satisfy (4); and therefore (4) represents a locus, which

passes through all the points where (1) and (2) intersect, as d and e vary. It does *not* follow that every point, whose co-ordinates satisfy equation (4), should be one of these points of intersection.

83. *The sides containing a given angle are in a given ratio, and the vertex is fixed; supposing the extremity of one of the sides to move in a given straight line, to find the locus of the extremity of the other.*

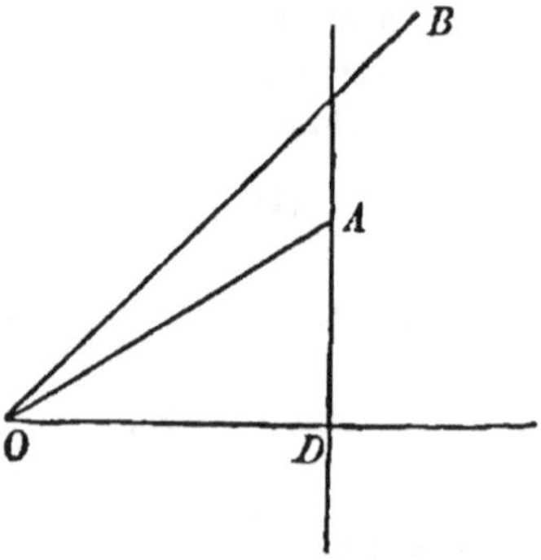

Let OA, OB be the two sides, where $OB=n\,.\,OA$, and let angle $AOB=\alpha$. Let A move on the straight line AD; it is required to find the locus of B. Take OD $(=p)$, the perpendicular from O on AD, as initial line, and let $OB=\rho$, angle $BOD=\theta$; then

$$OA=OD\sec AOD=p\sec(\theta-\alpha),$$

and $OB=n\,.\,OA$, or $\rho=np\sec(\theta-\alpha)$, which is therefore the polar equation to the locus of B. It represents (Art. 45) a straight line at a distance $=np$ from O, the inclination of this distance to OD being α.

84. *A straight line is drawn from a point O, cutting two other straight lines KA, KB in the points A, B, and in the straight line OAB a point P is taken, so that OA, OP, OB are in harmonic progression; find the locus of P.*

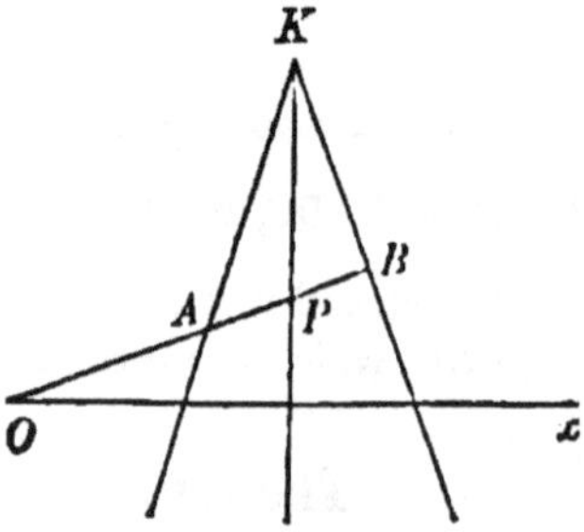

The condition to be satisfied is

$$\frac{2}{OP}=\frac{1}{OA}+\frac{1}{OB}.$$

Draw any initial line Ox, and let the equations to KA, KB be

$$\frac{1}{\rho}=a\cos\theta+b\sin\theta,\quad \frac{1}{\rho}=a'\cos\theta+b'\sin\theta,$$

since the general equation of Art. 44 may, by dividing and transposing, be written in this form. Then, if the angle $BOx=\theta$,

$$\frac{1}{OA}=a\cos\theta+b\sin\theta,\quad \frac{1}{OB}=a'\cos\theta+b'\sin\theta,$$

therefore $$\frac{2}{OP}=(a+a')\cos\theta+(b+b')\sin\theta,$$

and, writing ρ for OP, we have the polar equation to the locus of P. It is

(Art. 43) the equation to a straight line passing through the intersection of KA and KB; for it is equivalent to

$$\left(\frac{1}{\rho} - a\cos\theta - b\sin\theta\right) + \left(\frac{1}{\rho} - a'\cos\theta - b'\sin\theta\right) = 0.$$

EXAMPLES V.

1. In the fig. of Euc. I. 5, if BG, CF meet in H, shew that AH bisects the angle BAC.

2. In the fig. of Euc. I. 47, shew that AL, BK, FC intersect in one point, and that, if BG and CH be joined, the lines will be parallel.

3. Prove algebraically Euc. VI. 2.

4. Let Ax bisect any straight line CD in O; draw CB, cutting Ax in B and AD produced in E; join DB and let it meet AC produced in F, and join FE; FE shall be parallel to DC.

5. In two given straight lines, drawn from a point O, take points P, Q in one, and P', Q' in the other, so that OP, OQ, OP', OQ' are in harmonical progression; shew that the locus of the intersections of PQ' and $P'Q$ is a straight line bisecting the angle between the given lines.

6. Shew that the locus of a point, the algebraic sum of whose distances from the sides of a polygon is constant, is a straight line.

7. Taking the requisite data to fix a parallelogram in a plane, by equations to its sides, prove that the diagonals bisect each other.

8. $MANP$ is a parallelogram, having a given angle at A, and also its perimeter a given quantity: shew that the locus of P for all such parallelograms is a straight line.

9. A square is moved, so as to have the extremities of one of its diagonals upon two straight lines at right angles to one another, in its own plane; shew that the extremities of the other diagonal will move upon two other straight lines at right angles.

10. The hypotenuse of a right-angled triangle is made to slide between two perpendicular straight lines; find the locus of the right angle.

11. Given the base and the difference of the squares of the sides of a triangle; find the locus of the vertex.

12. Given the base and the sum of the sides of a triangle; if the perpendicular from the vertex on the base be produced through the vertex, till its whole length equals one of the sides, shew that the locus of the extremity of the perpendicular is a straight line.

13. If Ax, Ay be two straight lines, and through any point P there be drawn straight lines PP_1Q_1, PP_2Q_2, &c. meeting Ax in P_1, P_2, &c., and Ay in Q_1, Q_2, &c., then, if AP_1, AP_2, &c. are in harmonical progression, so also are AQ_1, AQ_2, &c.

14. AB, AC are two straight lines given in position; a straight line DE meets them in D, E, respectively, so that $AD + AE$ is a constant length; also DE is divided in the point P, so that DP bears a constant ratio to EP; the locus of P is a straight line.

15. If the angles of the triangle ABC are given, and A is fixed, while B moves along a fixed straight line, shew that the locus of C is a straight line.

*CHAPTER VI.

The Straight Line with Abridged Notation.

85. WE have shewn (Art. 43), that the equation

$$Ax + By + C + k(ax + by + c) = 0 \text{.........} (1)$$

can by varying k be made to represent *any* straight line passing through the intersection of the lines

$$Ax + By + C = 0,\ ax + by + c = 0 \text{.........} (2).$$

Let the symbols L and M stand for

$$Ax + By + C \text{ and } ax + by + c;$$

then equations (2) are written $L = 0$, $M = 0$, and equation (1) is written $L + kM = 0$. The lines $L = 0$, $M = 0$ are usually called 'the line (L),' 'the line (M),' and their point of intersection is called 'the point (L, M).'

Instead of the equation $L + kM = 0$, it is often more convenient to use the symmetrical form $lL + mM = 0$. This will obviously make no real difference in the equation, since we have simply substituted the arbitrary ratio $\frac{m}{l}$ for the arbitrary constant k.

86. Since (Art. 47) the equations to parallel straight lines may be written so as to differ only in their constant term, any straight line parallel to (L) can be written in the

form $L + c = 0$, where c is a constant. The lines $(L + c)$, $(L - c)$ are parallel to and equidistant from (L).

If any number of lines (L), (M), (N), &c. be parallel, then the equation

$$lL + mM + nN + \&c. = 0 \qquad (1)$$

will represent a straight line parallel to them; for (L), (M), (N), &c. will be of the forms

$$Ax + By + C = 0, \quad Ax + By + C' = 0, \quad Ax + By + C'' = 0, \text{ \&c.}$$

and equation (1) becomes

$$Ax + By + \frac{lC + mC' + nC'' + \&c.}{l + m + n + \&c.} = 0,$$

which represents a straight line parallel to the given lines.

Again, if (L), (M), (N), &c. pass through one point, (1) will *generally* represent a straight line passing through that point; for the co-ordinates of the point make $L = 0$, $M = 0$, $N = 0$ simultaneously, and therefore satisfy (1). We say *generally*, because l, m, n, *might* be such as to make the left-hand member of (1) vanish *identically*.

87. The following example will shew how equations to lines may be found in this system of Abridged Notation.

If (L), (M), (N) be the sides of a triangle ABC, opposite to A, B, C, and points D, E, F be taken in (L), (M), (N), respectively, to shew that the sides of the triangle DEF may be represented by

$$L + mM + \frac{N}{l} = 0, \qquad M + nN + \frac{L}{m} = 0, \qquad N + lL + \frac{M}{n} = 0,$$

where l, m, n are constants.

We may write the equations

to CF, $L + mM = 0$, to AD, $M + nN = 0$, to BE, $N + lL = 0$;

and we have for the points

$$D, \left.\begin{matrix} L = 0 \\ M + nN = 0 \end{matrix}\right\}; \qquad E, \left.\begin{matrix} M = 0 \\ N + lL = 0 \end{matrix}\right\}; \qquad F, \left.\begin{matrix} N = 0 \\ L + mM = 0 \end{matrix}\right\}.$$

Hence the equation to DE is, since it passes through D, of the form

$$M+nN+kL=0\text{..............................}(1);$$

and, since it passes through E, it is of the form

$$N+lL+k'M=0\text{..............................}(2);$$

and, comparing these forms, we have

$$\frac{k}{l}=\frac{1}{k'}=\frac{n}{1},$$

whence $k=nl$, $k'=\frac{1}{n}$, and (1) and (2) each become

$$N+lL+\frac{M}{n}=0.$$

Otherwise, after writing equation (1), we may proceed thus. Since (1) passes through E, the values $M=0$, $N+lL=0$ satisfy the equation, and therefore, substituting for N and M, we have, as before

$$nlL-kL=0 \text{ or } k=nl.$$

Similarly it may be shewn, that the equations to the other sides may be written as stated above.

88. If the equation to a straight line be written in the form

$$p-x\cos\alpha-y\sin\alpha=0,$$

so that (Art. 55) the origin is on the positive side of the line, it is usual to employ one of the Greek letters for its abbreviated form $\alpha=0$, so that all points, for which the expression α is positive, are on the origin side of the line. Let $\beta=0$, in like manner, be written for

$$q-x\cos\beta-y\sin\beta=0.$$

Then we have shewn (Art. 54) that the α and β of any point are its distances from the lines (α) and (β). We have shewn also (Art. 58) that the equations

$$\alpha-k\beta=0,\quad \alpha+k\beta=0,$$

are capable of a simple geometrical interpretation, and that they represent two straight lines drawn through the intersection of (α) and (β), so that the perpendiculars dropped from any point of either of them upon the lines (α), (β), are to one another as $k : 1$.

Particular attention must be paid to the position of the origin, and the rules laid down in Art. 59 for determining the sign of k. The positive and negative signs in the figure of Art. 58 are now reversed; but, exactly as in that article, for all lines which lie in the same angle as the origin and in the angle vertically opposite, the form is $\alpha - k\beta = 0$; for lines which lie in the other two angles the form is $\alpha + k\beta = 0$. The student must be careful to bear in mind, that these remarks apply to the particular form of equation only, which we denote by (α), (β), &c. Many problems occur, as above (Art. 87), where it is not necessary to introduce any limitation as to the forms of the equations or the position of the origin.

89. We have shewn in Art. 58, Cor., that, when $k = 1$, equations (7) and (8) of that article represent the bisectors of the supplementary angles between the lines; hence the equations to the bisectors of the angles between (α) and (β) are

$$\alpha - \beta = 0, \quad \alpha + \beta = 0.$$

A little consideration will shew that the lines ($\alpha - k\beta$), ($k\alpha - \beta$) are equally inclined to the line ($\alpha - \beta$); as are also ($\alpha + k\beta$), ($k\alpha + \beta$) to the line ($\alpha + \beta$).

90. The properties of the *Harmonic Pencil* will serve to illustrate this part of the subject.

DEF. Any four straight lines meeting in a point are called a *pencil* of four lines, and a straight line drawn across the pencil is called a *transversal*. A pencil is called *harmonic* if it divides any transversal *harmonically*, that is, so that *the whole line is to one extreme segment, as the other extreme segment is to the middle part.* The four points where a transversal meets the pencil are called a *Range.*

Let OA, OC, OB, OD be any pencil, and $ACBD$ a transversal; then we shall shew that the ratio $\frac{AD}{AC} \div \frac{BD}{BC}$ is constant, in whatever way AD be drawn across the system. If we denote the angles ABO, ACO, ADO, by B, C, D, we have

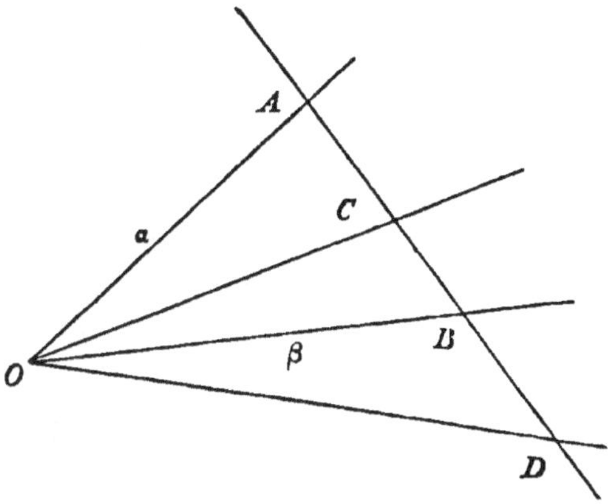

$$\frac{AD}{AC} = \frac{AD}{AO} \cdot \frac{AO}{AC} = \frac{\sin AOD}{\sin D} \cdot \frac{\sin C}{\sin AOC},$$

$$\frac{BD}{BC} = \frac{BD}{BO} \cdot \frac{BO}{BC} = \frac{\sin BOD}{\sin D} \cdot \frac{\sin C}{\sin BOC};$$

$$\therefore \frac{AD}{AC} \div \frac{BD}{BC} = \frac{\sin AOD}{\sin BOD} \div \frac{\sin AOC}{\sin BOC} \quad \text{.........(1)};$$

which is a ratio independent of the position of AD, and is called the *anharmonic* ratio of the pencil. When the right-hand member of equation (1) = 1, the transversal is *harmonically* divided, and the pencil is *harmonic.* We then have

$$AD : AC = BD : BC \quad \text{....................(2)},$$

and, if we consider AD, AB, AC as the first, second, and third quantities, respectively, equation (2) asserts that the first is to the third as the difference between the first and second is to the difference between the second and third, and the quantities are therefore in harmonical progression. The lines KO, KA, KP, KB, in the figure of Art. 84, form an harmonic pencil.

91. The lines

$$\alpha = 0, \quad \beta = 0, \quad \alpha - k\beta = 0, \quad \alpha + k\beta = 0,$$

form an harmonic pencil.

Let OA, OB, be the lines, $\alpha = 0$, $\beta = 0$, then, if we suppose the origin somewhere in the angle AOB, the lines $\alpha - k\beta = 0$, $\alpha + k\beta = 0$ will (Art. 88) lie as OC and OD respectively. Then we have

$$\frac{\sin AOC}{\sin BOC} = k = \frac{\sin AOD}{\sin BOD};$$

therefore, from Art. 90 (1),

$$AD : AC = BD : BC.$$

92. Art. 91 is equally true, if the lines are

$$L = 0, \quad M = 0, \quad L - kM = 0, \quad L + kM = 0,$$

for (Art. 27, Cor. 2) we have $L = \lambda\alpha$, $M = \mu\beta$, where λ and μ are constants; hence the equations

$$L - kM = 0 \text{ and } L + kM = 0,$$

may be written

$$\lambda\alpha - k\mu\beta = 0 \text{ and } \lambda\alpha + k\mu\beta = 0, \text{ or } \alpha - k'\beta = 0 \text{ and } \alpha + k'\beta = 0,$$

where k' is written for $\frac{k\mu}{\lambda}$. These lines therefore form an harmonic pencil with (α) and (β), that is with (L) and (M). The student must be careful not to assume that ($L - M$), ($L + M$) bisect the angles between (L) and (M); they do, however, form an harmonic pencil with them.

93. *If (L), (M), (N) be three straight lines, and we can find three constants l, m, n, such that*

$$lL + mM + nN = 0 \ldots\ldots\ldots\ldots\ldots\ldots (1),$$

identically, then all the three lines (L), (M), (N), pass through one point, or are parallel.

For, since (1) is true *identically*, that is, for all values of x and y, we have $L = -\frac{m}{l} M - \frac{n}{l} N$ *identically*, and therefore $L = 0$ is the equation to a straight line passing through the intersection of (M) and (N), if they meet, or (Art. 86) parallel to them, if they are parallel.

Ex. (1) *The three straight lines that bisect the angles of a triangle meet in a point.*

Taking the origin of co-ordinates within the triangle, let $\alpha = 0, \beta = 0, \gamma = 0$ be the equations to the three sides; then the equations to the straight lines bisecting the angles are

$$\beta - \gamma = 0, \quad \gamma - \alpha = 0, \quad \alpha - \beta = 0.$$

Here $l = m = n = 1$, and the lines therefore meet in a point.

Ex. (2) *If, through the angular points of a triangle, there be drawn any three straight lines meeting in a point, then three straight lines, drawn through the same angles, equally inclined to the bisectors of the angles, will also meet in a point.*

The first three lines may be represented by

$$m\beta - n\gamma = 0, \quad n\gamma - l\alpha = 0, \quad l\alpha - m\beta = 0,$$

putting $\frac{n}{m}$ for k &c.

Then (Art. 89), the equations to the other three will be

$$n\beta - m\gamma = 0, \quad l\gamma - n\alpha = 0, \quad m\alpha - l\beta = 0;$$

and multiplying these by $\frac{1}{mn}$, $\frac{1}{nl}$, and $\frac{1}{lm}$, respectively, they become

$$\frac{\beta}{m} - \frac{\gamma}{n} = 0, \quad \frac{\gamma}{n} - \frac{\alpha}{l} = 0, \quad \frac{\alpha}{l} - \frac{\beta}{m} = 0;$$

and therefore we see that these also pass through one point.

Ex. (3) *The straight lines joining the angular points of a triangle with the middle points of the opposite sides intersect in one point.*

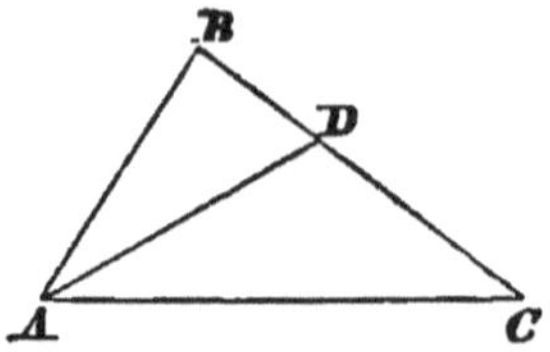

Let ABC be the triangle, and the origin as before within it: let $\alpha=0$, $\beta=0$, $\gamma=0$ be the equations to BC, CA, and AB. Then if D be the middle point of BC, the equation to AD is

$$\beta-\frac{\sin CAD}{\sin BAD}\gamma=0;$$

but $$\frac{\sin CAD}{\sin C}=\frac{CD}{AD}, \text{ and } \frac{\sin BAD}{\sin B}=\frac{BD \text{ or } CD}{AD},$$

therefore $$\frac{\sin CAD}{\sin BAD}=\frac{\sin C}{\sin B},$$

and the equation to AD becomes

$$\beta \sin B-\gamma \sin C=0.$$

Similarly the equations to the other two lines are

$$\gamma \sin C-\alpha \sin A=0, \quad \alpha \sin A-\beta \sin B=0,$$

and these three lines evidently pass through one point.

Ex. (4) It may be shewn in the same manner, that, if AD be perpendicular to BC, its equation is

$$\beta \cos B-\gamma \cos C=0,$$

and that those of the other perpendiculars are

$$\gamma \cos C-\alpha \cos A=0, \quad \alpha \cos A-\beta \cos B=0,$$

and these three pass through the same point.

94. *Trilinear Co-ordinates.*

The distances α, β, γ of a point from the three sides of a triangle, formed by the lines (α), (β), (γ), are called the *Trilinear Co-ordinates* of the point, and the triangle is called the *Triangle of reference.* By paying attention to the signs of α, β, γ, the position of a point may be defined by these distances, and the properties of lines investigated by means similar to those employed in the Cartesian System. In this system it is usual to consider α as positive, when P is on the *same* side of BC as A, and as negative when on the other side, so that the trilinear co-ordinates of a point within the triangle of reference are all positive. It will be seen that this is equivalent to considering the origin within the triangle, and the symbols α, &c. to stand for $p - x \cos \alpha - y \sin \alpha$, &c.

COR. By writing the above values for α, &c., any equation will be transformed from Trilinear to Cartesian co-ordinates.

95 (i). We proceed to shew the relation that must exist between the trilinear co-ordinates of a point. Let ABC be the triangle of reference, the lengths of whose sides are a, b, c. Take any point P within the triangle, and join it with the angular points; then the distances of P from BC, CA, AB are α, β, γ, the trilinear co-ordinates of the point; and the areas of the triangles PBC, PCA, PAB are

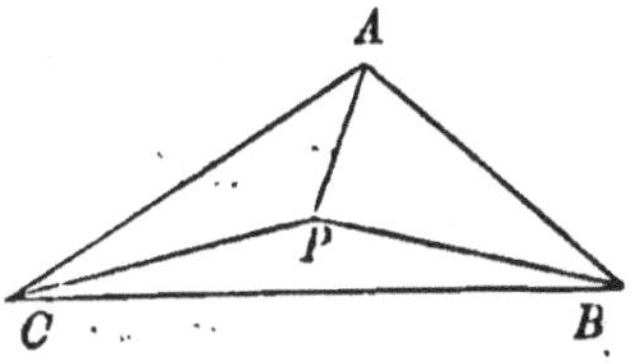

$$\frac{a\alpha}{2}, \frac{b\beta}{2}, \frac{c\gamma}{2}.$$

Hence, if Δ denote the area ABC, we have

$$a\alpha + b\beta + c\gamma = 2\Delta,$$

as a constant relation that must always subsist between the quantities α, β, γ. If the point be taken outside the triangle, the same relation will be seen to hold, by giving the proper sign to α, β, γ. It is clear that any point on BC gives $\alpha = 0$, which is therefore the equation to BC; similarly $\beta = 0$, $\gamma = 0$ are the equations to CA and AB. The point whose trilinear co-ordinates are α, β, γ may be called the point $(\alpha\beta\gamma)$: thus the point A is the point $\frac{2\Delta}{a}$, 0, 0.

95 (ii). It is important to observe, that a homogeneous equation in α, β, γ speaks, not of the actual values of these

quantities, but of their ratios; since every such equation may be written so as to involve the ratios only. Thus we may write

$$\alpha^3 - 3\alpha^2\beta + \gamma^3 = 0 \text{ or } \left(\frac{\alpha}{\gamma}\right)^3 - 3\left(\frac{\alpha}{\gamma}\right)^2 . \frac{\beta}{\gamma} + 1 = 0;$$

and we see that, if these equations are satisfied by α', β', γ', they are also satisfied by $k\alpha'$, $k\beta'$, $k\gamma'$. Such equations then involve *two* independent magnitudes only; and these can be at once eliminated from three simple equations.

If the ratios of the co-ordinates of any point be given, the actual values of the co-ordinates can be found from the relation

$$a\alpha + b\beta + c\gamma = 2\Delta,$$

as will be seen in the examples below. In practice, however, we rarely require the absolute values of the co-ordinates. It is usual to make all equations homogeneous by the method of Art. 96.

Ex. *To find the trilinear co-ordinates of the points of intersection in the Examples to Art.* 93.

In Ex. 1 we have at the point of intersection $\alpha = \beta = \gamma$;

therefore $$\frac{a\alpha}{a} = \frac{b\beta}{b} = \frac{c\gamma}{c} = \frac{a\alpha + b\beta + c\gamma}{a+b+c};$$

therefore $$\alpha = \beta = \gamma = \frac{2\Delta}{a+b+c}.$$

Compare this result with $r = \frac{S}{s}$, the formula for the radius of the circle inscribed in the triangle.

In Ex. 2 we have in like manner

$$\frac{a\alpha}{al} = \frac{b\beta}{bm} = \frac{c\gamma}{cn} = \frac{a\alpha + b\beta + c\gamma}{al + bm + cn};$$

therefore $$\frac{\alpha}{l} = \frac{\beta}{m} = \frac{\gamma}{n} = \frac{2\Delta}{al + bm + cn}.$$

The method is exactly the same for Ex. 3 and 4.

96. By means of the relation proved above (Art. 95) any equation in trilinear co-ordinates can be made homogeneous, by multiplying each term by $\frac{a\alpha + b\beta + c\gamma}{2\Delta}$ raised to a suitable power, since this quantity $= 1$. Thus, the equation to a straight line parallel to the line $l\alpha + m\beta + n\gamma = 0$,

is (Art. 86) $$l\alpha + m\beta + n\gamma + k = 0,$$

or as an equation homogeneous in α, β, γ,

$$2\Delta\,(l\alpha + m\beta + n\gamma) + k\,(a\alpha + b\beta + c\gamma) = 0.$$

Again, the equation $\alpha^2 + h\beta + k = 0$ becomes

$$4\Delta^2\alpha^2 + 2\Delta h\beta\,(a\alpha + b\beta + c\gamma) + k\,(a\alpha + b\beta + c\gamma)^2 = 0.$$

97. The equation to any locus can be transformed from Cartesian to Trilinear co-ordinates, without altering the degree of the equation. For (Art. 76 (i)) the axes may be made to coincide with two sides of the triangle of reference, CB, CA, without raising or depressing the equation; then, if x, y and α, β be the co-ordinates of any point P in the two systems, respectively, and the angle $ACB = \omega$, we have $x = \beta \operatorname{cosec} \omega$, $y = \alpha \operatorname{cosec} \omega$; hence any equation in x and y may be written as an equation of the same degree in α, β, and can be made homogeneous in α, β, γ, by the method of Art. 96. Hence the Cartesian equation to *any* straight line can be transformed into an equation of the form

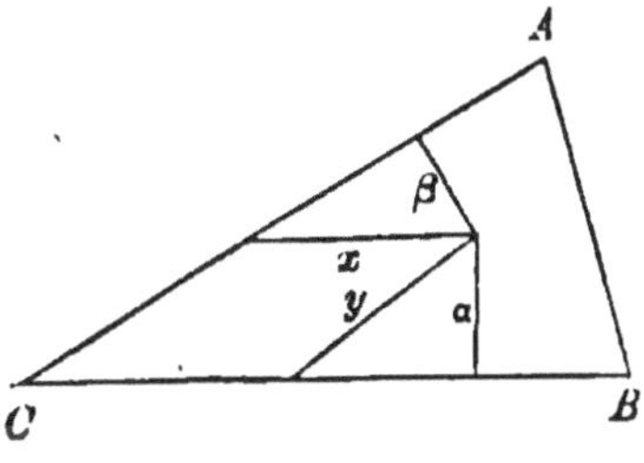

$$l\alpha + m\beta + n\gamma = 0,$$

that is to say, the general equation of the first degree in trilinear co-ordinates can be made to represent *any* straight line, by giving suitable values to l, m, n.

98. Conversely, with one exception, every equation of the form

$$l\alpha + m\beta + n\gamma = 0$$

will represent a straight line; for it may be written

$$\frac{l}{n}\alpha + \frac{m}{n}\beta + \frac{2\Delta - a\alpha - b\beta}{c} = 0,$$

or
$$\left(\frac{l}{n} - \frac{a}{c}\right)\alpha + \left(\frac{m}{n} - \frac{b}{c}\right)\beta + \frac{2\Delta}{c} = 0;$$

and by writing $\alpha = y \sin\omega$, $\beta = x \sin\omega$, as in Art. 97, we have an equation of the first degree in x and y, with CB, CA for axes. This will always represent a straight line, unless

$$\frac{l}{n} = \frac{a}{c}, \quad \frac{m}{n} = \frac{b}{c} \text{ or } l : m : n = a : b : c.$$

Cor. If the lines (L), (M), (N), are the sides of the triangle of reference, then, since (Art. 92) we may write

$$l\lambda\alpha + m\mu\beta + n\nu\gamma = 0 \text{ for } lL + mM + nN = 0,$$

it follows that any straight line can be represented by the latter equation, and that it will always represent a straight line, unless $l\lambda : m\mu : n\nu = a : b : c$.

99. We shall now examine the meaning of the equation

$$a\alpha + b\beta + c\gamma = 0, \text{ or } \alpha \sin A + \beta \sin B + \gamma \sin C = 0,$$

since these two are equivalent.

We saw (Art. 26) that the line $(Ax + By + C = 0)$ cuts off intercepts on the axes $-\frac{C}{A}$, $-\frac{C}{B}$; hence, if A and B become very small, the intercepts on the axes are very great. Let A and B each become indefinitely small, then the intercepts become indefinitely great, and the line is altogether at an infinite distance from the origin. The equation may now be written,

$$0 \cdot x + 0 \cdot y + C = 0,$$

which cannot be satisfied by any finite values of x and y, but may be satisfied by infinite values, since the product $0 \times \infty$ may be finite. We may express all this shortly by saying that the equation $C = 0$, that is, a constant $= 0$, *represents a straight line situated altogether at an infinite distance from the origin.* The direction of the line is wholly undetermined; and it must be clearly understood that the equation, impossible in itself and representing nothing, derives its meaning from the possible equation of which it is the limiting form. Similarly it would be absurd to say (Appendix IV.) that the equation $C = 0$ gave two infinite values of x, or that the equation $Bx + C = 0$ had one infinite and one finite root; but both these statements are intelligible, if these equations are the limiting forms of $Ax^2 + Bx + C = 0$, where A and B in the former and A in the latter are indefinitely small.

In the same way the equation

$$a\alpha + b\beta + c\gamma = 0$$

is in itself impossible, since we have proved that

$$a\alpha + b\beta + c\gamma$$

is a constant quantity, and cannot $= 0$; but the equation

$$l\alpha + m\beta + n\gamma = 0,$$

when the ratios $l : m : n$ approach indefinitely near to $a : b : c$, will represent *a straight line altogether at an infinite distance from the triangle of reference;* for when this equation is put in the form $Ax + By + C = 0$, as in Art. 98, the values of A and B become in this case indefinitely small, and the line, therefore, infinitely distant from the origin.

100 (i). *To find the condition that the two straight lines*

$$l\alpha + m\beta + n\gamma = 0 \text{..........................} (1),$$

$$l'\alpha + m'\beta + n'\gamma = 0 \text{..........................} (2),$$

should be parallel.

Suppose a third straight line

$$\lambda\alpha + \mu\beta + \nu\gamma = 0 \text{..........................} (3)$$

to pass through the point of intersection of (1) and (2). Then for this point (1), (2), (3) are true simultaneously; and a relation between the constants may be found, by solving (1) and (2) for $\frac{\alpha}{\gamma}$ and $\frac{\beta}{\gamma}$, and substituting the values so found in (3). Now suppose λ, μ, ν to become a, b, c; then the line (3), and therefore the point of intersection of (1) and (2), has moved off to an infinite distance, and the relation obtained is the condition of parallelism. It will be found to be

$$(mn' - m'n)\,a + (nl' - n'l)\,b + (lm' - l'm)\,c = 0.$$

100 (ii). *To find the condition that the two straight lines*

$$l\alpha + m\beta + n\gamma = 0 \text{......................} (1),$$

$$l'\alpha + m'\beta + n'\gamma = 0 \text{......................} (2),$$

should be perpendicular.

Suppose the origin of rectangular Cartesian co-ordinates to be somewhere within the triangle; then, if we transform the equations, as in Art. 94, Cor., using p, q, r and α, β, γ, (1) and (2) become

$$\left.\begin{array}{r} lp + mq + nr - l\cos\alpha \\ - m\cos\beta \\ - n\cos\gamma \end{array}\right| \left.\begin{array}{r} x - l\sin\alpha \\ - m\sin\beta \\ - n\sin\gamma \end{array}\right| y = 0 \text{.........} (3),$$

$$l'p+m'q+n'r - \left.\begin{array}{r} l'\cos\alpha \\ -m'\cos\beta \\ -n'\cos\gamma \end{array}\right| x - \left.\begin{array}{r} l'\sin\alpha \\ -m'\sin\beta \\ -n'\sin\gamma \end{array}\right| y = 0 \ldots\ldots\ldots(4).$$

Now (3) and (4) will be perpendicular, if the condition of Art. 47 ($Aa+Bb=0$) is fulfilled; that is, if

$$ll'+mm'+nn'+\left.\begin{array}{r} lm' \\ +l'm \end{array}\right| \cos(\alpha\sim\beta)+\left.\begin{array}{r} mn' \\ +m'n \end{array}\right| \cos(\beta\sim\gamma)+\left.\begin{array}{r} nl' \\ +n'l \end{array}\right| \cos(\gamma\sim\alpha)=0.$$

But it is easily seen from a figure, that

$$\cos(\alpha\sim\beta)=-\cos C,\ \cos(\beta\sim\gamma)=-\cos A,\ \cos(\gamma\sim\alpha)=-\cos B;$$

hence the condition becomes

$$ll'+mm'+nn'-\left.\begin{array}{r} lm' \\ -l'm \end{array}\right| \cos C-\left.\begin{array}{r} mn' \\ -m'n \end{array}\right| \cos A-\left.\begin{array}{r} nl' \\ -n'l \end{array}\right| \cos B=0.$$

101. *To find the perpendicular distance from a given point to a given straight line.*

Let $(\alpha'\beta'\gamma')$ be the given point, and

$$l\alpha+m\beta+n\gamma=0 \ldots\ldots\ldots\ldots\ldots\ldots\ldots(1)$$

the given straight line. Let (1) be transformed to equation (3), Art. 100 (ii), and let the point $(\alpha'\beta'\gamma')$ in (1) be the point $(x'y')$ in (3), so that $l\alpha'+m\beta'+n\gamma'$ would become the left-hand member of (3), written with x', y'. Then (Art. 52) the length of the perpendicular is

$$\frac{l\alpha'+m\beta'+n\gamma'}{\sqrt{(A^2+B^2)}},$$

where
$$-A=l\cos\alpha+m\cos\beta+n\cos\gamma,$$
$$-B=l\sin\alpha+m\sin\beta+n\sin\gamma.$$

Hence $A^2 + B^2 =$

$$l^2 + m^2 + n^2 + 2lm\cos(\alpha \sim \beta) + 2mn\cos(\beta \sim \gamma) + 2nl\cos(\gamma \sim \alpha)$$

$$= l^2 + m^2 + n^2 - 2lm\cos C - 2mn\cos A - 2nl\cos B.$$

102. *To find the equation to a straight line which passes through the points* $(\alpha'\beta'\gamma')$, $(\alpha''\beta''\gamma'')$.

Let the equation to the line be

$$l\alpha + m\beta + n\gamma = 0 \quad \text{.....................(1)};$$

then

$$l\alpha' + m\beta' + n\gamma' = 0 \quad \text{.....................(2)},$$

and

$$l\alpha'' + m\beta'' + n\gamma'' = 0 \quad \text{.....................(3)}.$$

Eliminating n and m successively from (2) and (3), we have

$$\frac{l}{\beta'\gamma'' - \beta''\gamma'} = \frac{m}{\gamma'\alpha'' - \gamma''\alpha'} = \frac{n}{\alpha'\beta'' - \alpha''\beta'},$$

and the required equation is

$$(\beta'\gamma'' - \beta''\gamma')\,\alpha + (\gamma'\alpha'' - \gamma''\alpha')\,\beta + (\alpha'\beta'' - \alpha''\beta')\,\gamma = 0.$$

103. *To find the equations to a straight line in the form*

$$\frac{\alpha - \alpha'}{\lambda} = \frac{\beta - \beta'}{\mu} = \frac{\gamma - \gamma'}{\nu} = \rho \quad \text{.................(1)},$$

where $(\alpha'\beta'\gamma')$ is a fixed point on the line, and l is the distance between $(\alpha\beta\gamma)$ and $(\alpha'\beta'\gamma')$, changing sign as in Art. 34.

Let O be the fixed point $(\alpha'\beta'\gamma')$ on the line OQ whose equation is required, and let P $(\alpha\beta\gamma)$ be any other point on it; draw Oa, Ob, Oc parallel to BC, CA, AB, respectively, and so that the angles bOc, cOa, aOb may be the supplements of the angles A, B, C, respectively. Let θ, ϕ, ψ denote the angles POa, POb, POc, all measured in the same direction from OQ as initial line, as in the figure. Draw PA', PB', PC' perpendicular to Oa, Ob, Oc, or those lines produced; then

$$\begin{aligned} \alpha-\alpha' &= PA' = OP \sin POA' = OP \sin\theta, \\ \beta-\beta' &= PB' = OP \sin POB' = OP \sin\phi, \\ \gamma-\gamma' &= -PC' = -OP \sin POC' = OP \sin\psi; \end{aligned}$$

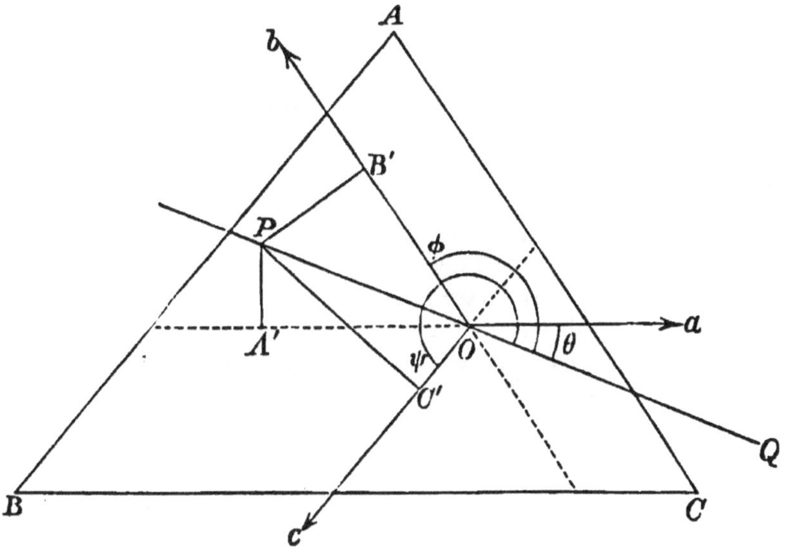

hence $$\frac{\alpha-\alpha'}{\sin\theta}=\frac{\beta-\beta'}{\sin\phi}=\frac{\gamma-\gamma'}{\sin\psi}=OP.$$

If we take P on the other side of O, each of these ratios $=-OP$; hence the required equations are

$$\frac{\alpha-\alpha'}{\sin\theta}=\frac{\beta-\beta'}{\sin\phi}=\frac{\gamma-\gamma'}{\sin\psi}=\rho,$$

where ρ is positive or negative. The denominators in these equations are called the *direction sines* of the line. The symbols λ, μ, ν, *when used with these equations*, will always mean these sines. If l, m, n are used, where $l : m : n = \lambda : \mu : \nu$, the ratios are still equal, but they are not then $=\rho$.

COR. Since each of the above ratios is equal to

$$\frac{a\alpha + b\beta + c\gamma - (a\alpha' + b\beta' + c\gamma')}{a\lambda + b\mu + c\nu},$$

the numerator of which is zero, it follows that

$$a\lambda + b\mu + c\nu = 0,$$

a relation between λ, μ, ν.

104 (i). The two following examples are good illustrations of the methods of this chapter. The latter deserves especial notice.

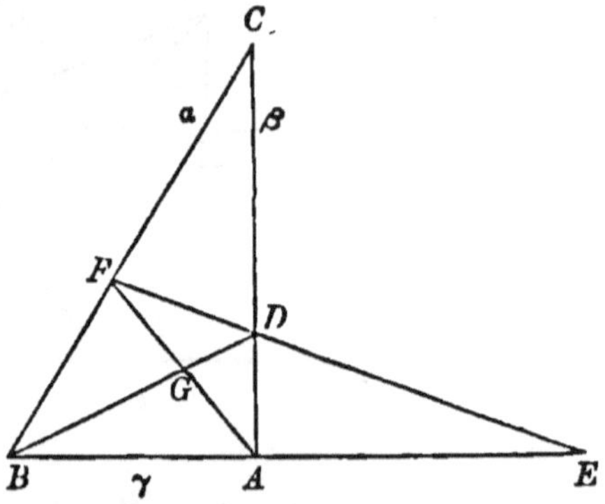

Ex. 1. Let A, B, F, D be any four points in a plane; then three pairs of straight lines can be drawn, so that each pair includes all four points. Let these pairs be BC, AC; BE, FE; BD, AF. The figure is now called a *complete* quadrilateral.

Let ABC be the triangle of reference; then we may write the equation to BD

$$l\alpha - n\gamma = 0 \text{............(1)},$$

since it passes through (α, γ), and any value may be given to the arbitrary ratio $\frac{n}{l}$. Similarly the equation to AF may be written

$$m\beta - n\gamma = 0 \text{............................(2)}.$$

Hence the equation to FE since it passes through F, the intersection of (α) and $(m\beta - n\gamma)$, may be written

$$m\beta - n\gamma + k\alpha = 0 \text{......................(3)},$$

and since it passes through D, the intersection of (β) and $(l\alpha - n\gamma)$, when $\beta = 0$ in (3), $l\alpha = n\gamma$: hence, substituting for β and γ, we have

$$-l\alpha + k\alpha = 0 \text{ or } k = l,$$

and the equation to FE is

$$l\alpha + m\beta - n\gamma = 0 \text{........................(4)}.$$

Similarly the equation to CG, which passes through C (α, β) and G $(l\alpha - n\gamma, m\beta - n\gamma)$, is

$$l\alpha - m\beta = 0 \qquad\qquad (5),$$

and the equation to CE, which passes through C (α, β) and E $(l\alpha + m\beta - n\gamma, \gamma)$, is

$$l\alpha + m\beta = 0 \qquad\qquad (6).$$

Hence (Art. 91) CB, CA, CG, CE form an harmonic pencil, for their equations are

$$\alpha = 0, \quad \beta = 0, \quad l\alpha - m\beta = 0, \quad l\alpha + m\beta = 0.$$

We leave it to the student to prove that EB, EF, EG, EC and GF, GD, GC, GE form harmonic pencils. We might with equal propriety have used the abbreviations L, M, N instead of the trilinear co-ordinates α, β, γ.

Ex. 2. *If there be two triangles ABC, abc, such that the intersections of the corresponding sides lie in a straight line, then the straight lines joining the corresponding angles will meet in a point, and conversely.*

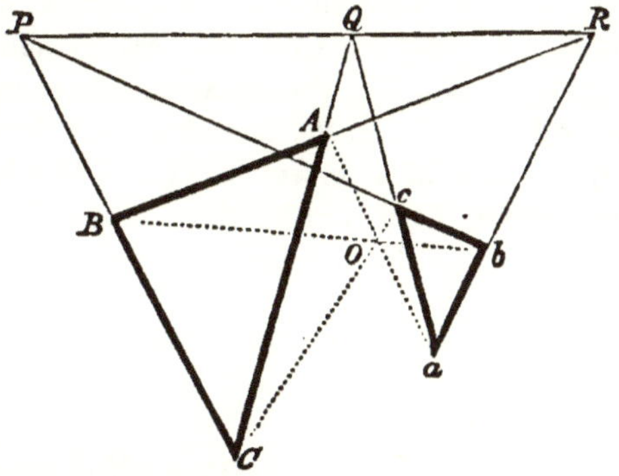

Let P, Q, R, lying in one straight line, be the intersections of the corresponding sides. Take ABC as the triangle of reference, and let the equation to PQR be

$$l\alpha + m\beta + n\gamma = 0 \ldots (PQR).$$

Then the equation to bc, since it passes through the intersection of PQR and BC, is of the form

$$l\alpha + m\beta + n\gamma + k\alpha = 0,$$

or

$$l'\alpha + m\beta + n\gamma = 0 \qquad\qquad (bc).$$

Similarly the equations to ca and ab are

$$l\alpha + m'\beta + n\gamma = 0 \qquad\qquad (ca),$$

$$l\alpha + m\beta + n'\gamma = 0 \qquad\qquad (ab).$$

From equations (bc), (ca) we obtain by subtraction

$$(l - l')\,\alpha - (m - m')\,\beta = 0 \qquad\qquad (Cc),$$

which therefore represents a straight line passing through c, the intersection of bc and ca. But it also represents a straight line passing through the intersection of (α) and (β), i.e. through C. Hence it is the equation to Cc.

Similarly the equations to Aa, Bb are

$$(m-m')\beta-(n-n')\gamma=0\text{.........}(Aa),$$

and
$$(n-n')\gamma-(l-l')\alpha=0\text{.........}(Bb),$$

and therefore (Art. 93) Aa, Bb, Cc meet in a point.

Conversely, suppose Aa, Bb, Cc to meet in O, and let the equation to PQ, the straight line joining the intersections of BC, bc and CA, ca, be

$$l\alpha+m\beta+n\gamma=0\text{....................}(PQ).$$

Then the equations to bc and ca will be (bc) and (ca) as above. We shall shew that the equation to ab is (ab). As above, the equation to COc is (Cc); and therefore at the point O we have

$$(l-l')\alpha=(m-m')\beta=(n-n')\gamma \text{ suppose.}$$

Hence the equations to AOa, BOb will be (Aa) and (Bb). Now the line represented by the equation (ab) passes through b, the intersection of BOb and bc; for it is obtained from the equations (Bb) and (bc) by subtraction; also it passes through a, the intersection of AOa and ca; for it is obtained from the equations (Aa) and (ca) by addition. Hence the equation (ab) represents ab. But the equation (ab) is evidently satisfied when $\gamma=0$ and $l\alpha+m\beta+n\gamma=0$, i.e. the line ab passes through the intersection of AB and PQ; or AB, ab intersect in PQ. The triangles ABC, abc are said to be *homologous; PQR* is called the *axis* and O the *centre of homology.*

104 (ii). *Areal Co-ordinates.*

The relation

$$a\alpha+b\beta+c\gamma=2\Delta,$$

between the trilinear co-ordinates of a point P, may be written

$$\frac{a\alpha}{2\Delta}+\frac{b\beta}{2\Delta}+\frac{c\gamma}{2\Delta}=1.$$

If therefore x, y, z denote the three ratios

$$\frac{a\alpha}{2\Delta},\ \frac{b\beta}{2\Delta},\ \frac{c\gamma}{2\Delta},$$

they will be subject to the simple relation

$$x+y+z=1\text{......................}(1).$$

Since x, y, z bear a constant ratio to α, β, γ, they may be used [Art. 95 (ii)] as the co-ordinates of P. They are called *areal* co-ordinates, since they represent the ratios of the triangles PBC, PCA, PAB, respectively, to the triangle ABC; and, on account of the simplicity of the relation (1), they may often be used with advantage.

In the use of these co-ordinates the same convention must be adopted with regard to algebraic sign, as in Art. 94. Thus the triangle PBC will be considered positive when it lies on the same side of BC as does the triangle of reference, and so for the other triangles.

The areal and trilinear co-ordinates of any point are connected by the relations

$$\frac{x}{a\alpha}=\frac{y}{b\beta}=\frac{z}{c\gamma}=\frac{1}{2\Delta},$$

so that we can transform any equation or expression from the one system to the other.

It follows from the above relation between the co-ordinates of the two systems, that an equation of the first degree in areal co-ordinates represents a straight line; also, if the same straight line be represented by

$$l\alpha+m\beta+n\gamma=0,$$

$$l'x+m'y+n'z=0,$$

we have
$$\frac{l}{l'a}=\frac{m}{m'b}=\frac{n}{n'c}.$$

Hence we may obtain the relation between the coefficients of the areal equation, which corresponds to any given relation between the coefficients in the trilinear equation.

Ex. 1. The straight lines drawn through the angular points of a triangle bisecting the opposite sides are (Art. 93, Ex. 3) represented by

$$y-z=0, \qquad z-x=0, \qquad x-y=0;$$

the internal bisectors of the angles (Art. 93, Ex. 1), by

$$\frac{y}{b}-\frac{z}{c}=0, \quad \frac{z}{c}-\frac{x}{a}=0, \quad \frac{x}{a}-\frac{y}{b}=0;$$

the straight line at infinity (Art. 99) by

$$x+y+z=0.$$

Ex. 2. The condition that the two straight lines

$$lx+my+nz=0,$$
$$l'x+m'y+n'z=0,$$

should be parallel (Art. 100 (i)), is

$$mn'-m'n+nl'-n'l+lm'-l'm=0.$$

Similarly, from Art. 100 (ii), may be obtained the condition of perpendicularity.

EXAMPLES VI.

The triangle ABC is supposed to be the triangle of reference.

1. Find the equation to a straight line through the vertex A of a triangle, parallel to the base BC.

2. Find the equation to the straight line joining the middle points of AB, AC.

3. Find the equation to the straight line passing through the point $(\alpha'\beta'\gamma')$, and parallel to $(l\alpha+m\beta+n\gamma=0)$.

4. Find the equation to the straight line joining the feet of the perpendiculars from A and B on BC and CA respectively.

5. If the lines represented by the equations

$$(a-b')(\alpha-b\beta)+(b-a)\gamma=0,$$
$$(b-b')(\alpha-a\beta)+(a-b)\gamma=0,$$

intersect in the line $\alpha-a'\beta=0$, shew that the following relation holds amongst the constants:

$$(a'+b')(a+b)=2(a'b'+ab).$$

6. Find the co-ordinates of the point of intersection of the lines $l\alpha+m\beta+n\gamma=0$ and $l'\alpha+m'\beta+n'\gamma=0$.

7. Find the condition that the three points

$$(\alpha'\beta'\gamma'),\ (\alpha''\beta''\gamma''),\ (\alpha'''\beta'''\gamma''')$$

may lie on one straight line.

8. Shew that the straight line, drawn through A parallel to BC, and the bisector of BC from A, form an harmonic pencil with AB and AC.

9. Interpret the equation $a\alpha + b\beta = 0$.

10. In Art. 87 find the condition that AD, BE, CF may pass through one point.

11. Find the equation to a straight line bisecting BC at right angles.

12. The straight lines bisecting the three sides of a triangle at right angles meet in a point.

13. Form the equation to a perpendicular to BC from C.

14. Find the condition that the equations

$$\frac{\alpha}{l} = \frac{\beta}{m},\quad \frac{\beta}{m} = \frac{\gamma}{n},\quad \frac{\gamma}{n} = \frac{\alpha}{l}$$

may represent three parallel straight lines.

15. The straight line joining the middle points of the sides of a triangle is parallel to the base.

16. Find the condition that the straight line,

$$l\alpha + m\beta + n\gamma = 0,$$

should be parallel to the bisector of the angle A of the triangle of reference.

17. If the line $l\alpha + m\beta + n\gamma = 0$ is drawn across the triangle of reference, shew that l, m, n cannot all have the same sign. If it meets AB, AC, not produced, in M, N, find the lengths of AM, AN.

18. From the angles A, B, C of any triangle are drawn three straight lines AA', BB', CC', bisecting the angles; through A, B, C are drawn three straight lines perpendicular to AA', BB', CC', to meet BC, CA, AB, produced, in G, H, K; G, H, K are in one straight line.

19. ABC is a triangle, D and E are points within the triangle, such that the angle $ABE = CBD$, and $BCD = ACE$; shew that $BAD = CAE$.

20. Interpret the equations

$$\alpha + \beta + \gamma = 0, \quad \beta + \gamma - \alpha = 0, \quad \gamma + \alpha - \beta = 0, \quad \alpha + \beta - \gamma = 0.$$

21. The four angles of a quadrilateral $ABCD$ are bisected by four straight lines; the bisectors of A, B meet in E, of B, C in F, of C, D in G, of D, A in H. Prove that the directions of EG and FH pass through the intersection of the directions of AD, BC, and AB, CD respectively.

22. From the angles of a triangle ABC straight lines are drawn through a given point O within the triangle, to meet the opposite sides in E, F, G; FG, GE, EF are produced to meet BC, CA, AB, in P, Q, R; shew that P, Q, R lie in one straight line.

23. If two similar triangles have their homologous sides parallel, the straight lines which join the equal angles meet in a point.

24. Shew that the result of Ex. 5, page 42, may be obtained by means of Art. 93.

25. Shew that the trilinear co-ordinates of the centre of the circle circumscribed about the triangle of reference, are given by the equations

$$\frac{\alpha}{\cos A} = \frac{\beta}{\cos B} = \frac{\gamma}{\cos C} = \frac{abc}{4\Delta}.$$

26. If O be the centre of the circle circumscribed about the triangle of reference, and if AO, BO, CO be produced to meet the opposite sides in A', B', C', shew that three of the four straight lines represented by the equations

$$\alpha \sec A \pm \beta \sec B \pm \gamma \sec C = 0$$

are the sides of the triangle $A'B'C'$. Shew that, if BC, $B'C'$ meet in P; CA, $C'A'$ in Q; AB, $A'B'$ in R; P, Q, R will lie on the fourth straight line.

CHAPTER VII.

The Circle.

105. We have seen (Art. 61), that the general equation of the second degree

$$Ax^2 + 2Hxy + By^2 + 2Gx + 2Fy + C = 0$$

may sometimes represent two straight lines. Before examining generally all the loci represented by it, we shall shew that certain particular forms of it are capable of being interpreted, and will represent circles. We shall afterwards see that the circle is a particular case of a class of curves represented by the general equation; and that the forms which we interpret are particular cases of one of the classes of equations into which we shall divide it. We adopt this plan on account of the simplicity of the circle, and because the reader is already familiar with its principal properties, geometrically treated.

106. *To find the equation to a circle whose centre and radius are given, the co-ordinates being rectangular.*

If C (ab) be the centre of the circle, P any point (xy) on the circumference, and CR be drawn parallel to Ox to meet the ordinate of P in R, we have

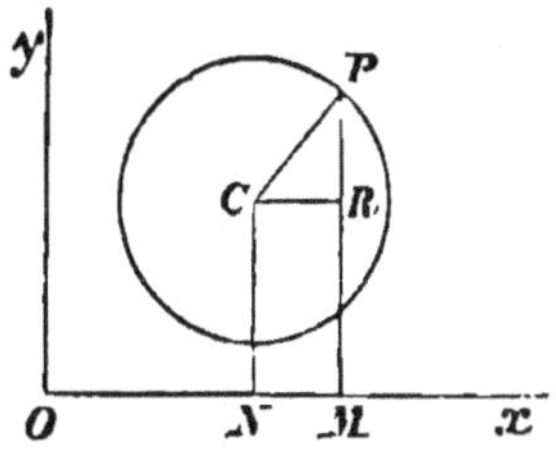

$$CR^2 + PR^2 = CP^2,$$

or $$(x - a)^2 + (y - b)^2 = r^2,$$

where r = the radius of the circle.

This follows directly from Art. 7; and in fact the equation only asserts, that the distance between the points (ab) and (xy) is constant and equal to r.

If the co-ordinates be oblique, and inclined to one another at an angle $= \omega$, we have, since angle CRP now $= 180^\circ - \omega$,

$$(x-a)^2 + (y-b)^2 + 2(x-a)(y-b)\cos\omega = r^2;$$

but we shall seldom have occasion to use this equation.

COR. Expanding the general equation to the circle referred to rectangular axes, we have

$$x^2 + y^2 - 2ax - 2by + a^2 + b^2 - r^2 = 0;$$

and hence it appears that the general equation to the circle is of the form

$$x^2 + y^2 + 2Gx + 2Fy + C = 0,$$

G, F, C being any constants. The equation

$$Ax^2 + Ay^2 + 2Gx + 2Fy + C = 0,$$

may be reduced to this form by dividing by A, and is therefore the most general form that the equation can assume, when the co-ordinates are rectangular.

107. Hence if we can reduce an equation to the form

$$x^2 + y^2 + 2Gx + 2Fy + C = 0,$$

we may always interpret it; for, adding $G^2 + F^2$ to both sides of the equation, we have

$$(x+G)^2 + (y+F)^2 = G^2 + F^2 - C.$$

I. If $G^2 + F^2 - C$ is positive, this is the equation to a circle, the co-ordinates of whose centre are

$x = -G$, $y = -F$, and whose radius $= (G^2 + F^2 - C)^{\frac{1}{2}}$.

II. If the quantity G^2+F^2-C be $=0$, the equation may be considered to represent a circle with an infinitely small radius, or (Art. 65) two imaginary straight lines

$$x+G+\sqrt{-1}\,(y+F)=0,\quad x+G-\sqrt{-1}\,(y+F)=0,$$

which intersect in the only real point for which the equation is satisfied, namely the point $(-G, -F)$.

III. If G^2+F^2-C is negative, there are (Art. 66) no values of x and y that can satisfy the equation, and the circle is imaginary.

Ex. 1. The equation $x^2+y^2-2x+4y+1=0$, may be written

$$(x-1)^2+(y+2)^2=4,$$

which represents a circle, the co-ordinates of whose centre are $x=1$, $y=-2$, and whose radius $=2$.

Ex. 2. The equation $x^2+y^2+2x-6y+10=0$, may be written

$$(x+1)^2+(y-3)^2=0,$$

a circle, the co-ordinates of whose centre are $x=-1$, $y=3$, and whose radius $=0$.

Ex. 3. $x^2+y^2+2x+6y+11=0$, or $(x+1)^2+(y+3)^2=-1$, represents an imaginary circle.

108 (i). If in the equation $(x-a)^2+(y-b)^2=r^2$, $a=0$, $b=0$, or the centre of the circle be origin, the equation becomes

$$x^2+y^2=r^2.$$

If $a=r$, $b=0$, or a diameter be chosen as axis of x, and its extremity as origin, the equation becomes

$$x^2-2rx+y^2=0;$$

and similarly, if the axis of y be a diameter, and the origin at its extremity, the equation is

$$x^2+y^2-2ry=0.$$

108 (ii). It may be observed here, that in the circle, as well as every other curve, if the origin is on the curve, there

will be no term which does not involve either x or y; for the equation must be satisfied by the values $x=0$, $y=0$, which cannot be the case, if there be a term which does not vanish when x and y vanish.

*109. From the figure of Art. 106, we see that, if C is origin, and the angle $PCR=\theta$,

$$x=r\cos\theta,\quad y=r\sin\theta;$$

or, if O is origin,

$$x=a+r\cos\theta,\quad y=b+r\sin\theta.$$

Thus the co-ordinates of the point P are expressed in terms of a single variable.

Ex. *To find the equation to the chord of a circle which passes through two points, defined by a single variable.*

Let the points be

$$r\cos\theta,\ r\sin\theta;\quad r\cos\phi,\ r\sin\phi.$$

The equation to the chord (Art. 35) is

$$\frac{x-r\cos\theta}{y-r\sin\theta}=\frac{r\cos\phi-r\cos\theta}{r\sin\phi-r\sin\theta},$$

$$=-\frac{\sin\frac{\theta+\phi}{2}}{\cos\frac{\theta+\phi}{2}},$$

whence $$\frac{x}{r}\cos\frac{\theta+\phi}{2}+\frac{y}{r}\sin\frac{\theta+\phi}{2}=\cos\theta\cos\frac{\theta+\phi}{2}+\sin\theta\sin\frac{\theta+\phi}{2},$$

$$=\cos\frac{\theta-\phi}{2}\ \ldots\ldots\ldots\ldots\ldots\ldots(1).$$

If $\theta=\phi$, the equation becomes

$$\frac{x}{r}\cos\phi+\frac{y}{r}\sin\phi=1\ \ldots\ldots\ldots\ldots\ldots\ldots(2),$$

the equation to a line which is defined below (Art. 115) as the *tangent* at the point $(r\cos\phi,\ r\sin\phi)$.

110. If we expand the equation to the circle (Art. 106), referred to oblique axes, we obtain

$$x^2 + 2\cos\omega \,.\, xy + y^2 - 2(a + b\cos\omega)\,x - 2(b + a\cos\omega)\,y$$
$$+ a^2 + b^2 + 2ab\cos\omega - r^2 = 0 \ldots\ldots (1).$$

Hence, if the inclination of the axes be ω, the general equation to the circle is

$$x^2 + 2xy\cos\omega + y^2 + 2Gx + 2Fy + C = 0 \ldots\ldots\ldots (2),$$

where G, F, C are constants.

COR. In order then that the general equation

$$Ax^2 + 2Hxy + By^2 + 2Gx + 2Fy + C = 0,$$

may represent a circle, we must have

$$A = B,\ \ H = A\cos\omega,$$

where ω is the angle between the axes; for then, by dividing by A, the equation can be reduced to the form of (2).

EX. *To determine the inclination of the co-ordinate axes, in order that the equation*

$$x^2 - xy + y^2 - kx - ky = 0$$

may represent a circle, and to find the magnitude of its radius.

Comparing the equation with equation (1), we have the equations

$$2\cos\omega = -1,\quad a^2 + b^2 + 2ab\cos\omega - r^2 = 0,$$
$$2(a + b\cos\omega) = k = 2(b + a\cos\omega),$$

from which we obtain

$$\omega = \frac{2}{3}\pi,\quad a = b = r = k.$$

111. The equation

$$x^2 + y^2 = r^2$$

will give us a well-known property of the circle; for it may be obtained by eliminating k, by means of multiplication, from the equations

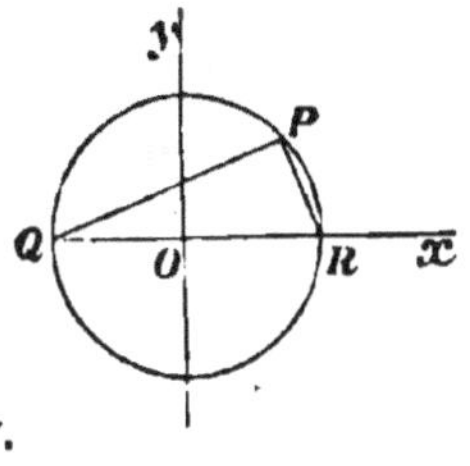

$$y = k(x - r) \ldots\ldots\ldots\ (1),$$
$$y = -\frac{1}{k}(x + r) \ \ldots\ldots\ (2),$$

where k is a constant and perfectly arbitrary.

But these equations evidently represent straight lines which

(i) pass through the extremities of the diameter QR, which is the axis of x; for (Art. 43) equation (1) represents a line passing through the intersection of the lines $y=0$, $x-r=0$, which is the point R; and (2) passes through the intersection of $y=0$, $x+r=0$, which is the point Q.

(ii) intersect in the circle, since by eliminating k between them we have the equation to the circle; and

(iii) are at right angles to one another by Art. 47; and they represent all lines which fulfil these three conditions. Hence we see that the locus of the vertices of all right-angled triangles on QR as base, is the semicircle QPR.

112. *To find the equation to the circle referred to polar co-ordinates.*

Let Ox be the initial line, O the pole; let the co-ordinates of the centre C be the known quantities ρ', θ', and of any point P in the circumference, ρ, θ; then

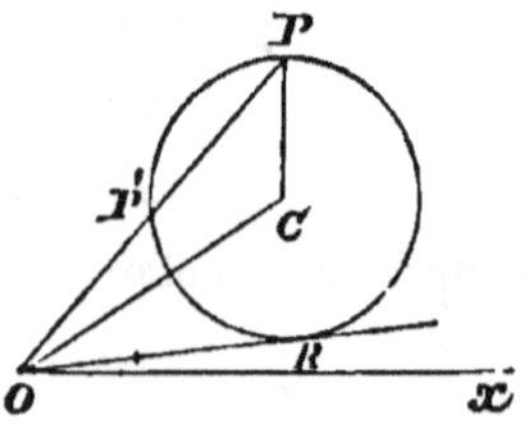

$$PO^2 + CO^2 - 2\,PO\,.\,CO\,.\cos POC = CP^2,$$

or $\rho^2 + \rho'^2 - 2\rho\rho' \cos(\theta - \theta') = r^2$,

which is the polar equation required.

It will be seen that this is the formula of Art. 14, and only asserts that the distance between the points, whose polar co-ordinates are ρ, θ, and ρ', θ', is constant and equal to r.

Cor. 1. The two values of ρ which may be found from the equation

$$\rho^2 - 2\rho' \cos(\theta - \theta')\,\rho + \rho'^2 - r^2 = 0,$$

are the two distances from the pole O of the points P, P', where the radius vector, which makes an angle θ with Ox, cuts the circle. The product of the roots of this equation (Appendix) $= \rho'^2 - r^2$, a quantity which does not change for different values of θ. Hence the rectangle $OP.OP'$ is constant for all positions of OP. When the roots are equal, or the line touches the circle, as OR, we have

$$\rho'^2 - r^2 = OR^2.$$

Hence $OP.OP' = OR^2$, as in Euc. III. 35, 36.

Cor. 2. If $\rho' = 0$, or the centre be pole, the equation becomes

$$\rho = r.$$

Cor. 3. If $\rho' = r$, and $\theta' = 0$, or a diameter be the initial line, and one extremity of it the pole, the equation becomes

$$\rho = 2r \cos\theta.$$

The reader will do well to verify by geometrical figures the results obtained here and in Art. 108.

113. In the foregoing articles we have assumed one only of the well-known geometrical properties of the circle, viz. that the distance from the centre to the circumference is constant, and from this property we have deduced the equation. Most of the following articles will admit of being proved in a very simple manner by those properties of the circle with which the reader is familiar; but we prefer to deduce our proofs from the equation alone, because this method is the same as that which we shall use in the case of other curves; and it is desirable that the student should perceive, that all the properties of the circle may be obtained from its equation, without any previous acquaintance with the curve. It will, however, be an exercise very profitable to the student, if he endeavour to deduce the equations of the following articles from any of the properties of tangents, &c. which he may find in Euclid.

114. *To find the length of a straight line drawn from a point $(x'y')$ to meet the circle.*

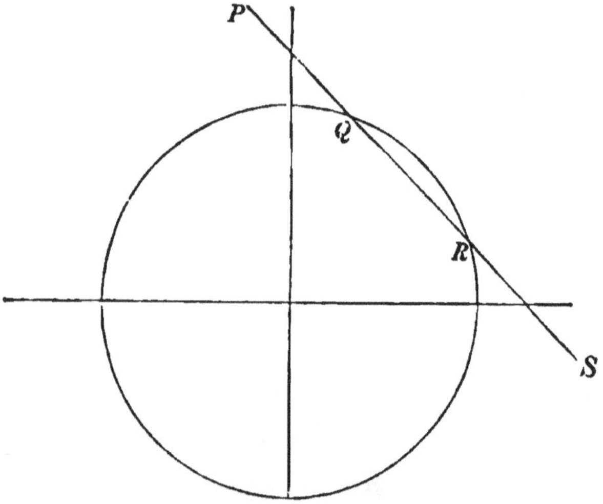

Let $PQRS$ be a line whose equation is

$$\frac{x-x'}{c}=\frac{y-y'}{s}=l \quad \text{.....................(1)},$$

drawn from the point $P(x'y')$ to cut the circle

$$x^2+y^2=r^2 \quad \text{..........................(2)}.$$

Now, if we substitute for x and y from (1) in (2), the result will be a quadratic in l, the two roots of which will be the distances PQ, PR of the point $(x'y')$ from the points where PS cuts the circle. As we have drawn the figure, the two values of l are (Art. 34, Cor.) negative; if P were to lie between Q and R, one root would be positive and the other negative: if P were to lie in RS, both roots would be positive. Making the substitution, we have

$$(cl+x')^2+(sl+y')^2=r^2,$$

therefore, since $$s^2+c^2=1,$$

$$l^2+2(cx'+sy')\,l+x'^2+y'^2-r^2=0 \quad \text{........(3)},$$

or $$l^2+Ql+R=0,$$

an equation which will always give *two* values for l; hence every straight line meets the circle in two real, coincident, or imaginary points, according as the roots of (3) are real and unequal, real and equal, or imaginary.

115. If the points of section Q, R, become coincident, by Q remaining fixed, till R, moving along the curve, approaches indefinitely near to Q, the line will be a tangent according to Euclid's definition; for it is of indefinite length, and meets the curve in *one point only*. We shall, however, find it convenient to adopt language similar to that used in Algebra, and to speak of *two coincident points*, just as we speak of *two equal* roots, and do not call them *one* root. We shall then take the following as our *definition* of a tangent, since it is found more convenient than Euclid's, when we treat curves by Algebraic methods.

DEF. If two points be taken on a curve, and a chord drawn through them; then, if the first point remains fixed, while the second, moving along the curve, approaches indefinitely near to the first, the chord in its limiting position is called the tangent to the curve at the first point.

116. We shall have occasion to consider the following particular forms which equation (3) of Art. 114 may assume.

If $R=0$, the point $(x'y')$ is on the circle, and one value of l becomes $=0$.

If $R=0$ and $Q=0$, both values of l become $=0$, and the line passes through two coincident points of the circle and is a tangent.

If $Q=0$, the roots of the equation are (Appendix) equal and of opposite signs, and $(x'y')$ is therefore the middle point of the chord.

117. *To find the equation to a straight line touching the circle at a point* $(x'y')$.

Let PS (fig. Art. 114) be the line

$$\frac{x-x'}{c}=\frac{y-y'}{s}=l \qquad \text{.....................(1)},$$

cutting the circle

$$x^2+y^2=r^2 \qquad \text{........................(2)},$$

in the point Q $(x'y')$. Then for the distances (l) between $(x'y')$ and the points of section of the line and circle, we obtain as in Art. 114, the equation

$$l^2+2(cx'+sy')\,l+x'^2+y'^2-r^2=0,$$

or

$$l^2+2(cx'+sy')\,l=0 \qquad \text{...................(3)},$$

since $(x'y')$ is on the circle, and therefore

$$x'^2+y'^2=r^2.$$

Equation (3) gives us $l=0$ (as it should, for $(x'y')$ coincides with one of the points of section), and also

$$l+2(cx'+sy')=0,$$

the value of l in which is the distance QR. But, if we suppose the point R to move up to Q, this distance vanishes, and the line becomes a tangent at Q $(x'y')$; and we have, as the condition that (1) should be tangent,

$$cx'+sy'=0 \qquad \text{......................(4)}.$$

Eliminating c and s by means of this equation and the equation to the line, we have

$$(x-x')\,x'+(y-y')\,y'=0 \qquad \text{.................(5)},$$

or

$$xx'+yy'-(x'^2+y'^2)=0,$$

whence

$$xx'+yy'=r^2,$$

which is the equation to the tangent at the point $(x'y')$.

We leave the reader to obtain this equation geometrically, by means of known properties of the tangent. For example, he will find it easy to shew, from Euc. III. 16, Cor. that

$$\frac{x-x'}{y-y'}=-\frac{y'}{x'},$$

which gives us equation (5) of this article.

118. If we transfer the origin to any point $(-a, -b)$, so that the co-ordinates of the centre are a and b, we must write in the equation to the tangent, by Art. 70,

$$x-a,\ x'-a,\ y-b,\ y'-b \text{ for } x,\ x',\ y,\ y',$$

respectively, and the equation becomes

$$(x-a)(x'-a)+(y-b)(y'-b)=r^2.$$

119. *To find the equation to the tangent in terms of its inclination to the axis of x.*

Let the equations to a straight line and a circle be

$$y=mx+b \quad \text{.......................(1)},$$

$$x^2+y^2=r^2 \quad \text{.......................(2)};$$

then, if we find values of x and y which satisfy both these equations, these will be the co-ordinates of the points, where line and circle intersect. Eliminating y between them, we obtain

$$x^2+(mx+b)^2=r^2,$$

$$(1+m^2)x^2+2bmx+b^2-r^2=0,$$

the roots of which equation are the abscissæ of the points of intersection. If these roots are equal, we have (Appendix)

$$b^2m^2=(b^2-r^2)(1+m^2),$$

whence $$b^2=r^2(1+m^2).$$

If the two values of x are equal to one another, the two

values of y must, from (1), be equal to one another. Hence the two points in which (2) is cut by (1) will be *coincident*, and (1) will be a tangent to (2). Hence, the equation to the tangent which makes an angle $\tan^{-1} m$ with the axis of x, is

$$y = mx \pm r\sqrt{1+m^2},$$

the double sign referring to the two tangents at the extremities of any diameter, which are parallel.

COR. 1. Hence the condition that any line ($y = mx + b$) should touch the circle ($x^2 + y^2 = r^2$), is

$$b^2 = r^2(1 + m^2).$$

If this condition be written

$$\frac{b}{(1+m^2)^{\frac{1}{2}}} = r,$$

it asserts (Art. 52) that the perpendicular from the centre (0, 0) on the line $y = mx + b$ is equal to the radius. Hence, more generally, the condition that the line

$$Ax + By + C = 0 \text{..................... (1)}$$

should touch the circle

$$(x-a)^2 + (y-b)^2 = r^2 \text{.................. (2)}$$

is that the perpendicular from the point (ab) on (1) should be equal to the radius of (2). That is (Art. 52)

$$\pm\frac{Aa + Bb + C}{(A^2+B^2)^{\frac{1}{2}}} = r.$$

COR. 2. Hence the equation to the tangent, in terms of its distance from the centre, and the angle α which that distance makes with the axis of x, is

$$x\cos\alpha + y\sin\alpha = r.$$

The same equation is obtained by writing $x' = r\cos\alpha$, $y' = r\sin\alpha$ (Art. 109) in the equation of Art. 117.

120. As a useful exercise, we shall obtain the equation of Art. 119 as follows, by the method of Art. 63 (ii).

The equation to the two straight lines joining the origin to the intersections of (1) and (2), will be

$$x^2 + y^2 - r^2 \frac{(y - mx)^2}{b^2} = 0,$$

or $$(b^2 - r^2m^2)\, x^2 + 2r^2m\, xy + (b^2 - r^2)\, y^2 = 0 \ldots\ldots (3).$$

If the lines represented by (3) coincide, we obtain

$$(b^2 - r^2m^2)(b^2 - r^2) = r^4m^2;$$

whence $$b^2 = r^2 (1 + m^2).$$

Ex. *To deduce each of the equations*

$$xx' + yy' = r^2 \ldots (1), \quad y = mx \pm r\sqrt{1 + m^2} \ldots (2)$$

from the other.

To deduce (2) from (1), suppose

$$xx' + yy' - r^2 = 0 \text{ and } y - mx - b = 0$$

to represent the same line; then we have

$$\frac{y'}{1} = -\frac{x'}{m} = \frac{r^2}{b};$$

and, since $x'^2 + y'^2 = r^2$,

$$r^2 = \frac{m^2r^4}{b^2} + \frac{r^4}{b^2}, \text{ or } b^2 = r^2 (1 + m^2).$$

To deduce (1) from (2). Let (2) be written in the form of equation (1), Art. 121; then, if we suppose $(x'y')$ to be *on* the circle, so that $x'^2 + y'^2 = r^2$, the equation becomes

$$y'^2m^2 + 2x'y'm + x'^2 = 0, \text{ or } my' + x' = 0;$$

hence μ and μ' are each equal to $-\frac{x'}{y'}$, and equations (2) of Art. 121 become

$$xx' + yy' = r^2,$$

since the tangents which can be drawn from $(x'y')$ now coincide.

121. *To determine the equations to the tangents drawn to a circle from any point $(x'y')$.*

Let the equation to the tangent be

$$y - mx = \pm r\sqrt{1 + m^2};$$

then, since it passes through $(x'y')$, the co-ordinates of that point satisfy the equation, and we have

$$(y' - mx')^2 = r^2 (1 + m^2),$$

or
$$(x'^2 - r^2) m^2 - 2x'y'm + y'^2 - r^2 = 0 \ldots\ldots\ldots (1),$$

a quadratic to determine the two values of m, in the equations to the two tangents which can be drawn to the circle from $(x'y')$. If μ and μ' be the two roots, the required equations will be (Art. 29, Cor. 1)

$$y - y' = \mu (x - x'), \quad y - y' = \mu' (x - x') \ldots\ldots\ldots\ldots (2).$$

COR. Solving equation (1), we have

$$m = \frac{x'y' \pm r\sqrt{x'^2 + y'^2 - r^2}}{x'^2 - r^2},$$

whence we see that the values of m are real, if $x'^2 + y'^2 > r^2$, or the point is outside the circle; they are equal, when $x'^2 + y'^2 = r^2$, or the point is on the circle; and they are imaginary, when $x'^2 + y'^2 < r^2$, or the point is within the circle. Hence we say, that from *any* point $(x'y')$, there can be drawn two real, coincident, or imaginary tangents to the circle. We do not attach any geometrical meaning to the term 'imaginary tangent.' We simply mean to say, that, even when μ and μ' are imaginary, equations (2) can be formed, which satisfy the conditions of tangency, and are satisfied by the co-ordinates of the point $(x'y')$. In Art. 62, Ex. 2, we have a numerical example of imaginary lines which pass through a real point.

The change in the sign of the expression, $x^2 + y^2 - r^2$, should be remarked, and compared with Art. 55.

122. The straight line drawn through any point in a curve, perpendicular to the tangent at that point, is called the *Normal*.

In the case of the circle, if the point be $(x'y')$, the equation to the normal is

$$y - y' = m(x - x');$$

and, since it is perpendicular to the tangent whose equation is

$$xx' + yy' = r^2,$$

we have (Art. 47, Cor. 2) $y' - mx' = 0$; hence the equation to the normal is

$$y - y' = \frac{y'}{x'}(x - x'),$$

which, after reduction, becomes

$$xy' - yx' = 0,$$

the equation to a straight line passing through the origin; hence every normal in the circle passes through the centre, as is proved in Euc. III. 19.

123. *To find the equation to the chord joining the points of contact of two tangents from any external point $(x'y')$.*

Let $P'(x'y')$ be the external point, and let PP'' be the line whose equation is required.

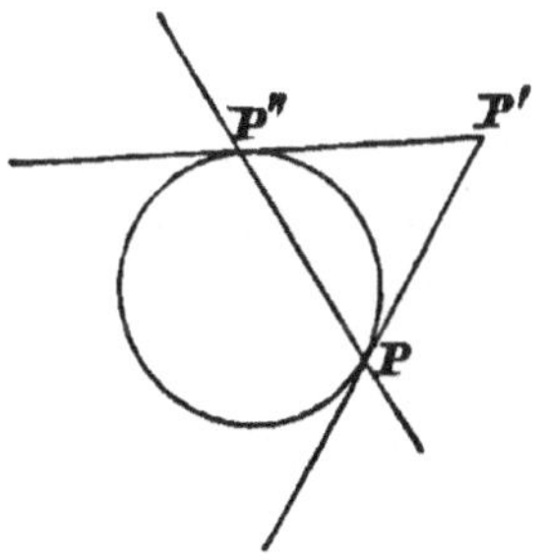

Now the equation to $P'P''$, the tangent at $P''(x''y'')$, is (Art. 117)

$$xx'' + yy'' = r^2,$$

and, since this line passes through $P'(x'y')$, the co-ordinates of P' satisfy the equation; hence

$$x'x'' + y'y'' = r^2;$$

hence the values $x = x''$, $y = y''$ satisfy the equation

$$xx' + yy' = r^2,$$

or $P''(x''y'')$ is a point on the line it represents; and by exactly similar reasoning, P, the other point of contact, is on this same straight line; hence

$$xx' + yy' = r^2$$

is the equation to the chord joining the points of contact of tangents drawn from the point $(x'y')$; for it is the equation to *some* straight line, and both P and P'' have been proved to lie in it.

Cor. Hence, to draw tangents to the circle from any external point $(x'y')$, we have the two equations

$$xx' + yy' = r^2, \quad x^2 + y^2 = r^2,$$

to determine the co-ordinates at the points of contact. These equations will always give two points, real or imaginary, corresponding to the points of intersection of the line and circle.

124. *A chord of a circle is drawn through a fixed point $(x'y')$, and tangents are drawn at the points where it cuts the circle; to find the equation to the locus of the intersection of these tangents, when the chord is turned about the point $(x'y')$.*

Let P' be the point $(x'y')$, $P'QR$ the chord, and let the tangents at Q and R intersect in P'' $(x''y'')$; it is required to find the locus of P'', as the chord turns about P'. Considering $P'R$ as the chord joining the points of contact of tangents drawn from the point $(x''y'')$, its equation is (Art. 123)

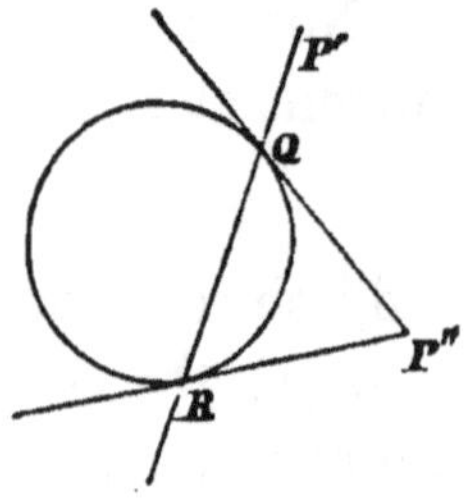

$$xx'' + yy'' = r^2 \quad\text{.........................(1);}$$

but, since it passes through P' $(x'y')$ we have

$$x'x'' + y'y'' = r^2 \quad\text{.......................(2);}$$

$$xx'' + yy'' = r^2 \dots\dots\dots\dots\dots\dots(1);$$

$$x'x'' + y'y'' = r^2 \dots\dots\dots\dots\dots\dots(2);$$

but $(x''y'')$ is *any* point in the required locus, and its co-ordinates satisfy the equation

$$xx' + yy' = r^2 \qquad\qquad (3);$$

hence the co-ordinates of every point in the locus satisfy this equation, which is therefore the equation required. The locus is therefore a straight line. It is[1] evidently perpendicular to the line $(xy' - yx' = 0)$, i.e. to the line joining the centre to the point $(x'y')$.

125. The reasoning of the preceding article is very often perplexing to the beginner. The difficulty commonly arises from the use of the co-ordinates of P'' $(x''y'')$ as *constants* in equation (1), and afterwards as *variables* in equation (3). We should bear in mind that, although P'' is a moveable point, we do not examine its position during its motion, but, taking it in any *one* of its several positions, we obtain a relation between its co-ordinates, *while in that position.* The relation so obtained is equally true for all positions, and is therefore the equation to the locus of the point. Thus, equation (2) is a relation obtained between the co-ordinates of a certain point P''. Equation (3) declares this relation to be true for *all* points determined by the same law as P''.

126. The line

$$xx' + yy' = r^2,$$

is called the *polar* of the point $(x'y')$ with regard to the circle

$$x^2 + y^2 = r^2,$$

and the point $(x'y')$ is called the *pole* of the line. These terms must not be supposed to have any such meaning, as

[1] This statement must be omitted, when we refer to this proof in the case of other curves.

they have in Art. 11. As the equation to the polar is one of the greatest importance, and will be frequently used in the following pages, we will define its exact meaning in all cases.

(1) The position of the point P' in Art. 124 is not subject to any limitation; hence, wherever the point $(x'y')$ may be, the equation $xx' + yy' = r^2$ represents the locus of the intersection of tangents drawn at the extremities of chords which all pass through $(x'y')$.

(2) If the point be *without* the circle, this locus is (Art. 123) identical with the chord joining the points of contact of tangents drawn from $(x'y')$.

(3) If the point be *on* the circle, the locus is also (Art. 117) the tangent at the point $(x'y')$.

In the following figures, P is the point $(x'y')$, and RQ is the line $(xx' + yy' = r^2)$.

Let P be *within* the circle; then, if ST, $S'T'$ are chords

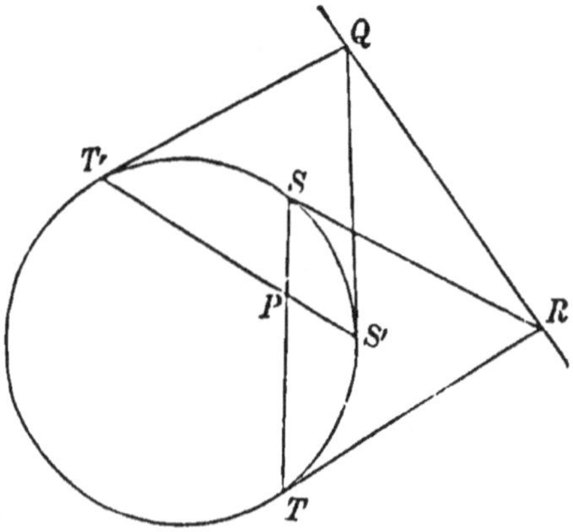

drawn through P, and the tangents at their extremities meet in R and Q, RQ is the locus, and the polar has the interpretation (1) only.

Let P be without the circle; then the polar has the interpretations (1) and (2); that is, RQ is the chord of contact of tangents drawn from P, and also, if any chord PST be drawn, the tangents at S and T intersect in RQ.

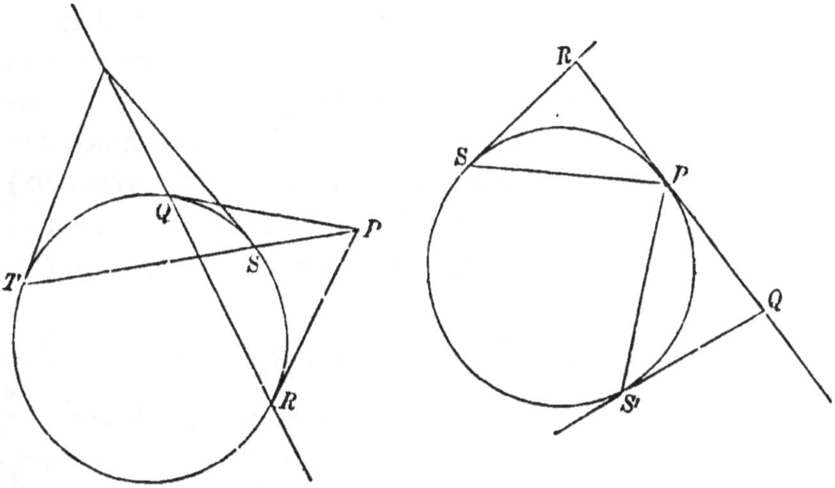

Let P be on the circle; then the polar is the tangent at P. It has also the interpretation (1); for, if any number of chords be drawn from P, as PS, PS', the tangents at the extremities of those chords intersect in the line RQ. This is evident, since RQ is itself one of the tangents in every case. It has also the interpretation (2), for the chord of contact now coincides with the tangent.

127. We saw (Art. 121) that from any point $(x'y')$ there could be drawn two real, coincident, or imaginary tangents to the circle; and we shewed in what sense we might say that imaginary lines passed through real points. In a similar sense we may say that real lines pass through imaginary points. Thus the equation to the polar of $(x'y')$ may be said to pass through the two real, coincident, or imaginary points of tangency, according as $(x'y')$ is *without*, *on*, or *within* the circle. We attach no geometrical meaning to the

term 'imaginary point;' we simply mean that the imaginary values of x and y, obtained from the equations

$$xx' + yy' = r^2, \quad x^2 + y^2 = r^2 \ldots\ldots\ldots\ldots\ldots(1),$$

satisfy the real equation $xx' + yy' = r^2$, which is obviously true. We will examine this a little more closely, that we may see the conditions, under which two imaginary points may lie upon and determine a real straight line, as it is evident that not *every* two such points will do so. Since the imaginary roots of a quadratic equation assume (Appendix) the form $\alpha + \beta\sqrt{-1}$, $\alpha - \beta\sqrt{-1}$, it follows that the two values of x, obtained by eliminating y between equations (1), may be assumed to be

$$x_1 = \alpha + \beta\sqrt{-1}, \quad x_2 = \alpha - \beta\sqrt{-1};$$

and the corresponding values of y will be in like manner

$$y_1 = \gamma + \delta\sqrt{-1}, \quad y_2 = \gamma - \delta\sqrt{-1}.$$

The equation to the straight line through (x_1y_1), (x_2y_2) will be (Art. 30)

$$\frac{y - \gamma - \delta\sqrt{-1}}{x - \alpha - \beta\sqrt{-1}} = \frac{-2\delta\sqrt{-1}}{-2\beta\sqrt{-1}},$$

or $$(y - \gamma)\beta = (x - \alpha)\delta,$$

which is a real straight line. Thus we see that a *pair* of imaginary points, such as the above, will always determine a real straight line; also the middle point between them is a real point on the line, whose co-ordinates are

$$x = \frac{x_1 + x_2}{2} = \alpha, \quad y = \frac{y_1 + y_2}{2} = \gamma.$$

Any real straight line will contain any number of imaginary points, since any imaginary value of one variable will give us a corresponding imaginary value of the other; but an imaginary straight line will pass through one real point only, namely that for which (Art. 65) the real and the imaginary parts of the equation vanish simultaneously.

128. We saw (Art. 124) that the polar of a point $(x'y')$ is perpendicular to the line joining $(x'y')$ with the centre. Also, if P be the point $(x'y')$, QR the polar $(xx' + yy' = r^2)$, and CLR the perpendicular from C, we have (Art. 52)

$$CR = \frac{r^2}{(x'^2 + y'^2)^{\frac{1}{2}}};$$

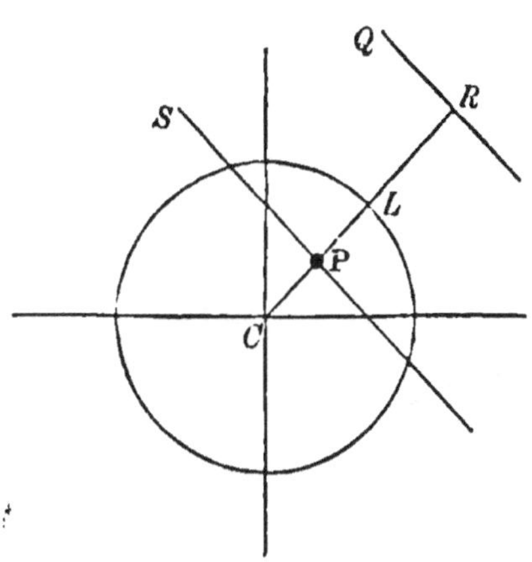

but $CP = (x'^2 + y'^2)^{\frac{1}{2}}$, $CL = r$,

$$\therefore CR \, . \, CP = CL^2;$$

hence we have an easy geometrical construction for the polar; for, if in CP or CP produced (according as $CP >$ or $< CL$) we take a point R, so that

$$CR : CL = CL : CP,$$

and through R draw RQ perpendicular to CR, RQ will be the polar of P. If SP is perpendicular to CR, SP is the polar of R. This is a particular case of the property proved in the next article.

129. *If any point $(x'y')$ be taken on the polar of the point $(x''y'')$, then the polar of $(x'y')$ shall pass through $(x''y'')$.*

For the equation to the polar of $(x''y'')$ is

$$xx'' + yy'' = r^2,$$

and the condition that $(x'y')$ should lie on this line, is

$$x'x'' + y'y'' = r^2;$$

but this is also the condition to be fulfilled in order that the polar of $(x'y')$, whose equation is

$$xx' + yy' = r^2,$$

should pass through the point $(x''y'')$.

DEF. Two points are said to be *conjugate* with respect to a curve, when each lies upon the polar of the other; and two straight lines are said to be *conjugate*, when each passes through the pole of the other.

COR. Any straight line $(Ax + By + C = 0)$ can be written in the form

$$-\frac{Ar^2}{C}x - \frac{Br^2}{C}y - r^2 = 0,$$

which is of the form $xx' + yy' - r^2 = 0$, and is therefore the polar of a point $-\frac{Ar^2}{C}, -\frac{Br^2}{C}$; hence the preceding proposition may be thus enunciated: *If points be taken in any straight line, the polars of these points will all pass through a fixed point.* This fixed point is, from above, the pole of the straight line.

EX. *To find the pole of the line $3x + 5y - 6 = 0$ with respect to the circle $x^2 + y^2 = 30$.*

The equation to the line may be written

$$\frac{3 \times 30}{6}x + \frac{5 \times 30}{6}y - 30 = 0,$$

the pole of which is the point $x = 15$, $y = 25$.

130. *The polar of the intersection of two straight lines is the straight line which joins their poles.*

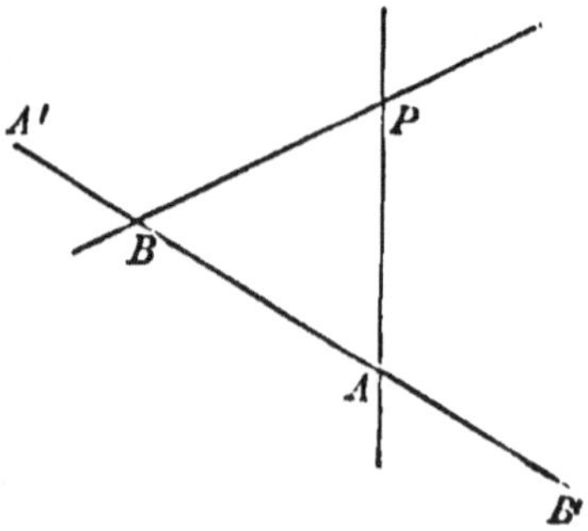

Let AP, BP be the two lines intersecting in P, and let A', B' be their poles. Then, since the point P is taken upon AP, the polar of A', therefore the polar of P passes through A'. Similarly the polar of P passes through B'; therefore $A'B'$ is

133. *The equations to two circles are given; to find the equation of the straight line which joins their points of intersection.*

Let the equation to the two circles be

$$x^2 + y^2 + 2Gx + 2Fy + C = 0 \text{............} (1),$$

$$x^2 + y^2 + 2G'x + 2F'y + C' = 0 \text{............} (2);$$

then, by Art. 43, if k be any constant,

$$x^2 + y^2 + 2Gx + 2Fy + C + k(x^2 + y^2 + 2G'x + 2F'y + C') = 0 \text{...} (3)$$

is the equation to some locus passing through the points where (1) and (2) intersect, since it is satisfied by those values of x and y which satisfy (1) and (2). The equation may be written

$$(1 + k)x^2 + (1 + k)y^2 + 2(G + kG')x + 2(F + kF')y + C + kC' = 0,$$

which evidently represents a circle (Art. 107) since it can be reduced to the form

$$x^2 + y^2 + Px + Qy + R = 0,$$

by dividing by $1 + k$.

If $k = -1$, or (2) be simply subtracted from (1), the equation becomes

$$2(G - G')x + 2(F - F')y + C - C' = 0,$$

which is therefore the equation to a *straight line* passing through the intersections of (1) and (2).

Cor. Equation (3) may be made to represent *any* particular circle which passes through the intersection of (1) and (2). For suppose that it passes through some other point $(x'y')$; then, if the values x' and y' are substituted for x and y in (3), we shall have an equation to determine the value of k in the circle which we wish to represent. The circle is now completely determined, as we should expect, since only one circle can be described through three points, i.e. about a triangle.

*134. Let the equation to *any* circle (Art. 106, Cor.), referred to rectangular axes, be

$$x^2 + y^2 + 2Gx + 2Fy + C = 0 \quad\text{..............(1)};$$

then (Art. 107) it can be written in the form

$$(x-a)^2 + (y-b)^2 - r^2 = 0 \quad\text{..............(2)},$$

where the point (ab) is the centre, and r the radius. Let S denote the left-hand member of (1) and (2); then, since $(x-a)^2 + (y-b)^2$ is the square of the distance of any point (xy) in the plane from the centre (ab), it is evident (Euc. I. 47) that S is the square of the length of the tangent drawn from (xy) to the circle. It is easily seen that S is positive, zero, or negative, according as (xy) is without, on, or within the circle.

Cor. If the co-ordinates of any point (xy) satisfy the equation $S = kS'$, then the square of the tangent drawn from (xy) to the circle $S=0$ is equal to k times the square of the tangent drawn from it to the circle $S'=0$. Hence Art. 133 proves the following proposition: *If a point moves, so that the tangents from it to two given circles are in a constant ratio, its locus is another circle, which passes through the points of intersection of the given circles.*

*135. We have seen, by Art. 133, that, if S and S' are symbols standing for the expressions

$$x^2 + y^2 + 2Gx + 2Fy + C,\ x^2 + y^2 + 2G'x + 2F'y + C',$$

so that

$$S = 0 \quad\text{..............(1)}, \qquad S' = 0 \quad\text{..............(2)}$$

are the equations to two circles, then the equation

$$S - S' = 0 \quad\text{........................(3)}$$

is the equation to a straight line, and that, if the circles intersect, the points of intersection will lie on (3). But (3) is a real line, whether the circles intersect or not, and in the latter case (Art. 127) is satisfied by the imaginary values of the co-ordinates of intersection. In either case it possesses the following important property with regard to the circles: *if, from any point of it, straight lines be drawn to touch both circles, the lengths of those lines are equal;* for the squares of the lengths of the tangents to (1) and (2) from any point (xy) on (3) are (Art. 134) equal to S and S' respectively, and (3) asserts that these are equal. The line (3) is called the *Radical Axis* of the two circles.

*136. *The three radical axes belonging to three given circles meet in a point.*

Let the equations to the three circles be

$$S=0, \quad S'=0, \quad S''=0;$$

then the equations to the radical axes of these circles taken two and two together, are

$$S-S'=0, \quad S'-S''=0, \quad S''-S=0.$$

At the point where the straight lines represented by the first two of these equations intersect, we have

$$S-S'=0, \quad S'-S''=0,$$

hence $$S''-S=0;$$

that is, the third straight line passes through the intersection of the first two.

*137. If in the equation to one of the circles,

$$(x-a)^2+(y-b)^2-r^2=0,$$

we suppose r to vanish, the circle is reduced to the point (ab); hence, if we subtract this from the other equation, the radical axis is now a straight line, such that the tangents from any point of it to the circle are equal in length to the distance of that point from a given point (*ab*).

If both radii vanish, and both circles become points, the radical axis becomes a straight line, every point in which is equally distant from two given points.

The student will find it easy to prove, that the radical axis is perpendicular to the line joining the centres of the two circles, and is consequently the common tangent, when the circles touch.

*138. If there be two unequal circles which do not intersect, there will evidently be two points, on the indefinite straight line joining their centres, from each of which a pair of tangents common to the two circles may be drawn. The points will lie, one between the centres, and the other exterior to the smaller circle. They are called the *Centres of Similitude*[1].

EXAMPLES VII.

1. To find the centre and radius of the circle

$$x^2 + y^2 - 6x + 4y + 4 = 0.$$

2. Investigate the line or lines represented by the equation

$$x^3 + xy^2 - x^2r - xr^2 - ry^2 + r^3 = 0.$$

3. Find the common chord of two circles

$$(x-1)^2 + (y-2)^2 = 6, \ (x-2)^2 + (y-3)^2 = 8.$$

4. To find the equation to a straight line which passes through the centres of the two circles

$$x^2 + 2x + y^2 = 0, \ y^2 + 2y + x^2 = 0.$$

[1] The properties of these points are discussed at length in Salmon's *Conic Sections*.

5. To find the equation to a circle having for its diameter the straight line joining the points of intersection of the line, $y = mx$, and the circle, $y^2 = 2rx - x^2$.

6. Find the equation to the circle, the diameter of which is the common chord of the circles

$$x^2 + y^2 = r^2, \quad (x - a)^2 + y^2 = r^2.$$

7. What is represented by the equation

$$x(x - 2) + y(y - 4) + 8 = 0\,?$$

8. Find a relation between the coefficients of the equation

$$A(x^2 + y^2) + 2Gx + 2Fy + C = 0,$$

in order that (1) the axis of x, and (2) the axis of y, may be tangents to the circle.

9. To find the inclination to the axis of x of the tangents drawn from any point $(x'y')$ to the circle whose equation is

$$(x - a)^2 + (y - b)^2 - r^2 = 0.$$

10. To find the relation between the quantities a, b, r, in order that the line $\frac{x}{a} + \frac{y}{b} = 1$ may touch the circle $x^2 + y^2 = r^2$.

11. To find the equation to a circle, the centre of which is at the origin of co-ordinates, and which is touched by the line

$$y = 2x + 3.$$

12. To find the intercepts on the axes of co-ordinates of the tangent to a circle $(x^2 + y^2 = r^2)$, drawn parallel to a given straight line, $(x \cos \alpha + y \sin \alpha = p)$.

13. If $2\alpha'$, $2\alpha''$ be the inclination of two radii of a circle, $x^2 + y^2 = r^2$, to the axis of x, to find the equation to the chord joining the extremities of the radii.

14. If the pole always lie on a line

$$\frac{x}{a}+\frac{y}{b}=1,$$

and the equation to the circle is $x^2+y^2=r^2$, the equation to the polar is of the form

$$(ax-r^2)+k(by-r^2)=0,$$

where k is any constant.

15. If the pole of a straight line with regard to the circle $x^2+y^2=r^2$ lie on the circle $x^2+y^2=4r^2$, the polar will touch the circle

$$x^2+y^2=\frac{r^2}{4}.$$

16. Find the equation to the circle which has each of the co-ordinates of the centre $=-\frac{1}{3}$, and its radius $=\frac{2}{\sqrt{3}}$, the axes being inclined at an angle of 60°.

17. Prove that the circles

$$x^2+y^2=(c+a)^2,\quad (x-a)^2+y^2=c^2,$$

have only one common tangent, and find its equation.

18. Find the locus of the middle points of chords drawn from the extremity of the diameter of any circle.

19. Shew that the polar of the point $(x'y')$ with regard to the circle $(x-a)^2+(y-b)^2=r^2$ is

$$(x-a)(x'-a)+(y-b)(y'-b)=r^2.$$

20. Find the locus of the vertices of all triangles which have a given base, and a given vertical angle.

21. Prove Euc. III. 31, from the resulting equation.

22. Tangents are drawn to a circle $x^2+y^2=r^2$, at two points $(x'y')$, $(x''y'')$; to find the distance of a point (hk) from a straight line passing through the centre and the intersection of the two tangents.

23. To find the equations to straight lines touching a circle

$$x^2 + y^2 = 10,$$

at points, the common abscissa of which is unity.

24. Find the equation to a straight line touching the circle

$$(x-a)^2 + (y-b)^2 = r^2,$$

and parallel to a given line $y = mx + c$.

25. To find the equation to the straight line passing through the origin of co-ordinates, and touching the circle

$$x^2 + y^2 - 3x + 4y = 0.$$

26. To find the length of the common chord of the circles

$$(x-a)^2 + (y-b)^2 = r^2, \ (x-b)^2 + (y-a)^2 = r^2.$$

27. Find the area between the two circles

$$x^2 + y^2 + 2x + 4y = 0, \ x^2 + y^2 + 2x + 4y = 1.$$

28. To find the length of the chord of a circle $x^2 + y^2 = r^2$, made by the straight line $\frac{x}{a} + \frac{y}{b} = 1$.

29. If from a given point S, a perpendicular be drawn to the tangent PY at any point P of a circle, of which the centre is C, and, in the line MP at right angles to CS and produced if necessary, a point Q be taken, such that $QM = SY$, to find the locus of Q.

30. Given the equation to a circle, and the chord of a circle; shew that a perpendicular let fall upon the chord from the centre bisects the chord.

31. Find the diameter of the circle

$$x^2 + 2xy \cos \omega + y^2 - ax - by = 0.$$

32. In the equation $Ax + By + C = 0$, if C is constant, and A and B vary, subject to the condition $A^2 + B^2 =$ a constant, the equation represents a series of tangents to a given circle.

33. Find the equation to the circle which passes through the points (0, 0), $(-8a, 0)$, $(0, 6a)$, the axes being rectangular.

34. To find the locus of middle points of chords which pass through a given point.

35. If on any radius vector through a fixed point O, OQ be taken in a constant ratio to OP, where P is a point on a given circle, find the locus of Q.

36. The circles represented by the equation

$$(n+1)(x^2+y^2) = ax + nby,$$

where n is arbitrary, have a common chord.

37. Prove algebraically that the angles in the same segment of a circle are equal, and that the angle in a semicircle is a right angle.

38. Two sides of a triangle are b and c, and they include an angle A; if these sides be taken as axes, the equation to the circumscribed circle is

$$x^2 + 2xy \cos A + y^2 - bx - cy = 0.$$

39. Given the base and vertical angle, to shew that the locus of the point of intersection of the perpendiculars from the angles on the sides is a circle.

40. Given base and ratio of sides of a triangle; shew that the locus of the vertex is a circle.

41. When will the locus of a point be a circle, if the square of its distance from the base of a triangle be in a constant ratio to the product of its distances from the sides?

42. When will the locus of a point be a circle, if the sum of the squares of the three perpendiculars from it on the sides of a triangle be constant?

43. Find the locus of a point, the square of whose distance from a given point is proportional to its distance from a given right line.

44. Given the base of a triangle, and m times the square of one of its sides $\pm n$ times the square of the other = a constant; find locus of the vertex; find centre and radius of resulting circle, and where it cuts base.

45. Find the equations to the circles which touch the three lines, referred to rectangular axes,

$$x = a, \; y = 2b, \; y = 2b'.$$

46. The locus of the centres of all circles inscribed in all right-angled triangles on the same hypotenuse is the quadrant described on the hypotenuse.

47. The equation to a circle is $y^2 + x^2 = a(y + x)$; what is the equation to that diameter which passes through the origin of co-ordinates?

48. To find the equation to a circle referred to two tangents at right angles, as axes.

49. If through any point of a quadrant whose radius is R, two circles be drawn, touching the bounding radii of the quadrant, and r, r' be the radii of these circles, $rr' = R^2$.

50. To find the equations to the straight lines which touch both the circles

$$x^2 + y^2 = r^2, \; (x - a)^2 + y^2 = r'^2.$$

51. To find the equation to the circle which touches the three straight lines, referred to rectangular axes,

$$x = 0, \; y = 0, \; \frac{x}{a} + \frac{y}{b} = 1.$$

52. To find the equations to two circles, which touch rectangular axes of x and y, and pass through a given point (ab).

53. The straight lines joining the angles of a triangle with the points in which the escribed circles touch the opposite sides, meet in a point.

54. In any circle draw a chord AB; from the middle point E of the lesser segment draw any straight line cutting AB in C,

and meeting the circumference in D; join AD, and in AD take $AP = AC$; find locus of P.

55. The axes Ox, Oy cut a circle in points A, A', B, B' respectively; to compare the values of x, y, at the intersection of the chords AB', $A'B$.

56. Determine the magnitude and position of the circle

$$\rho^2 - 2\rho(\cos\theta + \sqrt{3}\sin\theta) = 5.$$

57. Find the locus of a point, such that, if straight lines be drawn to it from the four corners of a given square, the sum of their squares may be invariable.

58. ACB is the segment of a circle, and any chord AC is produced to a point P, so that $AC : CP$ is a given ratio; required to find the locus of P.

59. Find the equation to a straight line which cuts a given circle, when the straight lines drawn from the points of intersection to the centre contain a right angle, and one of them is inclined to the axis of x at a given angle.

60. Two straight lines revolve uniformly in one plane about one extremity, the one moving twice as fast as the other. Find the locus of their point of intersection, supposing them to begin to move together in the same direction, from the straight line joining their fixed extremities.

61. If any number of circles touch one another in one point, all their polars, which correspond to a common pole, pass through a single point.

62. To find the locus of the pole, when the polar of a given circle always passes through a given point.

63. In the sides AB, AC of a given triangle ABC, take two points M, N, such that $\frac{BM}{MA} = \frac{AN}{NC}$, and shew that the circle described about the triangle AMN will always pass through a given point.

CHAPTER VIII.

General Equation of the Second Degree.

139. THE equation of the first degree has given us but one species of line, viz. the straight line. We saw, in the case of the circle, that an equation of the second degree may represent a curve limited in every direction; and the case where it represents two straight lines, shews us that it may represent loci extending to infinity. These are but particular cases of the equation of this degree, of which the most general form is (Art. 22)

$$Ax^2 + 2Hxy + By^2 + 2Gx + 2Fy + C = 0.$$

Our object in the present chapter is to interpret this equation, which for brevity we shall write $\phi(xy) = 0$; and we shall shew how, by a proper selection of origin and axes, we may in every case reduce it to a form, from which we shall be able to trace the locus, and deduce its most remarkable properties. It will be shewn in a future chapter, that much of this can be accomplished without reduction of the equation; and properties so established have the advantage of applying to every particular form that the locus may assume; but the proofs are of necessity more cumbrous and difficult to a beginner, than those which we are about to use.

140. If a point be so situated with regard to a locus, that all chords of the locus, drawn through the point, are bisected in it, the point is called the *centre* of the locus.

Suppose C to be the centre of a locus and the chord PCQ to meet the locus in P and Q; then $CP = CQ$, and, if we take C as the origin of co-ordinates, it is evident that the co-ordinates of Q are the same as those of P, but with opposite signs; *i.e.* if the co-ordinates of P be x', y', the co-ordinates of

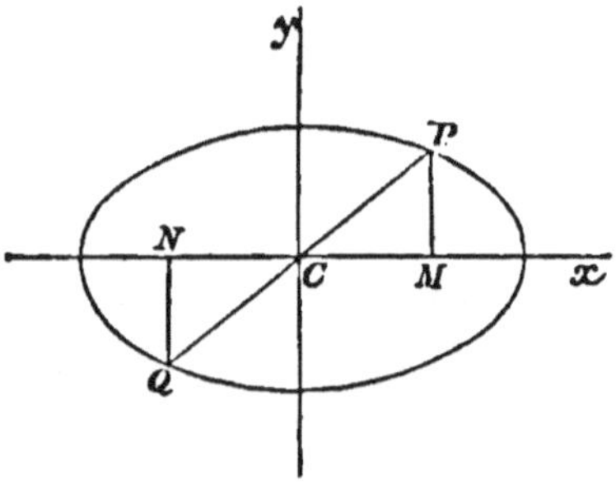

Q are $-x'$, $-y'$; and this is true, whatever be the position of P. Hence, the centre being origin, for every point $(x'y')$, whose co-ordinates satisfy the equation, there will be a corresponding point $(-x' \; -y')$, whose co-ordinates also satisfy the equation. From this it follows, that, *when the centre is the origin, the equation will not be altered by writing* $-x$, $-y$ *for* x *and* y.

Also, if for every point $P(x'y')$ there is a corresponding point $Q(-x', -y')$, the origin is the centre; for, since the triangles PCM, QCN will be equal in every respect, $CP = CQ$; and, since the angles PCM, QCN are equal, CP and CQ are in the same straight line; therefore every chord passing through C is bisected. Hence, conversely, *if an equation is not altered by writing* $-x$ *and* $-y$ *for* x *and* y, *the origin is the centre of the locus.*

Now the equation $\phi(xy) = 0$ cannot remain unaltered, as above, unless the terms $2Gx$ and $2Fy$ vanish from it; hence, if we select the centre of the locus for origin, we shall simplify the equation by getting rid of these two terms.

141. In order, then, to find the centre of the locus, we must transfer the origin to a point $(x'y')$, and then observe what values of x', y' make the new coefficients of x and y vanish. These values will be the co-ordinates of the centre with reference to the original axes.

Writing (Art. 70) $x+x'$ for x, $y+y'$ for y in the general equation, we have

$$A(x+x')^2+2H(x+x')(y+y')+B(y+y')^2$$
$$+2G(x+x')+2F(y+y')+C=0,$$

$$\text{or}\quad Ax^2+2Hxy+By^2+\left.\begin{matrix}2Ax'\\+2Hy'\\+2G\end{matrix}\right|x+\left.\begin{matrix}2Hx'\\+2By'\\+2F\end{matrix}\right|y+\phi(x'y')=0,$$

where $\phi(x'y')=Ax'^2+2Hx'y'+By'^2+2Gx'+2Fy'+C.$

Hence the co-ordinates of the centre, with reference to the original axes, will be determined by the equations

$$Ax'+Hy'+G=0\ldots\ldots\ldots\ldots\ldots\ldots(1),$$
$$Hx'+By'+F=0\ldots\ldots\ldots\ldots\ldots\ldots(2).$$

These two equations will, in general, give one and only one value of x' and y'; hence, *loci of the second degree have in general one, and only one, centre.*

Its co-ordinates are found, by solving the above equations, to be

$$x'=-\frac{HF-BG}{H^2-AB},\quad y'=-\frac{HG-AF}{H^2-AB}.$$

Equations (1) and (2) may be thus remembered[1]: For (1), take those terms only of the general equation, which involve

[1] The reader of the Differential Calculus will see, that the equations of the centre are obtained by differentiating the equation $\phi(xy)=0$ with respect to x and y. The equation $\phi(xy)=0$, when the origin is transferred to a point $(x'y')$, is $\phi(x'+x,\ y'+y)=0$, or, by a familiar expansion,

$$0=\phi(x'y')+\frac{d\phi}{dx'}x+\frac{d\phi}{dy'}y+\tfrac{1}{2}\left\{\frac{d^2\phi}{dx'^2}x^2+2\frac{d^2\phi}{dx'dy'}xy+\frac{d^2\phi}{dy'^2}y^2\right\}$$

+terms which involve higher differential coefficients, and therefore vanish; hence, in order that the terms of the first degree in x and y may vanish, we must have $\dfrac{d\phi}{dx'}=0$, $\dfrac{d\phi}{dy'}=0$.

x; multiply each term by the index of x in it, and diminish that index by unity. Equation (2) may be obtained similarly, by substituting y for x in the above rule.

Ex. The equations for the centre of the locus represented by

$$3x^2+2xy+3y^2-16y+23=0,$$

are $6x+2y=0$, and $2x+6y-16=0$, which become (1) and (2), when divided by two.

Cor. Since the equations for the centre do not involve the constant term C, it follows that all central loci, whose equations can be written so as to differ in the constant term only, are concentric.

*142. The calculation of the value of $\phi(x'y')$, when $(x'y')$ is the centre, may be thus facilitated;

$$\phi(x'y')=(Ax'+Hy'+G)x'+(Hx'+By'+F)y'+Gx'+Fy'+C;$$

but these first two terms are each $=0$, when $(x'y')$ is the centre; therefore

$$\phi(x'y')=Gx'+Fy'+C;$$

hence the equation $\phi(xy)=0$, becomes, when the origin is transferred to the centre,

$$Ax^2+2Hxy+By^2+2G\frac{x'}{2}+2F\frac{y'}{2}+C=0;$$

that is, we can transfer the origin to the centre, by substituting the halves of the co-ordinates of the centre, for x and y respectively, in the terms $2Gx$ and $2Fy$.

Ex. The equations for the centre of

$$3x^2+2xy+3y^2-16y+23=0,$$

are $6x+2y=0$, and $2x+6y-16=0$; whence $x=-1$, $y=3$ are the co-ordinates of the centre; and writing $\frac{3}{2}$ for y in the term $-16y$, we have for the reduced equation,

$$3x^2+2xy+3y^2-1=0.$$

143. We see then, that it will be always possible to find one, and only one, pair of values for the co-ordinates of the centre, except when $H^2 - AB = 0$. Hence the loci of the second degree may be divided into two classes: (i) loci which *have a centre*, where $H^2 - AB$ is not zero; and (ii) loci which in general *have not a centre*, or rather, whose centre is infinitely distant, where $H^2 - AB = 0$. It will be seen in Art. 157, why we say 'in general.'

144. We shall first consider the case of Central Loci.

We see by Art. 141, that, in the case of central loci, the general equation may, by taking the centre $(x'y')$ of the locus as origin, be reduced to

$$Ax^2 + 2Hxy + By^2 + \phi(x'y') = 0.$$

We next proceed to inquire, whether, by any change in the direction of the axes, we can get rid of the term involving xy, as it will be seen hereafter, that this will greatly facilitate our inquiries into the form and properties of the curve. Now it is manifest, that, if we can so transform the axes as to get rid of the term involving xy, the equation will be left in the form

$$Px^2 + Qy^2 + R = 0,$$

where, if any value be given to one of the variables, the other will have two equal values with opposite signs; hence, in this case, each axis will bisect all chords parallel to the other.

145. We have not hitherto supposed the axes necessarily rectangular; but the generality of our reasoning would not have been affected by such a supposition, since, if they had been oblique, by transforming them to rectangular axes, we should (Art. 76 (i)) have obtained an equation of the same degree, and of the same form as the one we have assumed.

We shall now, for the sake of simplicity, suppose such a change, if necessary, to have been made, and the axes to be rectangular; then, if we turn them round through any angle θ, we must (Art. 71) substitute in the equation,

$$Ax^2 + 2Hxy + By^2 + \phi(x'y') = 0 \ldots\ldots\ldots\ldots(1),$$

for x, $x\cos\theta - y\sin\theta$; for y, $x\sin\theta + y\cos\theta$.

Substituting, and arranging the terms, we have

$$\left.\begin{array}{l} A\cos^2\theta \\ +2H\sin\theta\cos\theta \\ +B\sin^2\theta \end{array}\right| x^2 - \left.\begin{array}{l} 2A\sin\theta\cos\theta \\ +2H\cos^2\theta \\ -2H\sin^2\theta \\ +2B\sin\theta\cos\theta \end{array}\right| xy + \left.\begin{array}{l} A\sin^2\theta \\ -2H\sin\theta\cos\theta \\ +B\cos^2\theta \end{array}\right| y^2 + \phi(x'y') = 0. \ldots\ldots\ldots (2).$$

If now we put the new coefficient of $xy = 0$, we obtain

$$\tan 2\theta = \frac{2H}{A - B} \ldots\ldots\ldots\ldots\ldots\ldots\ldots(3),$$

an equation from which we may determine the angle θ, through which the axes must be moved, in order that the term involving xy may vanish.

As the tangent of an angle may have any magnitude, it follows that this equation will always give real values for 2θ. There will be an infinite number of solutions; for, if 2α be *one* value of 2θ which satisfies the equation, then the equation is satisfied, if $2\theta = n\pi + 2\alpha$, where n is any integer. Hence the values of θ are expressed by $\frac{n\pi}{2} + \alpha$, and we obtain a series of angles

$$\alpha,\ \frac{\pi}{2} + \alpha,\ \pi + \alpha,\ \frac{3\pi}{2} + \alpha,\ \&c.$$

The only difference, that will be made by selecting different values of θ, will be, that the axis of x and y in one case may occupy the position of the axis of y and x in another, or the positive and negative directions of the axes may change places. Hence, when the centre is origin, there exists one system of rectangular axes, and one only, about which the curve is so situated, that each axis bisects all chords parallel to the other. Also there can be no such system with any other origin; for a straight line bisecting a system of parallel chords must pass through the centre, which is the bisection of one of them. These axes are called *the axes* of the curve. We shall see hereafter that their position is that of the figures.

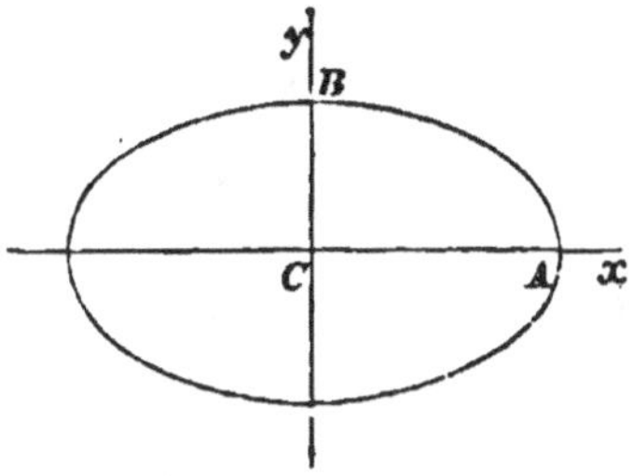

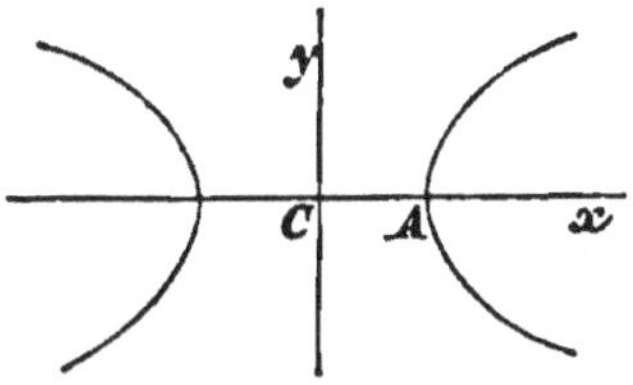

Cor. 1. Since the direction of the axes depends upon the quantities A, H, B only, it follows that loci, whose equations can be written so as to have the first three terms the same, have their axes parallel.

Cor. 2. If we have $H=0$, and $A=B$, the new coefficient of xy vanishes for every value of θ, and $\tan 2\theta$ becomes indeterminate. Hence in this case we may take any rectangular axes whatsoever, without introducing the term xy into the equation. This agrees with Art. 106, for the curve is then a circle.

*146. The axes of the locus, then, make angles θ and $90^\circ+\theta$ with the axis of x, where θ is determined from the equation

$$\tan 2\theta = \frac{2H}{A-B};$$

hence, the centre being origin, the equations to the axes are

$$y - x\tan\theta = 0, \quad y + x\cot\theta = 0,$$

or, as one locus,

$$y^2 + (\cot\theta - \tan\theta)\,xy - x^2 = 0,$$

or, since $$\cot\theta - \tan\theta = \frac{2}{\tan 2\theta} = \frac{A-B}{H},$$

$$x^2 - \frac{A-B}{H}\,xy - y^2 = 0.$$

The student will observe that this is (Art. 68) the equation to the straight lines which bisect the angles between the straight lines represented by

$$Ax^2 + 2Hxy + By^2 = 0,$$

a coincidence which will be hereafter (Art. 186) explained.

147. We have shewn (Art. 141) that the coefficients of the first three terms of the equation $\phi(xy) = 0$, are not altered by a transfer of the origin; we shall now shew, that when the axes are moved through an angle θ, and the new coefficients denoted by A', $2H'$, B', we have the relations,

$$A' + B' = A + B,\quad H'^2 - A'B' = H^2 - AB.$$

From equation (2) of Art. 145 we have

$$A' = A\cos^2\theta + 2H\sin\theta\cos\theta + B\sin^2\theta,$$

$$B' = A\sin^2\theta - 2H\sin\theta\cos\theta + B\cos^2\theta;$$

therefore $$A' + B' = A + B,$$

$$A' - B' = 2H\sin 2\theta + (A-B)\cos 2\theta;$$

therefore

$$-4A'B' = (A'-B')^2 - (A'+B')^2$$

$$= \{2H\sin 2\theta + (A-B)\cos 2\theta\}^2 - (A+B)^2;$$

also $$4H'^2 = \{2H\cos 2\theta - (A-B)\sin 2\theta\}^2;$$

hence $$4(H'^2 - A'B') = 4H^2 + (A-B)^2 - (A+B)^2,$$

or $$H'^2 - A'B' = H^2 - AB.$$

These results are very important, because we shall shew that the nature of· the curve depends upon the sign of $H^2 - AB$. We have on this account proved them independently, instead of taking them as a particular case of Art. 76 (ii). It may be there seen that the *sign* of $H^2 - AB$ does not change for any transformation, since $\sin^2 \omega$ and $\sin^2 \phi$ are positive quantities.

COR. Since the axis of x of any rectangular system can be made to coincide with the axis of x of any oblique system with the same origin, without changing $A + B$ and $H^2 - AB$, it follows that this article taken with Art. 73, Ex. proves Art. 76 (ii).

*148. The preceding article furnishes us with an easy method of arriving at the actual values of A' and B'. For, since the equation can be reduced to the form

$$A'x^2 + B'y^2 + \phi (x'y') = 0,$$

H' is $= 0$, and we have

$$A' + B' = A + B\ldots(1),\quad - A'B' = H^2 - AB\ldots(2);$$

hence

$$A' - B' = \pm \sqrt{(A - B)^2 + 4H^2}\ldots\ldots\ldots\ldots(3),$$

and from (1) and (3) A' and B' may be determined, the values of A' and B' with the upper sign corresponding to the values of B' and A' with the lower.

The two results

$$A'x^2 + B'y^2 + \phi (x'y') = 0,\quad B'x^2 + A'y^2 + \phi (x'y') = 0,$$

will represent the same locus, the axes of x and y respectively in the one being called the axes of y and x in the other.

*149. If we simply wish to find the nature of the locus, without fixing its· position with regard to the new axes, Art.

148 is sufficient. To find the appropriate values for A' and B', we may proceed as follows. As in Art. 147, we have

$$A' - B' = 2H \sin 2\theta + (A - B) \cos 2\theta \text{.........(1)}.$$

And from the value of tan 2θ in Art. 145,

$$0 = (A - B) \sin 2\theta - 2H \cos 2\theta \text{..............(2)}.$$

Eliminating cos 2θ between equations (1) and (2), we have

$$2H (A' - B') = \{(A - B)^2 + 4H^2\} \sin 2\theta \text{......(3)}.$$

If we determine to take for 2θ the smallest positive angle that satisfies equation (3), Art. 145, 2θ must be less than 180°, since for angles between 0° and 180° the tangent passes through every possible value; hence sin 2θ is positive, and so also is $(A - B)^2 + 4H^2$. It follows then from (3) that $A' - B'$ is of the same sign as H. This will enable us to select the proper sign in equation (3), Art. 148.

150. We see then, that the equation $\phi (xy) = 0$ may, in the case of central loci, always be reduced to

$$A'x'^2 + B'y'^2 + \phi (x'y') = 0.$$

We shall now divide these loci into two classes.

Class I. $H^2 - AB$ negative. Since the sign of this quantity remains the same after transformation, and the coefficient of xy is now $= 0$, $-A'B'$ must be negative, i.e. A' and B' must have the same sign, and the equation may be written in the form

$$Px^2 + Qy^2 = R,$$

where P and Q are positive quantities. We shall now have three cases.

(i) If R is positive, we have a locus not yet investigated, which is called an *Ellipse*.

(ii) If $R = 0$, the locus (Art. 65) is the two imaginary straight lines

$$\sqrt{P}x + \sqrt{-Q}y = 0, \quad \sqrt{P}x - \sqrt{-Q}y = 0,$$

which meet in the real point $x = 0$, $y = 0$; or, as will appear hereafter, is the Ellipse indefinitely diminished.

(iii) If R is negative, the equation (Art. 66) can be satisfied by no real value of x and y, and the locus may be called an *imaginary Ellipse*.

The circle belongs to this class; for its most general equation may (Art. 110) be written, so that $A = B = 1$, and $H = \cos\omega$; hence $H^2 - AB = \cos^2\omega - 1$, which is always negative. It is in fact a particular case of locus (i), when $P = Q$.

Class II. $H^2 - AB$ positive. Here $-A'B'$ must be positive, i.e. A' and B' must have different signs, and the equation may be written in the form

$$Px^2 - Qy^2 = R,$$

where P and Q are positive. In this case the sign of R will make no difference in the *nature* of the locus, since the equation $Qy^2 - Px^2 = R$ will represent one of the same form, the axis of x in the former equation having the same position with reference to the locus, as the axis of y in the latter. Hence we have two cases.

(i) When R is not $= 0$, we have a locus not yet investigated, which is called an *Hyperbola*.

(ii) When $R = 0$ we have the two intersecting straight lines

$$\sqrt{P}x + \sqrt{Q}y = 0, \quad \sqrt{P}x - \sqrt{Q}y = 0.$$

*151. *To sum up briefly.* In order to reduce the equation to a Central Locus,

(i) Find the Class (Art. 150) to which the locus belongs; and then, if $H^2 - AB$ is not $= 0$,

(ii) Write the equations of the centre (Art. 141), and obtain the values of its co-ordinates.

(iii) Substitute the halves of these co-ordinates (Art. 142) in the terms $2Gx$ and $2Fy$, and so reduce the equation to the form

$$Ax^2 + 2Hxy + By^2 + \phi(x'y') = 0.$$

The origin is now transferred to the centre of the locus.

(iv) To reduce the equation to the form

$$A'x^2 + B'y^2 + \phi(x'y') = 0,$$

we have (Art. 148)

$$A' + B' = A + B, \quad -A'B' = H^2 - AB,$$

and the value of $A' - B'$ hence derived must (Art. 149) be taken with the same sign as H. The axes have now been turned through an angle determined by the equation

$$\tan 2\theta = \frac{2H}{A - B}.$$

*Ex. Let the proposed equation be

$$5x^2 + 2xy + 5y^2 - 12x - 12y = 0 \qquad\qquad (1).$$

Here $H^2 - AB = -24$, or the locus belongs to Class I. The equations of the centre are

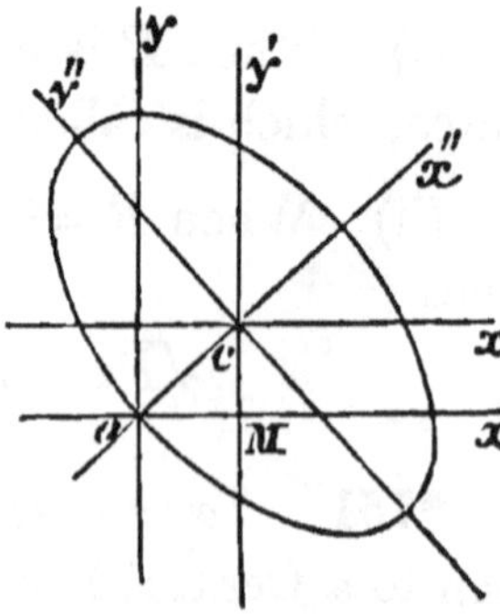

$$5x + y - 6 = 0,$$
$$x + 5y - 6 = 0,$$

whence $x = y = 1$. If then we transfer the origin to a point C, so that $OM = CM = 1$, C will be the centre, and the equation with the new axes Cx', Cy', parallel to the old ones, is found by writing $x = \frac{1}{2}$, $y = \frac{1}{2}$ in the last two terms of (1), to be

$$5x^2 + 2xy + 5y^2 - 12 = 0 \qquad\qquad (2).$$

Next, if we take $\tan 2\theta = \frac{2H}{A-B} = \frac{2}{5-5} = \infty$, we have $2\theta = 90^\circ$, $\theta = 45^\circ$; hence, if axes are turned through 45°, equation (2) becomes

$$A'x^2 + B'y^2 - 12 = 0 \ldots\ldots\ldots\ldots(3),$$

where $A' + B' = A + B = 10$, $-A'B' = H^2 - AB = -24$; whence $A' - B' = 2$ (Art. 149), since H is positive. Hence $A' = 6$, $B' = 4$, and (3) becomes

$$3x^2 + 2y^2 - 6 = 0.$$

This is now seen to be the locus which (Art. 150) we have called an *Ellipse;* and its position with regard to the new axes will be seen hereafter to be that of the figure. As the original equation contained no absolute term, it is evident (Art. 108 (ii)) that O is a point on the locus. Art. 288, applied to the original axes, will be of use in tracing the curve.

153. *Class* III. We shall now consider the case where $H^2 - AB = 0$. We saw (Art. 143) that in this case the coordinates of the centre become generally infinite; hence we cannot destroy the terms $2Gx$ and $2Fy$, by removing the origin to the centre; but, if we proceed as in Art. 145, and move the axes through an angle θ, the new coefficient of xy will vanish, when

$$\tan 2\theta = \frac{2H}{A-B},$$

for the introduction of the terms $2Gx$ and $2Fy$ will not affect the proof. But, since (Art. 147)

$$H'^2 - A'B' = H^2 - AB = 0,$$

if $H' = 0$, either A' or B' must $= 0$, and the equation will assume the form

$$B'y^2 + 2G'x + 2F'y + C' = 0 \ldots\ldots\ldots\ldots(1),$$

where we have supposed A' to vanish. If we supposed B' to vanish, the equation would not represent a different form of locus, but one having a situation with regard to the axes of x and y, similar to the situation of the supposed locus, with regard to the axes of y and x respectively.

If we transform the origin to a point $(x'y')$, equation (1) becomes

$$B'y^2 + 2G'x + 2(B'y' + F')y + B'y'^2 + 2G'x' + 2F'y' + C' = 0;$$

and, if we take for x' and y' the only pair of values that can be obtained from the equations

$$B'y' + F' = 0, \quad B'y'^2 + 2G'x' + 2F'y' + C' = 0,$$

the equation assumes its simplest form

$$B'y^2 + 2G'x = 0, \text{ or } y^2 = Lx.$$

Precisely as in Art. 145, it will be seen that there is one system of rectangular axes, and one only, which gives the equation to the locus in this form. The axis of x is now called *the axis* of the locus, and the origin of co-ordinates is called *the vertex*.

If $G' = 0$ in (1), the equation becomes

$$B'y^2 + 2F'y + C' = 0,$$

which represents (Art. 64) two straight lines parallel to the new axis of x, which are real and different, real and coincident, or imaginary, according as

$$F'^2 - B'C' > = < 0.$$

Hence, when $H^2 - AB = 0$, we have

(i) A locus not yet investigated, which is called a *Parabola*.

(ii) Two parallel straight lines.

(iii) Two coincident straight lines.

(iv) A locus, of which no geometrical conception can be formed, called two imaginary parallel straight lines.

*154. The preceding article is valuable, chiefly on account of the general conclusions to which it leads us. The values of A' and B' may be obtained as in Arts. 148, 149, and the student may prove that B' or A' vanish, according as H is positive or negative, and that $C' = C$. But the calculation of the values of G' and F' is tedious; and the student will find the method of Art. 156 more convenient for the working of numerical examples.

155. The reduction of Art. 153 may be made by a method suggested by the form of the equation itself. For, since $H^2 - AB = 0$, the first three terms in the general equation form a perfect square, and it may be written

$$(ax + by)^2 + 2Gx + 2Fy + C = 0 \ldots\ldots\ldots\ldots\ldots (1),$$

where a and b are written for $\sqrt{A}$ and $\sqrt{B}$. Suppose now (Art. 52, Cor. 1), that the lengths of straight lines drawn from (xy) to meet

$$ax + by = 0 \ldots\ldots\ldots\ldots (2), \quad 2Gx + 2Fy + C = 0 \ldots\ldots\ldots\ldots (3),$$

respectively, and parallel to (3) and (2) respectively, are

$$P(ax + by), \quad Q\,(2Gx + 2Fy + C),$$

where P and Q are constants. Now equation (1) may be written

$$P^2\,(ax + by)^2 = -\frac{P^2}{Q}\,Q\,(2Gx + 2Fy + C)\,;$$

this equation asserts, that the square of a straight line drawn from any point (xy) on the locus, parallel to (3) to meet (2), varies as the straight line drawn from the same point parallel to (2) to meet (3); hence, if we take (2) for the axis of x and (3) for the axis of y, the equation will assume the form

$$y^2 = Lx.$$

*156[1]. The new axes used in the last article are in general not rectangular. We shall now shew how to transform equation (1) to the same form, the new axes being

[1] From Salmon's *Conic Sections*.

rectangular. We shall suppose the co-ordinates in equation (1) to be rectangular.

If k be any constant, (1) may be written

$$(ax + by + k)^2 + 2(G - ak)\,x + 2\,(F - bk)\,y + C - k^2 = 0 \ldots (2).$$

The condition (Art. 47, Cor.) that the two lines

$$ax + by + k = 0 \ldots\ldots\ldots\ldots\ldots\ldots\ldots\ldots (3),$$

$$2\,(G - ak)\,x + 2\,(F - bk)\,y + C - k^2 = 0 \ldots\ldots (4),$$

should be at right angles, is

$$a\,(G - ak) + b\,(F - bk) = 0 \ldots\ldots\ldots\ldots\ldots (5);$$

whence $$k = \frac{Ga + Fb}{a^2 + b^2} = \kappa \text{ say.}$$

Substitute this value for k in (2), and for brevity write it

$$(ax + by + \kappa)^2 + Px + Qy + R = 0 \ldots\ldots\ldots\ldots (6),$$

where the two lines

$$ax + by + \kappa = 0 \;\ldots\ldots\; (7), \qquad Px + Qy + R = 0 \;\ldots\ldots\; (8),$$

are at right angles to one another. Now (6) may be written

$$\frac{(ax + by + \kappa)^2}{a^2 + b^2} = -\frac{\sqrt{P^2 + Q^2}}{a^2 + b^2} \times \frac{Px + Qy + R}{\sqrt{P^2 + Q^2}} \;\ldots\ldots\; (9),$$

and we know from Art. 52, that

$$\frac{ax + by + \kappa}{\sqrt{a^2 + b^2}} \text{ and } \frac{Px + Qy + R}{\sqrt{P^2 + Q^2}},$$

are the lengths of the perpendiculars from a point (xy) on (7) and (8); hence, if we construct the two lines (7) and (8), the equation (9) asserts, that the square of the perpendicular from any point (xy) of the locus on the first line, varies as the perpendicular on the second line.

Hence, if we transform our axes, and make the line (7)

our new axis of x, and (8) our new axis of y, then our new y will be the perpendicular on (7), and our new x will be the perpendicular on (8), and equation (9) is reduced to the form

$$y^2 = Lx.$$

If the lines $(ax + by = 0)$ and $(2Gx + 2Fy + C = 0)$ are at right angles to one another, equation (1) may be written in the form of equation (9) at once.

Since the left-hand member of equation (9) is always positive, the points on the locus must be such as to make the right-hand member positive also. This will be seen in working the examples at the end of the article.

The lines (3) and (4), with κ substituted for k in them, are, as will be seen hereafter, the axis of the parabola and the tangent at its vertex.

COR. Since the axis of the parabola $(ax + by + \kappa = 0)$ is always parallel to $(ax + by = 0)$, it follows that, if the equations to two parabolas can be written so as to have the first three terms the same, their axes will be parallel. Also, since κ and L do not depend upon the constant term C, if the equations can be written so as to differ in the constant term only, the parabolas will have their axes coincident, and will be equal.

*Ex. 1. Let the proposed equation be

$$x^2 - 2xy + y^2 - 8x + 16 = 0 \text{ ....(1)},$$

therefore $$(x - y + k)^2 - (8 + 2k)\, x + 2ky + 16 - k^2 = 0 \text{.......(2)}.$$

Then, that the lines may be at right angles,

$$-(8 + 2k) - 2k = 0, \text{ whence } k = -2.$$

Equation (2) then becomes

$$(x - y - 2)^2 - 4\,(x + y - 3) = 0,$$

or $$\frac{(x - y - 2)^2}{2} = 2\sqrt{2}\,\frac{x + y - 3}{\sqrt{2}} \text{ ....(3)},$$

where (3) is written in terms of the perpendiculars on the lines

$$x - y - 2 = 0 \text{............(4)}, \qquad x + y - 3 = 0 \text{............(5)}.$$

Let CX and CY be the lines (4) and (5); then, in order to make the expression $x+y-3$ positive, the point (xy) must be on the positive side of the line (5); that is, every point on the locus is (Art. 55, Ex.) on the side of CY remote from the origin; hence equation (3) asserts that

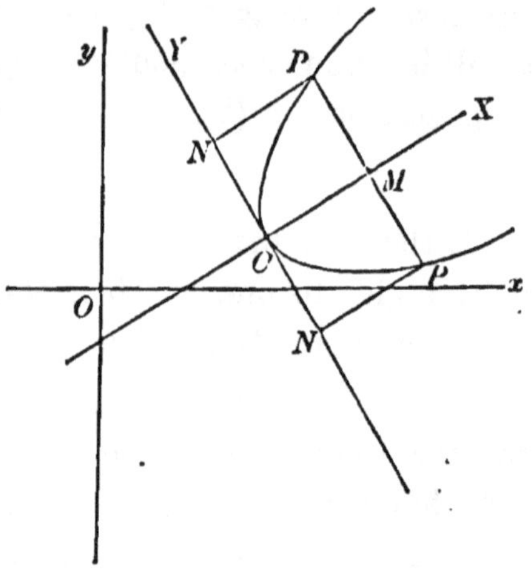

$$PM^2 = 2\sqrt{2}\, PN,$$

where PM may be drawn on either side of CX, but PN is always on the side of CY remote from the origin. If we take CX and CY as the positive directions of the axes of x and y respectively, equation (3) becomes

$$y^2 = 2\sqrt{2}x.$$

The form of the locus will be seen hereafter to be that of the figure.

Ex. 2. Let the equation be

$$x^2 - 2xy + y^2 + 8y - 8 = 0 \ldots\ldots\ldots\ldots\ldots\ldots(1);$$

then, exactly as in Ex. (1), the equation reduces to

$$\frac{(x-y-2)^2}{2} = -2\sqrt{2}\,\frac{x+y-3}{\sqrt{2}} \ldots\ldots\ldots\ldots(2).$$

Here the lines CX, CY are the same as in Ex. (1); and, in order to make $x+y-3$ *negative*, every point of the locus must be on the negative or origin side of CY; hence the curve is the same as that in the figure, but is drawn in the opposite direction; and, if we take axes as in Ex. (1), equation (2) becomes

$$y^2 = -2\sqrt{2}x.$$

Ex. 3. The equation

$$4x^2 - 4xy + y^2 - 10x - 20y = 0,$$

may at once be written

$$\frac{(2x-y)^2}{5} = \sqrt{5}\,\frac{x+2y}{\sqrt{5}},$$

since $(2x-y=0)$ and $(x+2y=0)$ are at right angles. The perpendicular on the left hand may be on either side of the line $(2x-y=0)$, but that on the right must (Art. 56) be *above* the line $(x+2y=0)$. The figure will resemble that of Ex. (1), the point C coinciding with O. Art. 288, applied to the original axes, will be of use in tracing the curves.

157. We said, Art. 143, that, when $H^2 - AB = 0$, the co-ordinates of the centre become *generally* infinite.

If however $H^2 - AB = 0$ *and* $HF - BG = 0$;

or
$$\frac{A}{H} = \frac{H}{B} = \frac{G}{F} \text{.........................} (1),$$

then also $HG - AF = 0$, and both the co-ordinates of the centre in Art. 141 become $= \frac{0}{0}$, and are therefore indeterminate. The two equations

$$Ax + Hy + G = 0, \quad Hx + By + F = 0,$$

will now, from (1), represent two straight lines that coincide; hence there are an indefinite number of centres, all situated in that line. The proposed equation, with the above relation between its coefficients, no longer represents a parabola; for the solution of Art. 62 becomes

$$Hx + By + F \pm \sqrt{F^2 - BC} = 0,$$

which represents two parallel straight lines equidistant from the line $Hx + By + F = 0$, which is therefore a line of centres.

*EXAMPLES VIII.

Transform the following equations, illustrating each transformation by a figure, as in Arts. 151, 156.

1. $3y^2 + 2x + 1 = 0$ to $y^2 = -\frac{2}{3}x$.

2. $3x^2 + 2y^2 - 2x + y - 1 = 0$ to $72x^2 + 48y^2 = 35$.

3. $3x^2 + 2xy + 3y^2 - 16y + 23 = 0$ to $4x^2 + 2y^2 = 1$.

4. $x^2 - 10xy + y^2 + x + y + 1 = 0$ to $32x^2 - 48y^2 = 9$.

5. $x^2 - 2xy + y^2 - 6x - 6y + 9 = 0$ to $y^2 = 3\sqrt{2}x$.

6. $x^2 + xy + y^2 + x + y - 5 = 0$ to $9x^2 + 3y^2 = 32$.

7. $x^2 + 2xy - y^2 - 2 = 0$ to $x^2 - y^2 - \sqrt{2} = 0$.

8. $x^2 - y^2 + y = 0$ to $4x^2 - 4y^2 + 1 = 0$.

9. Shew by transformation that the equation

$$12xy + 8x - 27y - 18 = 0,$$

represents two straight lines parallel to the axes.

10. Shew by transformation that the equation

$$3x^2 - 2xy + y^2 - 10x - 2y + 19 = 0,$$

represents two imaginary straight lines passing through the point (3, 4).

11. Shew by transformation that the equation

$$5x^2 - 4xy + y^2 - 4x + 2y + 2 = 0,$$

represents an imaginary ellipse.

12. Shew that any point on the line $(y = x + 1)$ is a centre of the locus

$$x^2 - 2xy + y^2 + 2x - 2y = 0.$$

13. Shew by transformation that the equation

$$x^2 + 2xy + y^2 + 1 = 0,$$

represents two imaginary parallel straight lines.

14. What is the equation to the axis in Ex. 5, and to the axes in Art. 151 Ex.?

15. Transform $16x^2 + 16xy + 7y^2 + 64x + 32y + 28 = 0$, the axes being inclined at an angle of 60°, to $4x^2 + y^2 = 9$, the axes being rectangular, and the axis of x remaining the same.

CHAPTER IX.

Central Conic Sections[1], *referred to their axes.*

158. In this chapter we shall consider the equations, whose loci (Art. 150) we have called the Ellipse and Hyperbola, namely,

$$Px^2 + Qy^2 = R, \quad Px^2 - Qy^2 = R,$$

where P, Q, R are supposed positive.

The equation to the ellipse may be written

$$\frac{x^2}{\frac{R}{P}} + \frac{y^2}{\frac{R}{Q}} = 1.$$

Put $\dfrac{R}{P} = a^2$, $\dfrac{R}{Q} = b^2$; then the equation becomes

$$\frac{x^2}{a^2} + \frac{y^2}{b^2} = 1,$$

where a and b are evidently the intercepts of the curve on the axes of x and y respectively.

The equation to the hyperbola, which differs from that

[1] The term 'conic section,' or 'conic,' must be understood to mean 'locus of the second degree,' and to embrace every variety of locus mentioned in Chap. VIII. It can be proved (Art. 323) that the section, made by any plane in a cone standing on a circular base, is a locus of the second degree. It was as sections of the cone, that the properties of these curves were first examined. Hence the name.

of the ellipse in the sign of the coefficient of y^2 only, may be written in the corresponding form

$$\frac{x^2}{a^2} - \frac{y^2}{b^2} = 1;$$

but, in this case, when $x = 0$, $y^2 = -b^2$, so that the intercept on the axis of y is an imaginary quantity, or the curve does not meet that axis.

159. The figure of the curves may now be deduced from the simple form to which we have reduced their equations. We will begin with the ellipse, and, since we may choose whichever axis we please for axis of x, we shall suppose that we have so chosen the axes, that a may be greater than b.

160. *To find the polar equation to the ellipse; the centre being the pole.*

Writing (Art. 13) $\rho \cos\theta$ for x, and $\rho \sin\theta$ for y in the equation $\frac{x^2}{a^2} + \frac{y^2}{b^2} = 1$, we have

$$\frac{1}{\rho^2} = \frac{\cos^2\theta}{a^2} + \frac{\sin^2\theta}{b^2},$$

or
$$\rho^2 = \frac{a^2b^2}{a^2\sin^2\theta + b^2\cos^2\theta},$$

which is the required equation. It may be written

$$\rho^2 = \frac{a^2b^2}{a^2 - (a^2 - b^2)\cos^2\theta} = \frac{a^2b^2}{b^2 + (a^2 - b^2)\sin^2\theta},$$

and it will appear hereafter, that it is convenient to use the abbreviation $\frac{a^2 - b^2}{a^2} = e^2$. Hence, dividing numerator and denominator by a^2, we have

$$\rho^2 = \frac{b^2}{1 - e^2\cos^2\theta},$$

the form most commonly used.

161. This equation will be found the most convenient for tracing the ellipse[1]. The least value that $b^2 + (a^2 - b^2)\sin^2\theta$ can have, is when $\theta = 0°$ or $180°$; therefore, since

$$\rho^2 = \frac{a^2b^2}{b^2 + (a^2 - b^2)\sin^2\theta},$$

the *greatest* values of ρ are the intercepts on the axis of x, which are each $= a$.

Again, the greatest values of $b^2 + (a^2 - b^2)\sin^2\theta$ are, when $\sin^2\theta = 1$, i.e., when $\theta = 90°$ or $270°$; hence, the *least* values of ρ are the intercepts on the axis of y, which are each $= b$. The greatest chord then that can be drawn through the centre is the axis of x, and the least chord, the axis of y.

From this property, these lines, $A'A$ $(= 2a)$ and $B'B$ $(= 2b)$ are called the axis *major* and the axis *minor* of the curve. It is plain, that the smaller $\sin^2\theta$ is, the greater ρ will be; hence, *the nearer any radius vector is to the axis major, the greater it will be.* By taking the two values of ρ, positive and negative, for each value of θ, we shall, as θ varies from $0°$ to $90°$, trace the portions AB, $A'B'$; and, as θ varies from $90°$ to $180°$, the portions BA', $B'A$. The form of the curve will therefore be that of the figure. The points A', A are called the vertices.

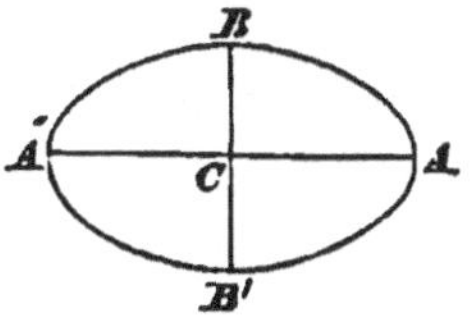

162. We obtain the same value for ρ, whether we suppose $\theta = \alpha$, or $\theta = -\alpha$. Hence, *Two radii vectores which make equal angles with the axis will be equal.* And it is easy to shew that the converse of this theorem is true.

163. The figure of the ellipse may also be seen from the following construction.

[1] The Articles on the figure of the curves are taken chiefly from Salmon's *Conic Sections.*

Solving the equation to the ellipse for y, we have

$$y = \pm \frac{b}{a}\sqrt{a^2 - x^2}.$$

Now, if we describe a concentric circle with the radius a, its equation will be

$$y = \pm \sqrt{a^2 - x^2}.$$

Hence, if a circle be described on the axis major, and on each ordinate MQ a point P be taken, such that MP may be to MQ in the constant ratio $b : a$, then the locus of P will be the required ellipse; hence the circle described on the axis major lies wholly *without* the curve.

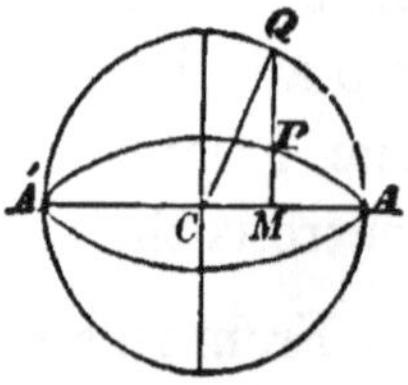

We might, in like manner, construct the ellipse, by describing a circle on the axis minor, and *increasing* each ordinate in the constant ratio $a : b$; hence the circle described on the axis minor lies wholly *within* the curve.

We see also, that the equation to the circle is the particular form which the equation to the ellipse assumes, when $b = a$.

*164. Let CQ be joined, and let the angle $QCM = \phi$; then, if x and y are the co-ordinates of P,

$$x = CQ \cos\phi = a \cos\phi; \quad y = \frac{b}{a}\, QM = \frac{b}{a}\, a \sin\phi = b \sin\phi.$$

Thus the co-ordinates of any point may be expressed in terms of a single variable ϕ. These values of x and y evidently satisfy the equation to the ellipse; for, when substituted in it, they produce the equation

$$\cos^2\phi + \sin^2\phi = 1,$$

which is always true. The angle ϕ is called the *eccentric angle* of the point P.

Ex. 1. *To find the equation to the chord of an ellipse, which passes through two points whose eccentric angles are given.*

Let the points be

$$a\cos\theta,\ b\sin\theta; \qquad a\cos\phi,\ b\sin\phi;$$

then the equation to the chord is

$$\frac{x-a\cos\theta}{y-b\sin\theta}=\frac{a(\cos\phi-\cos\theta)}{b(\sin\phi-\sin\theta)},$$

whence, exactly as in Art. 109, Ex. 1,

$$\frac{x}{a}\cos\frac{\theta+\phi}{2}+\frac{y}{b}\sin\frac{\theta+\phi}{2}=\cos\frac{\theta-\phi}{2}.$$

If the chord becomes a tangent, $\theta=\phi$, and the equation to the tangent is

$$\frac{x}{a}\cos\phi+\frac{y}{b}\sin\phi=1.$$

Ex. 2. *To find the equation to the normal at a point of an ellipse, whose eccentric angle is given.*

Let ϕ be the eccentric angle of the point; then the equation to the tangent is

$$\frac{x}{a}\cos\phi+\frac{y}{b}\sin\phi=1 \text{.......................}(1),$$

and the equation to a line passing through ($a\cos\phi$, $b\sin\phi$) perpendicular to (1) is (Art. 50, Cor. 2)

$$\frac{y-b\sin\phi}{x-a\cos\phi}=\frac{a\sin\phi}{b\cos\phi},$$

or

$$\frac{ax}{\cos\phi}-\frac{by}{\sin\phi}=a^2-b^2.$$

165. *To find the equation to the ellipse, when one of the vertices is origin, the direction of the axes being the same as before.*

The problem is, to transfer the origin of co-ordinates to the point A' $(-a, 0)$; hence, writing $x-a$ for x in the equation to the ellipse,

$$y^2=\frac{b^2}{a^2}(a^2-x^2),$$

we have $$y^2 = \frac{b^2}{a^2}(2ax - x^2),$$

for the equation to the ellipse when the origin is the vertex A'.

166. We shall now investigate the figure of the hyperbola from its equation

$$\frac{x^2}{a^2} - \frac{y^2}{b^2} = 1.$$

The intercept on the axis of x is evidently $= \pm a$, but that on the axis of y, being found from the equation $\frac{y^2}{b^2} = -1$, is imaginary; the axis of y then does not meet the curve in real points.

If, however, we take an hyperbola whose equation is

$$\frac{y^2}{b^2} - \frac{x^2}{a^2} = 1,$$

the axis of y will meet this curve at a distance $= \pm b$ from the origin, and the axis of x will not meet it in real points. This (for reasons evident hereafter) is called the *conjugate hyperbola*, and possesses properties which will be of use to us in considering those of the original curve. We shall then call the distance $A'A$ $(=2a)$, between the two vertices, or points where the curve meets the axis of x, the *transverse axis*, and we shall call the distance $B'B$ between the two points where the conjugate hyperbola meets the axis of y, the *conjugate* axis. For we have chosen as axis of x that which meets the hyperbola in real points, and are therefore not entitled to assume a greater than b, so that the terms axis major and axis minor would not here be applicable.

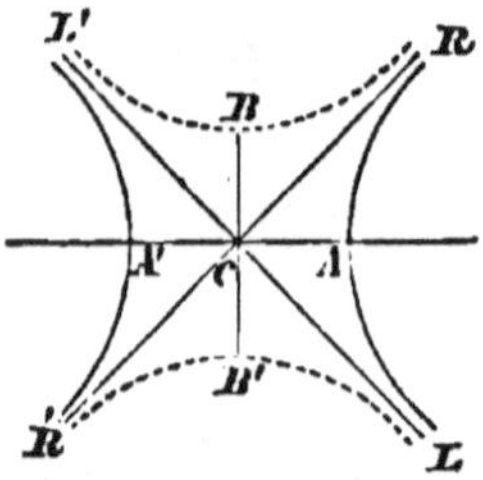

The conjugate hyperbola will evidently be an hyperbola, whose transverse and conjugate axes correspond to the conjugate and transverse axes of the original curve.

167. *To find the polar equation to the hyperbola, centre pole.*

Transforming the equation to the hyperbola to polar co-ordinates as in the case of the ellipse, we get

$$\rho^2 = \frac{a^2b^2}{b^2\cos^2\theta - a^2\sin^2\theta} = \frac{a^2b^2}{b^2 - (a^2+b^2)\sin^2\theta} = \frac{a^2b^2}{(a^2+b^2)\cos^2\theta - a^2}.$$

Since formulæ concerning the ellipse are altered to the corresponding formulæ for the hyperbola by changing the sign of b^2, we must, in this case, use the abbreviation e^2 for $\frac{a^2+b^2}{a^2}$. Dividing numerator and denominator by a^2, we have for the polar equation to the hyperbola, the centre being pole,

$$\rho^2 = \frac{b^2}{e^2\cos^2\theta - 1},$$

the form most commonly used.

168. The hyperbola may be conveniently traced from this equation. The denominator $b^2 - (a^2+b^2)\sin^2\theta$ will plainly be greatest when $\theta = 0°$ or $180°$, therefore, for these values, ρ will be least; or, *the transverse axis is the least chord which can be drawn through the centre to the curve.*

As θ increases from $0°$, ρ continually increases, until

$$\sin\theta = \frac{b}{\sqrt{a^2+b^2}}\left(\text{or } \tan\theta = \frac{b}{a}\right),$$

when the denominator of the value of ρ becomes $= 0$, and ρ becomes infinite. After this value of θ, ρ^2 becomes negative, and the radii vectores cease to meet the curve in real points, until again

$$\sin\theta = \frac{b}{\sqrt{a^2+b^2}} \left(\text{or } \tan\theta = -\frac{b}{a}\right),$$

when ρ again becomes infinite. It then decreases regularly as θ increases, until θ becomes $= 180°$, when it again receives its minimum value $= a$. The lower part of the curve evidently corresponds exactly with the upper.

The form of the hyperbola, therefore, is that represented by the dark line in the figure, where the branches LAR, $L'A'R'$ extend to infinity. We have drawn the curves always concave to the axis of x. We shall hereafter (Art. 188) prove the correctness of the figures in this respect.

169. It was shewn that the radii vectores answering to $\tan\theta = \pm\frac{b}{a}$ meet the curve at infinity. These radii vectores, indefinitely produced, are, for reasons given in Article 171, called the *asymptotes* of the curve. They are the lines $R'R$, $L'L$ of the figure, and evidently separate the lines which meet the curve in real points, from those which meet it in imaginary points; i.e. the whole of the curve is included in the angles RCL, $R'CL'$. Hence the equations to the asymptotes are

$$y = \pm\frac{b}{a}x, \text{ or, as one locus, } \frac{x^2}{a^2} - \frac{y^2}{b^2} = 0.$$

Similarly for the conjugate hyperbola, the equations to the asymptotes are

$$x = \pm\frac{a}{b}y, \text{ or, as one locus, } \frac{x^2}{a^2} - \frac{y^2}{b^2} = 0.$$

Hence the asymptotes of the conjugate hyperbola coincide with those of the original curve, and it lies wholly within the angles RCL', LCR', corresponding to the dotted line in the figure.

170. If $a=b$, the hyperbola and its conjugate become equal in every respect, and, since (Art. 169)

$$\frac{b}{a}=1=\tan RCA,$$

RCA will $=45^\circ$, or the angle between the asymptotes is a right angle. This is called the rectangular or equilateral hyperbola, and is to the hyperbola what the circle is to the ellipse. Its equation is

$$x^2-y^2=a^2 \quad\text{.......................(1).}$$

See Art. 186 and Cors.

171. DEF. *An asymptote is a straight line, the distance of which from a point of a curve diminishes without limit, as the point in the curve moves to an infinite distance from the origin.*

It must not be assumed, if the value of any radius vector becomes infinite, as in Art. 169, that it is therefore an asymptote to the curve. In fact, it will be proved hereafter, that any lines drawn parallel to RR' and LL' have *one* point of intersection with the curve at an infinite distance, just as in Art. 42 we found the co-ordinates of the intersection of parallel lines to be infinite: but in neither case do the loci approach indefinitely near to one another. We shall therefore shew that the lines RR' and LL' correspond to the above definition.

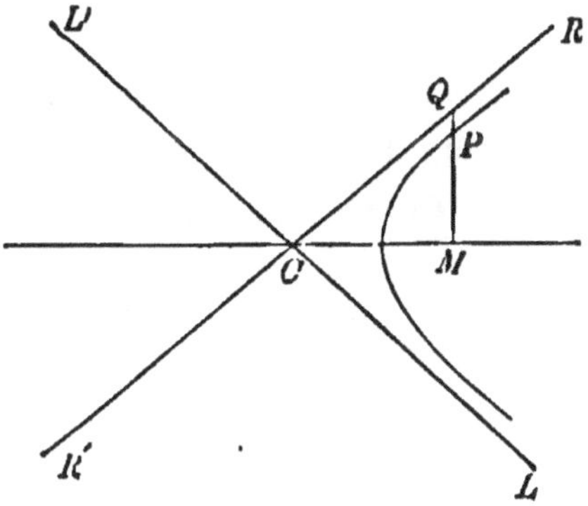

If the ordinate MP be produced to meet RR' in Q, the distance of the point P from $RR'=PQ\sin PQC$, and therefore varies as PQ. Now if $CM=x'$, $PM=y'$, $QM=y_1$, we have from the equations to the curve and asymptote,

$$\frac{y'^2}{b^2} = \frac{x'^2}{a^2} - 1, \quad \frac{y_1^2}{b^2} = \frac{x'^2}{a^2},$$

therefore $\quad y_1^2 - y'^2 = b^2$, or $y_1 - y' = \dfrac{b^2}{y_1 + y'}$.

But this value of $y_1 - y'$, which $= PQ$, becomes indefinitely small, when y_1 and y' become indefinitely large, and therefore RR' is an asymptote. Similarly LL' is an asymptote.

*172. The co-ordinates of any point in the hyperbola may, as in the ellipse, be expressed by a single variable by means of the *eccentric angle;* for we may put

$$x = a \sec \phi, \quad y = b \tan \phi,$$

since these values, when substituted in the equation

$$\frac{x^2}{a^2} - \frac{y^2}{b^2} = 1 \qquad (1),$$

will give

$$\sec^2 \phi - \tan^2 \phi = 1 \qquad (2),$$

which is always true. If a tangent MQ be drawn from the foot of the ordinate MP, to the circle on the transverse axis, the angle QCM is the eccentric angle of the point P; for

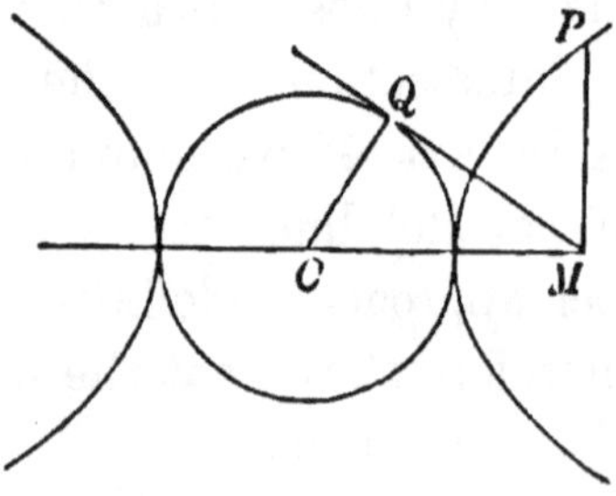

$$CM = CQ \sec QCM,$$

or

$$x = a \sec \phi,$$

and therefore, from (1) and (2), $y = b \tan \phi$.

173. As in the case of the ellipse, we shall find the equation to the hyperbola, when the vertex A is taken for origin, by writing $x + a$ for x in the equation

$$y^2 = \frac{b^2}{a^2}(x^2 - a^2).$$

The result is

$$y^2 = \frac{b^2}{a^2}(2ax + x^2).$$

174. The quantity e is called the *eccentricity* of the curve;

in the ellipse, $$e^2 = \frac{a^2 - b^2}{a^2};$$

in the hyperbola, $$e^2 = \frac{a^2 + b^2}{a^2}.$$

Hence in the ellipse e is less than unity; and in the hyperbola greater than unity. When $a = b$ in the ellipse, or the curve becomes a circle, $e = 0$, and, if a remain the same, e increases as b diminishes, or as the curve passes from a circular to an oval form. If θ is the angle between the asymptotes of the hyperbola, in which the transverse axis lies, we have

$$\tan\frac{\theta}{2} = \frac{b}{a}; \quad \text{therefore } \sec^2\frac{\theta}{2} = \frac{a^2 + b^2}{a^2} = e^2.$$

175. In the following investigations we shall in most cases consider both the ellipse and hyperbola together, as they have many properties in common, resulting from the similarity of their equations; and the properties of the one may be deduced from those of the other by changing the sign of b^2. The properties of the conjugate hyperbola may be deduced from those of the ellipse by changing the sign of a^2, or from those of the given hyperbola by changing the signs of a^2 and b^2. We shall, whenever we speak of the ellipse and hyperbola together, use the signs which apply to the ellipse.

176. *To find the length of a straight line drawn to the curve*

$$\frac{x^2}{a^2} + \frac{y^2}{b^2} = 1,$$

from the point $(x'y')$.

Let the equation to the line be

Then, as in Art. 114, for the distances (l) between $(x'y')$ and the points of section of the line and curve, we have

$$\frac{(cl+x')^2}{a^2}+\frac{(sl+y')^2}{b^2}=1;$$

$$\therefore \left(\frac{c^2}{a^2}+\frac{s^2}{b^2}\right)l^2+2\left(\frac{cx'}{a^2}+\frac{sy'}{b^2}\right)l+\frac{x'^2}{a^2}+\frac{y'^2}{b^2}-1=0 \ldots\ldots(2),$$

or $$Pl^2+Ql+R=0.$$

Now this equation will always give two values for l; hence every straight line meets the curve in two real, coincident, or imaginary points, according as the roots of (2) are real and unequal, real and equal, or imaginary.

We shall hereafter have occasion to consider the following particular forms that this equation may assume:

If $P=0$, one value of l (Appendix) becomes infinite.

If $P=0$ and $Q=0$, both values of l become infinite.

If $R=0$, the point $(x'y')$ is on the curve, and one value of l becomes $=0$.

If $R=0$ and $Q=0$, both values of l become $=0$, and the line passes through two coincident points of the curve, and is a tangent.

If $Q=0$, the roots of the equation are equal and of opposite signs, and $(x'y')$ is therefore the middle point of the chord.

177. If $P=0$ in equation (2), we have, according as the curve is an ellipse or hyperbola,

$$\frac{c^2}{a^2}+\frac{s^2}{b^2}=0, \text{ or } \frac{c^2}{a^2}-\frac{s^2}{b^2}=0.$$

The former equation gives imaginary values for $\frac{s}{c}$; the latter gives

$$\frac{s}{c} = \pm \frac{b}{a};$$

but $\frac{s}{c}$ is the tangent of the angle that the line (1) makes with the axis of x; hence (Art. 169) *one* value of l becomes infinite, if the line (1) is parallel to either asymptote, i.e. *a straight line drawn parallel to an asymptote meets the curve in one point only at a finite distance from the origin, and in one point at an infinite distance.*

178. If $P=0$ and $Q=0$, we have in the hyperbola

$$\frac{s}{c} = \pm \frac{b}{a}, \quad \frac{y'}{x'} = \frac{cb^2}{sa^2} = \pm \frac{b}{a}.$$

Hence by the first condition the line (1) is parallel to an asymptote, and by the second the point $(x'y')$ is on the same asymptote; i.e. the line (1) is one of the asymptotes, and both the values of l are infinite.

179. We have said that a line parallel to an asymptote meets the curve in one point at an infinite distance. This is a short way of enunciating the property, which may be stated more clearly as follows. Let K be the point $(x'y')$, and PKD a straight line cutting the hyperbola in P and D, and let KL be a straight line parallel to the asymptote CV; then, if the point D moves along the curve to an infinite distance, and the line PKD turns about K, it

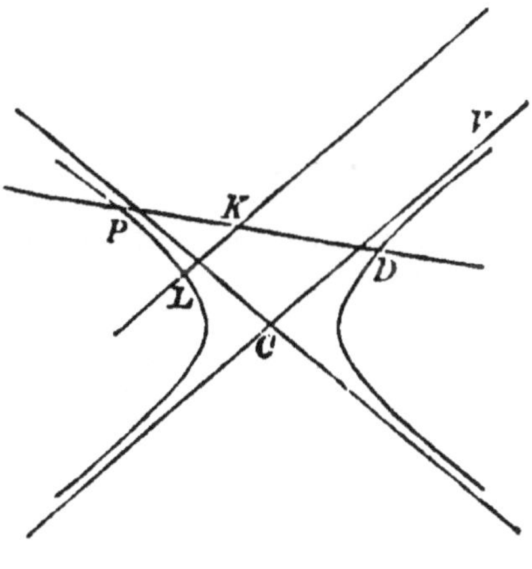

will tend to coincide with KL, as its limiting position. Without this explanation, the statement would probably perplex the student, inasmuch as he would see that KL does not actually meet the curve at infinity, for the distance between it and the curve can never become less than the distance between it and the asymptote. In the same way we have said (Art. 42) that parallel straight lines meet at infinity; and our statement here simply asserts, that the curve tends to become a straight line parallel to KL, i.e. the straight line CV.

180. *To find the equation to a straight line touching the curve*

$$\left(\frac{x^2}{a^2}+\frac{y^2}{b^2}=1\right),$$

at the point $(x'y')$.

The method employed is precisely the same as that used in the case of the tangent to the circle (Art. 117), to which the student may refer for the figure. We shall simply indicate the steps.

Let the equation to a straight line cutting the curve in the point $Q\,(x'y')$, be

$$\frac{x-x'}{c}=\frac{y-y'}{s}=l \text{.....................(1)},$$

then, for the distances (l) between $(x'y')$ and the points of section of the line and curve, we obtain equation (2) of Art. 176, which becomes

$$\left(\frac{c^2}{a^2}+\frac{s^2}{b^2}\right)l^2+2\left(\frac{cx'}{a^2}+\frac{sy'}{b^2}\right)l=0 \text{...........(2)},$$

since $\frac{x'^2}{a^2}+\frac{y'^2}{b^2}=1$; for $(x'y')$ is on the curve.

Equation (2) gives us $l=0$, as it should, for $(x'y')$ coincides with one of the points of section, and also

$$\left(\frac{c^2}{a^2}+\frac{s^2}{b^2}\right)l+2\left(\frac{cx'}{a^2}+\frac{sy'}{b^2}\right)=0,$$

the value of l in which equation is the distance (QR) of $(x'y')$ from the other point of section R. If the line be a tangent at Q $(x'y')$, this distance vanishes, and we have

$$\frac{cx'}{a^2}+\frac{sy'}{b^2}=0.$$

Eliminating c and s, by means of this equation and equation (1), we have

$$(x-x')\frac{x'}{a^2}+(y-y')\frac{y'}{b^2}=0,$$

or
$$\frac{xx'}{a^2}+\frac{yy'}{b^2}-\left(\frac{x'^2}{a^2}+\frac{y'^2}{b^2}\right)=0,$$

whence
$$\frac{xx'}{a^2}+\frac{yy'}{b^2}=1,$$

which is the equation to the tangent at the point $(x'y')$, and is easily remembered from its similarity to the equation to the curve. The form of the equation, when the origin is transferred, may be seen from Art. 118.

181. If m be the trigonometrical tangent of the angle which the tangent at $(x'y')$ makes with the axis of x, we have

$$m=\frac{s}{c}=-\frac{b^2x'}{a^2y'}.$$

Since this value of m does not change, when x', y' are replaced by $-x'$, $-y'$, we see that tangents at the extremities of chords through the centre are parallel. Also at the points B, B', for which $x'=0$, $y'=\pm b$, $m=0$, or the tangents are parallel to the axis of x; and at A, A', for which $x=\pm a$, $y=0$, $m=\infty$, or the tangents are perpendicular to the axis of x.

182. *To find the equation to the tangent in terms of its inclination to the axis of x.*

The reader will have no difficulty in using the methods of Arts. 119, 120 for the purpose of finding this equation independently, or in deducing it from Art. 180. The resulting equation to the tangent is

$$y = mx \pm \sqrt{m^2a^2 + b^2}.$$

The double sign refers to the two tangents at the extremities of chords through the centre, since, by the last article, these tangents have the same inclination to the axis of x.

Cor. Hence if a straight line and a conic be represented by

$$y = mx + \beta, \text{ and } \frac{x^2}{a^2} + \frac{y^2}{b^2} = 1,$$

the condition of tangency is

$$\beta^2 = m^2a^2 + b^2.$$

183. If the tangent pass through the centre, we have $\sqrt{m^2a^2 + b^2} = 0$. (Art. 28, Cor. 3.) This gives an imaginary value for m in the case of the ellipse, but for the hyperbola we have

$$m^2a^2 - b^2 = 0, \text{ or } m = \pm\frac{b}{a};$$

hence a tangent to the hyperbola drawn from the centre coincides (Art. 169) with the asymptote, and meets the curve at an infinite distance only. The asymptotes may therefore be considered as straight lines *touching* the curve at an infinite distance.

184. The result of the last article may be obtained from the equation to the tangent at any point $(x'y')$, as follows:

The equation to the tangent to the hyperbola at $(x'y')$, is (Art. 180),

$$\frac{xx'}{a^2} - \frac{yy'}{b^2} = 1 \ldots\ldots (1), \qquad \text{or } \frac{x}{a^2} \cdot \frac{x'}{y'} - \frac{y}{b^2} = \frac{1}{y'} \ldots\ldots (2);$$

hence, when y' is indefinitely large, or the point of contact is indefinitely distant, the right-hand member of (2) vanishes; also from the equation to the hyperbola, since $(x'y')$ is a point on the curve,

$$\frac{x'^2}{y'^2} = \frac{a^2}{b^2} + \frac{a^2}{y'^2};$$

and therefore, when x' and y' become infinitely great, ultimately

$$\frac{x'^2}{y'^2} = \frac{a^2}{b^2} \text{ or } \frac{x'}{y'} = \pm \frac{a}{b}.$$

Hence equation (2) becomes, after reducing and transposing,

$$\frac{x}{a} = \pm \frac{y}{b}, \text{ or as one locus, } \frac{x^2}{a^2} - \frac{y^2}{b^2} = 0 \ldots\ldots\ldots\ldots (2),$$

which represents the two asymptotes.

In the ellipse the same process would give ultimately

$$\frac{x'^2}{y'^2} = -\frac{a^2}{b^2},$$

which gives for the asymptotes of the ellipse the two imaginary straight lines

$$\frac{x}{a} = \pm \sqrt{-1}\, \frac{y}{b}, \text{ or as one locus, } \frac{x^2}{a^2} + \frac{y^2}{b^2} = 0.$$

It will perhaps be observed, that in the above proof we began with the equation to *one* straight line (1), and, on endeavouring to trace its limiting position, consequent on the indefinite increase of x' and y', we have obtained an equation which represents *two* straight lines. Can each of these be the limiting position of (1)? The difficulty vanishes, if we take into consideration that (1) represents, not one, but a class of lines, indefinite in their number, and subject only to the limitation that $(x'y')$ should be on the curve. The suppositions we have made with regard to x' and y', are equally applicable to the whole class. Hence the resulting equation (2) represents the limiting position, not of one tangent, but of all of them.

185. Since the asymptote touches the branch RAL (fig. Art. 166) at infinity, it must also, by the symmetry of the figure, touch the branch $R'A'L'$ at infinity. But we saw (Art. 176) that, when we combine the equations to a straight line and the curve, the result is a quadratic, which will determine two points of *intersection*, and only two. That is, a straight line can only intersect a curve of the second order in two points, and, when these two points coincide, the line is said to touch the curve; hence the two points of contact of the asymptote would seem to coincide. This difficulty will be removed, when we consider that the asymptote RR' does not really *touch the hyperbola at infinity*, and that this expression is only a short way of saying, that, as the distance of the point of contact becomes very great, whether on the branch AR or $A'R'$, the tangent in each case tends to coincide with the line RR'.

186. We see then that, if a central conic is referred to its axes, the equations to the conic and its asymptotes are

$$Px^2 + Qy^2 = R, \qquad Px^2 + Qy^2 = 0,$$

the asymptotes being real, only when P and Q are of opposite signs. Now let any transformation of axes be made, the centre remaining the origin, so that we write (Art. 72) $ax + by$ for x, and $a'x + b'y$ for y, where a, a', b, b' represent constant quantities; then, if the equation to the conic becomes

$$Ax^2 + 2Hxy + By^2 = R,$$

(for R will evidently not be altered), the equation to the asymptotes will become

$$Ax^2 + 2Hxy + By^2 = 0.$$

Hence, with any axes whose origin is the centre, the equation to the asymptotes is found by equating to zero the terms of the second degree in the equation to the conic. Also, since (Art. 141) these terms are not altered, when the origin is transformed from the centre to any point whatsoever, the direction of the axes remaining the same, we shall, by equating them to zero in *any* equation, find a pair of straight lines drawn through the origin parallel to the asymptotes. This article explains the coincidence mentioned in Art. 146, since the axes bisect the angles between the asymptotes.

Cor. 1. The equations of a conic and its asymptotes differ by a constant only; and this remains true after any transformation.

Cor. 2. If P and Q are of different signs, the equations

$$Px^2 + Qy^2 = R, \quad Px^2 + Qy^2 = 0, \quad Px^2 + Qy^2 = -R,$$

represent the hyperbola, the asymptotes, and the conjugate. Hence the equations of two conjugate hyperbolas differ from the equation of their asymptotes by constants which are equal and of opposite signs; and this also remains true after any transformation.

Cor. 3. The angle between the asymptotes is found by Art. 67. If the asymptotes are at right angles, the hyperbola is (Art. 170) rectangular, and we have

$$A + B = 0, \text{ or } A + B - 2H\cos\omega = 0,$$

according as the axes of co-ordinates are rectangular or oblique.

COR. 4. Since the equation $Ax^2 + 2Hxy + By^2 = 0$ determines the direction of the asymptotes, it follows that, if $A = 0$ in the general equation of the second degree, one asymptote will be parallel to the axis of x, for its direction is given by the equation $y = 0$. Similarly, if $B = 0$, one asymptote will be parallel to the axis of y. The locus will in either case (Art. 150) belong to the hyperbola class, since $H^2 - AB$ becomes H^2, and must be positive.

187. *To determine the equations to the tangents to an ellipse or hyperbola, which pass through a given point* $(x'y')$.

The equation to the tangent is (Art. 182)

$$y - mx = \pm\sqrt{a^2m^2 + b^2}.$$

Since it passes through $(x'y')$, we have

$$(y' - mx')^2 = a^2m^2 + b^2;$$

or

$$m^2 + \frac{2x'y'}{a^2 - x'^2}m + \frac{b^2 - y'^2}{a^2 - x'^2} = 0 \;\ldots\ldots\ldots\ldots(1),$$

which equation gives two values of m; let them be μ and μ'; then the equations of the tangents required are

$$y - y' = \mu(x - x'),\; y - y' = \mu'(x - x') \;\ldots\ldots\ldots(2).$$

188. The roots of equation (1) of Art. 187 are real and different, real and equal, or imaginary, according as

$$x'^2y'^2 >=< (b^2 - y'^2)(a^2 - x'^2) \;\ldots\ldots\ldots\ldots(3),$$

or as

$$\frac{x'^2}{a^2} + \frac{y'^2}{b^2} - 1 >=< 0 \;\ldots\ldots\ldots\ldots\ldots\ldots(4),$$

and a little consideration will shew, that inequality (4) gives the conditions that $(x'y')$ shall be *without*, *on*, or *within* the ellipse. Hence no real tangent can be drawn to the ellipse from within the curve.

To prove the same property for the hyperbola, we must write $-b^2$ for b^2 in (3), and we shall then obtain, instead of inequality (4),

$$\frac{x'^2}{a^2} - \frac{y'^2}{b^2} - 1 < = > 0 \text{.....................} (5).$$

Condition (5) will evidently *always* give possible roots, when x' does not exceed $\pm a$; for then $\frac{x'^2}{a^2}$ is less than unity, and consequently the whole expression is negative; but, when x' exceeds those values, $\frac{x'^2}{a^2}$ becomes greater than unity, and, in order that the whole expression may be negative, the value of y' must not be less than that of the ordinate of the curve corresponding to x', since it is that value which makes

$$\frac{x'^2}{a^2} - \frac{y'^2}{b^2} - 1 = 0.$$

Hence inequality (5) gives the condition that $(x'y')$ should be *without*, *on*, or *within* the curve (where the foci are considered as *within*); and no real tangent can be drawn to the hyperbola from within the curve. The above reasoning shews that the curves are, as we have drawn them, always concave to the axis of x; for, if this were not the case, it would evidently be possible to draw a straight line from within, which should pass through two coincident points of the curve.

The reader will observe, that the expressions

$$\frac{x^2}{a^2} + \frac{y^2}{b^2} - 1, \quad \frac{x^2}{a^2} - \frac{y^2}{b^2} - 1,$$

are negative when (xy) is on the origin side of the curves, and change sign when (xy) crosses the curves. This result should be compared with Arts. 55, 121.

189. *To determine the locus of the intersection of two tangents at right angles to one another.*

If, in Article 187, the two tangents are at right angles, $\mu\mu' = -1$. Hence, from equation (1), since $\mu\mu'$ is the product of the roots,

$$\frac{b^2 - y'^2}{a^2 - x'^2} = \mu\mu' = -1;$$

therefore $$x'^2 + y'^2 = a^2 + b^2;$$

or the locus of the point $(x'y')$, where the tangents intersect, is a circle, whose centre is C and radius $= \sqrt{a^2 + b^2}$.

The corresponding equation for the hyperbola is, of course,

$$x^2 + y^2 = a^2 - b^2,$$

which is a circle, unless b^2 be greater than a^2, in which case the locus is impossible; i.e. two tangents cannot be drawn at right angles to one another, when b^2 is greater than a^2. In the equilateral hyperbola $b = a$, and the circle is reduced to a point, namely the origin; hence only one pair of tangents at right angles to one another can be drawn to the equilateral hyperbola. These pass through the centre, and are the asymptotes, which are tangents at an infinite distance.

190. *To find the perpendicular from the centre on the tangent, in terms of the angle which it makes with the axis of x.*

If the perpendicular p make an angle α with the axis of x, the equation to the tangent is (Art. 27)

$$x \cos \alpha + y \sin \alpha = p,$$

and hence, by the condition given in Art. 182, Cor.,

$$p^2 = a^2 \cos^2 \alpha + b^2 \sin^2 \alpha,$$

or since (Art. 174) $$b^2 = a^2 (1 - e^2),$$

$$p^2 = a^2 (1 - e^2 \sin^2 \alpha).$$

Hence the equation to the tangent in the form given in Art. 27 is

$$x \cos \alpha + y \sin \alpha - a\sqrt{1 - e^2 \sin^2 \alpha} = 0.$$

If we had used the signs suitable to the hyperbola, we should have obtained the same result for that curve.

191. *To find the equation to the normal at any point $(x'y')$.*

By reasoning similar to that used in the case of the circle (Art. 122), since the normal passes through the point $(x'y')$ and is perpendicular to the tangent, whose equation is

$$\frac{xx'}{a^2} + \frac{yy'}{b^2} = 1,$$

its equation is
$$y - y' = \frac{a^2y'}{b^2x'}(x - x'),$$

or
$$\frac{a^2x}{x'} - \frac{b^2y}{y'} = a^2 - b^2.$$

192. To find the equation to the normal in terms of its inclination to the major axis, we may write it

$$y - y' = m(x - x') \ldots\ldots (1), \quad \text{where } m = \frac{a^2y'}{b^2x'} \ldots\ldots (2),$$

and is the tangent of the angle which the normal makes with the axis of x. We have then to express x' and y' in terms of m; that is, we must eliminate x' and y' between the three equations, (1), (2), and the equation to the curve.

From (2)
$$\frac{bm}{a} = \frac{ay'}{bx'},$$

whence
$$\frac{b^2m^2 + a^2}{a^2} = \frac{a^2y'^2 + b^2x'^2}{b^2x'^2} = \frac{a^2}{x'^2},$$

since
$$a^2y'^2 + b^2x'^2 = a^2b^2;$$

therefore $x' = \dfrac{a^2}{\sqrt{b^2m^2 + a^2}}$; and similarly $y' = \dfrac{b^2m}{\sqrt{b^2m^2 + a^2}}$.

Substituting these values in (1), we have

$$y = mx - \frac{(a^2 - b^2)\,m}{\sqrt{b^2m^2 + a^2}}.$$

193. The intercepts of the tangent and normal on the axis of x may be found by putting $y = 0$ in their equations.

Let them meet the axis in T and G; then, putting $y = 0$ in the equation to the tangent, we have

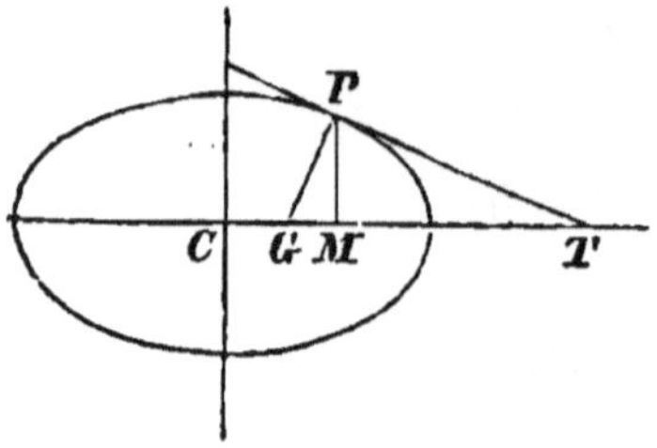

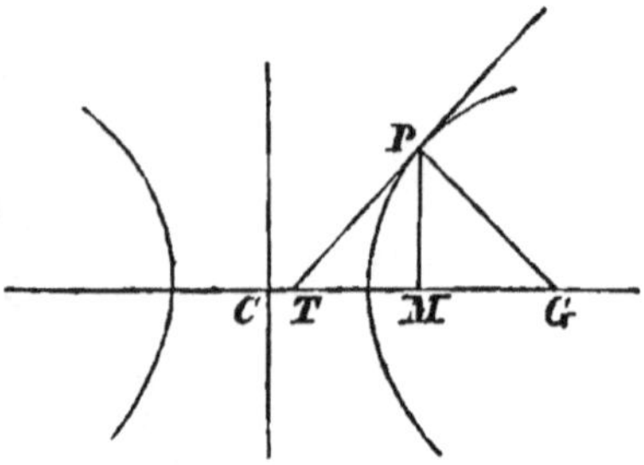

$$\frac{xx'}{a^2} = 1, \quad \text{or } CT = \frac{a^2}{x'};$$

and in the same manner from the equation to the normal, when $y = 0$, we have

$$x = x'\left(1 - \frac{b^2}{a^2}\right), \text{ or } CG = e^2x'.$$

The portion MT, intercepted on the axis between the tangent and the ordinate of the point of contact, is called the *subtangent*, and MG is called the *subnormal*. The length of the subtangent is,

in the ellipse, $\qquad CT - CM = \dfrac{a^2}{x'} - x' = \dfrac{a^2 - x'^2}{x'}$;

and in the hyperbola, $\qquad CM - CT = x' - \dfrac{a^2}{x'} = \dfrac{x'^2 - a^2}{x'}$.

In the same manner the length of the subnormal is,

in the ellipse, $$MG = x' - e^2x' = \frac{b^2}{a^2}x',$$

in the hyperbola, $$MG = e^2x' - x' = \frac{b^2}{a^2}x'.$$

194. If in the equation to the tangent

$$\frac{xx'}{a^2} + \frac{yy'}{b^2} = 1,$$

we write x', x for x, x', and y', y for y, y' respectively, the equation remains unchanged. Hence (Art. 131) all the theories of poles and polars proved for the circle in Arts. 123—127, 129, 130 are equally true for the ellipse and hyperbola, and the proofs will require no alteration, except that we must write the equations to the ellipse or hyperbola and its tangent, in the place of the equations to the circle and its tangent. These properties are so important, that the student is recommended to convince himself thoroughly of the truth of the above assertion, by writing out the articles with the requisite changes.

195. If, in the equation to the polar of the point $(x'y')$, we make $y' = 0$, the equation becomes $\frac{xx'}{a^2} = 1$, the equation to a straight line parallel to the axis of y; hence the polar of any point on the axis of x, is parallel to the axis of y; and simi-

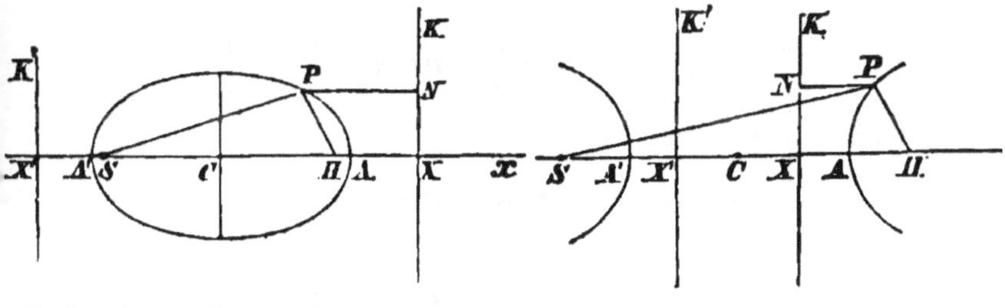

larly the polar of any point on the axis of y, is parallel to the axis of x.

If we take two points, H, S, on the axis of x, so that

in the ellipse $$CS = CH = \sqrt{a^2 - b^2} = ae,$$

and in the hyperbola $$CS = CH = \sqrt{a^2 + b^2} = ae,$$

the equations to the polars of these points will be, writing 0, ae and 0, $-ae$ for y', x',

$$x = \frac{a}{e} \text{ for the polar of } H, \quad x = -\frac{a}{e} \text{ for the polar of } S.$$

These points are called the *foci*, and their polars are called the *directrices* of the curve.

In the ellipse, since e is less than unity, the foci lie between the centre and the vertex, and the directrices KX, $K'X'$, the polars, respectively, of H and S, lie beyond the vertex. In the hyperbola, since e is greater than unity, the reverse will be the case. In the circle $a = b$, and the foci coincide with the centre; also (Art. 174) $e = 0$, so that the directrices are infinitely distant. It will be seen hereafter, that the focus and its polar, the directrix, possess many remarkable properties in connexion with the curve.

Cor. From the results of this article we have in the ellipse,

$$AH = AC - CH = a - ae = a(1 - e),$$

$$AX = CX - CA = \frac{a}{e} - a = \frac{a(1 - e)}{e},$$

$$HX = CX - CH = \frac{a}{e} - ae = \frac{a(1 - e^2)}{e}.$$

In the hyperbola

$$AH = CH - CA = ae - a = a(e - 1),$$

$$AX = CA - CX = a - \frac{a}{e} = \frac{a(e - 1)}{e},$$

$$HX = CH - CX = ae - \frac{a}{e} = \frac{a(e^2 - 1)}{e}.$$

196. *To find the distance of any point in the ellipse from the focus, in terms of the abscissa of the point.*

Since the co-ordinates of the point H are $(x = ae, y = 0)$, the square of the distance of any point P $(x'y')$ from it, is (Art. 7)

$$(x' - ae)^2 + y'^2 = x'^2 - 2aex' + a^2e^2 + y'^2;$$

and, if P be a point in the curve,

$$y'^2 = \frac{b^2}{a^2}(a^2 - x'^2) = (1 - e^2)(a^2 - x'^2),$$

since $b^2 = a^2 - a^2e^2$. Hence

$$HP^2 = a^2 - 2aex' + e^2x'^2,$$

or $$HP = a - ex'.$$

We do not notice the value $(ex' - a)$, obtained by giving the negative sign to the square root. For e is less than 1, and x' less than a, hence $ex' - a$ is constantly negative, and need not be taken into consideration, since we are now examining the *magnitude*, not the direction, of the radius vector HP.

Writing $-ae$ for ae in the preceding proof, we have, for the distance of P from the other focus,

$$SP = a + ex';$$

hence $$SP + HP = 2a,$$

or *the sum of the distances of any point in an ellipse from the foci is constant, and equal to the axis major.*

197. In the case of the hyperbola, we obtain the same value for HP^2, but, in extracting the root, we must take the value

$$HP = ex' - a,$$

since in the hyperbola x' is greater than a, and e is greater than unity, and consequently $a - ex'$ is constantly negative.

In like manner, we have

$$SP = ex' + a\,;$$

hence
$$SP - HP = 2a,$$

or, *in the hyperbola the difference of the focal radii is constant, and equal to the transverse axis.*

198. The property proved in the two last articles will enable us to describe an ellipse or hyperbola mechanically; for, evidently, if a string SPH be fastened to two points S and H, a pencil P, moved so as to keep the string always stretched, will describe an ellipse of which S and H are the foci, since $SP + HP$ will be a constant quantity.

Also, any portion of an hyperbola may be described by a ruler and cord; for let a ruler SR revolve round S in the plane of the paper, and let a cord be fastened to H, shorter than SR by a given difference c; then a pencil P, which should always keep the string stretched against SR, would describe an hyperbola; for the difference $SP - PH$ would always equal a constant quantity c, the difference in the lengths of the ruler and cord.

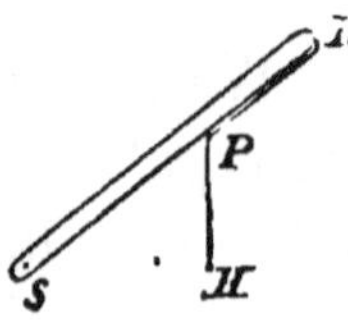

199. *The distance of any point on the curve from the focus, is in a constant ratio to its distance from the directrix.*

The equation to the directrix KX is (Art. 195)

$$ex - a = 0 \quad\text{.........................}(1),$$

and the length of the perpendicular (fig. Art. 195) from any point $P(x'y')$ in the ellipse on (1) is, (Art. 54), since P *is* on the origin side of KX,

$$= -\frac{ex' - a}{e} = \frac{1}{e}\,HP.$$

In the hyperbola, since P *is not* on the origin side of

KX, the perpendicular is

$$=\frac{ex'-a}{e}=\frac{1}{e}HP.$$

Hence in both curves the distance of any point P from the focus is to its distance from the directrix in the constant ratio of e to 1.

In the ellipse $e=\frac{\sqrt{a^2-b^2}}{a}$, and is less than 1, or the distance from the focus is *less* than the distance from the directrix. In the hyperbola $e=\frac{\sqrt{a^2+b^2}}{a}$, and is greater than 1, or the distance from the focus is *greater* than the distance from the directrix. It will afterwards be seen, that the parabola has a focus, and that the distance of any point from the focus is *equal* to the distance from the directrix.

Hence it is often given as a definition of a conic section, that *it is the locus of a point, whose distance from a given point has a fixed ratio to its distance from a given straight line.*

200. The double ordinate through the focus is called the *Latus Rectum.* Putting $x=ae$ in the equation to the curve, we have

$$y^2=b^2(1-e^2)=\frac{b^4}{a^2},$$

or $y=\pm\frac{b^2}{a}$; hence the latus rectum $=\frac{2b^2}{a}=2a(1-e^2)$ in the ellipse, and $=2a(e^2-1)$ in the hyperbola.

*201. *In the ellipse the normal bisects the interior angle between the focal distances, and in the hyperbola the exterior angle; and the focal radii make equal angles with the tangent.*

The equation to HP, since it passes through $(x'y')$, $(ae, 0)$ is

$$\frac{y}{x-ae}=\frac{y'}{x'-ae} \quad \text{..............(1)},$$

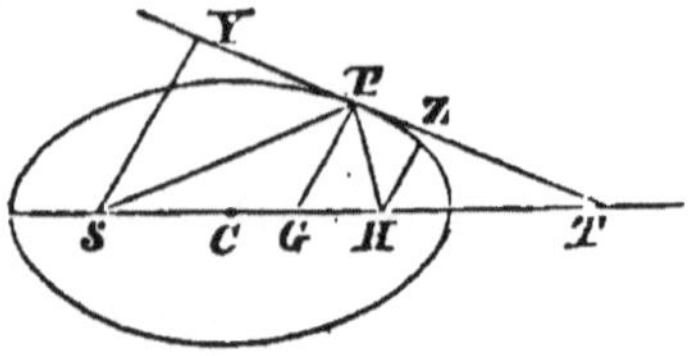

and similarly, the equation to SP is

$$\frac{y}{x+ae}=\frac{y'}{x'+ae} \quad \text{..............(2)}.$$

Hence, forming the equation to the bisector PG of the angle between (1) and (2), we have, by the rules of Art. 59,

$$\frac{-(x'-ae)\,y+y'x-aey'}{\{y'^2+(x'-ae)^2\}^{\frac{1}{2}}}=\frac{(x'+ae)\,y-y'x-aey'}{\{y'^2+(x'+ae)^2\}^{\frac{1}{2}}} \quad \text{......(3)},$$

which we might shew by reduction to be the equation to the normal at $(x'y')$. This may however be proved briefly as follows. The denominators of (3) are evidently HP and SP, and therefore reduce, for the ellipse, to $a-ex'$ and $a+ex'$, as in Art. 196; hence, to find where (3) meets the axis of x, we have, making $y=0$,

$$\frac{x-ae}{a-ex'}=\frac{-(x+ae)}{a+ex'};$$

hence $x=CG=e^2x'$, and therefore (Art. 193) the bisector PG is the normal.

The reasoning for the hyperbola is precisely the same; but the denominators of (3) will now become $ex'-a$ and $ex'+a$, and we shall have, to determine the point T where the bisector PT meets the axis,

$$\frac{x-ae}{ex'-a}=\frac{-(x+ae)}{ex'+a};$$

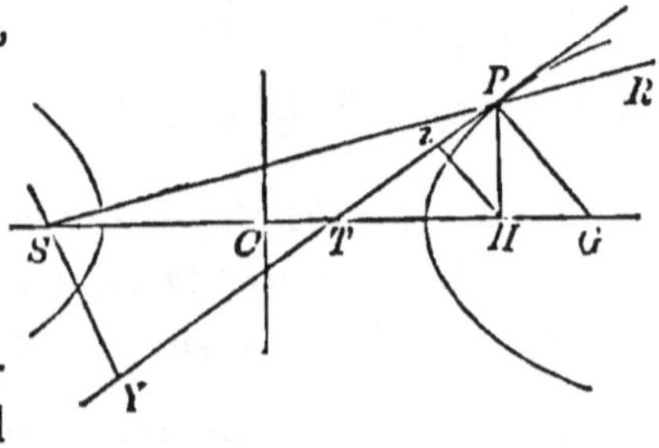

hence $x=CT=\dfrac{a^2}{x'}$,

and therefore (Art. 193) the bisector PT is the tangent, and consequently the normal PG bisects the exterior angle HPR.

202. The following is an easy geometrical proof of the preceding results. In the ellipse and hyperbola $CG = e^2x'$; hence in the ellipse,

$$SG = ae + e^2x', \quad HG = ae - e^2x';$$

also $$SP = a + ex', \quad HP = a - ex';$$

therefore $$SG : HG = SP : HP,$$

and therefore, by Euc. VI. 3, PG bisects the angle SPH. In the hyperbola,

$$SG = e^2x' + ae, \quad HG = e^2x' - ae;$$

also $$SP = ex' + a, \quad HP = ex' - a;$$

therefore $$SG : HG = SP : HP,$$

and therefore, by Euc. VI. A, PG bisects the exterior angle HPR.

203. *To find the locus of the extremities of perpendiculars dropped from the foci upon the tangent.*

Let SY, HZ be these perpendiculars; then the equation to PT is, taking the fig. of the ellipse,

$$y = mx + \sqrt{m^2a^2 + b^2},$$

and, since HZ passes through H $(\sqrt{a^2 - b^2}, 0)$, and is perpendicular to PT, its equation is

$$y = -\frac{1}{m}(x - \sqrt{a^2 - b^2}).$$

If we eliminate m between these equations, we shall obtain the equation to the locus required. The equations may be written

$$y - mx = \sqrt{m^2a^2 + b^2}, \quad my + x = \sqrt{a^2 - b^2};$$

adding their squares,

$$(x^2+y^2)(m^2+1)=a^2(m^2+1),$$

or

$$x^2+y^2=a^2,$$

the equation to the locus of Z, which represents a circle on the axis major as diameter.

204. If HZ is the perpendicular from H $(ae, 0)$ on the tangent, whose equation is

$$\frac{xx'}{a^2}+\frac{yy'}{b^2}=1 \quad\text{.....................(1)},$$

we have for the ellipse (Art. 54), since H *is* on the origin side of (1),

$$HZ=-\frac{\frac{ex'}{a}-1}{\sqrt{\frac{x'^2}{a^4}+\frac{y'^2}{b^4}}}=\frac{ab^2(a-ex')}{\sqrt{a^4y'^2+b^4x'^2}}.$$

But
$$a^4y'^2+b^4x'^2=a^2b^2\left(\frac{a^2}{b^2}y'^2+\frac{b^2}{a^2}x'^2\right)$$
$$=a^2b^2\{a^2-x'^2+(1-e^2)x'^2\}$$
$$=a^2b^2(a^2-e^2x'^2),$$

and therefore
$$HZ=b\sqrt{\frac{a-ex'}{a+ex'}} \quad\text{......................... (2)}.$$

Similarly
$$SY=b\sqrt{\frac{a+ex'}{a-ex'}} \quad\text{......................... (3)}.$$

Hence
$$SY.HZ=b^2 \quad\text{......................... (4)}.$$

In working this problem for the hyperbola, we must remember that, in the case of HZ (fig. Art. 201), H *is not* on the origin side of (1). We mention this, because the student would otherwise probably be puzzled, by getting $-b^2$ as the right-hand member of (4).

If $SY = p$, $SP = \rho$, equation (3) may be written

$$p^2 = b^2 \frac{\rho}{2a \mp \rho},$$

the upper or lower sign being taken, according as the curve is an ellipse or hyperbola.

205. *Any focal chord is perpendicular to the line joining its pole with the focus.*

By definition, the directrix is the polar of the focus, and conversely, by Art. 129, the polar of any point in the directrix passes through the focus. Hence assume the pole of any focal chord PHQ to be a point $K\left(\frac{a}{e}, y'\right)$ on the directrix,

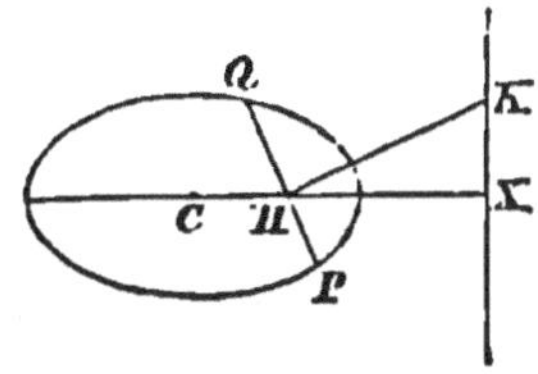

then the equation to the chord is (Art. 123)

$$\frac{x}{ae} + \frac{yy'}{b^2} = 1,$$

and the equation to a perpendicular HK to this chord, drawn through the focus, is

$$y = \frac{aey'}{b^2}(x - ae);$$

and, when $x = CX = \frac{a}{e}$ in this equation, $y = y' = KX$, or the line which it represents passes through K, the pole of the chord.

206. We have deduced all the above properties of the ellipse and hyperbola from their equations alone; we shall now shew how, conversely, their equations may be deduced from a knowledge of their properties; for example, let it be required *to find the locus of a point* (P), *the distance of which*

from a given point (S) has a constant ratio to its distance from a given straight line (KX).

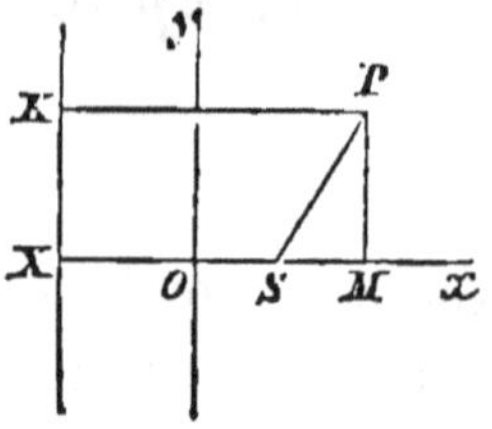

Suppose $SP : PK = e : 1$; draw SX perpendicular to KX, and divide SX in O, so that $SO : OX = e : 1$; then O will be a point in the locus; take OSx, Oy as axes, and let OM $(= x)$, MP $(= y)$ be the co-ordinates of P; let $OS = d$, and, therefore, $OX = \frac{d}{e}$. Then

$$SM^2 + PM^2 = SP^2,$$

or

$$(x-d)^2 + y^2 = e^2 . PK^2;$$

$$= e^2 \left(\frac{d}{e} + x\right)^2;$$

whence we obtain

$$(1-e^2)\,x^2 + y^2 - 2\,(1+e)\,dx = 0 \text{..........}(1),$$

an equation of the second degree between x and y.

207. We proceed to interpret this equation by the methods of Chap. VIII. The locus belongs to the Ellipse, Hyperbola, or Parabola class, according as $H^2 - AB$ is negative, positive, or zero, where A, $2H$, B are the coefficients of x^2, xy, y^2, respectively; that is, according as

$$4\,(e^2-1) <>= 0, \text{ or as } e <>= 1.$$

I. Suppose e to be less than unity, or the locus to belong to Class I. The equation may be written

$$x^2 - \frac{2dx}{1-e} + \frac{y^2}{1-e^2} = 0,$$

or

$$\left(x - \frac{d}{1-e}\right)^2 + \frac{y^2}{1-e^2} = \frac{d^2}{(1-e)^2}.$$

Transferring the origin to a point $x = \frac{d}{1-e}$, $y = 0$, we obtain

$$x^2 + \frac{y^2}{1-e^2} = \frac{d^2}{(1-e)^2},$$

which represents an ellipse with the centre for origin, unless $d = 0$, or the given point is on the given line, in which case the equation (Art. 107. II.) represents two imaginary straight lines, and is satisfied by the real values $x = 0$, $y = 0$ only. If we replace the known quantity $\frac{d}{1-e}$ by a, and put $a^2(1-e^2) = b^2$, we obtain the equation

$$\frac{x^2}{a^2} + \frac{y^2}{b^2} = 1,$$

which has been already discussed.

II. Suppose e to be greater than unity, or the locus to belong to Class II. In this case $1 - e$ and $1 - e^2$ are negative; we shall therefore write the equation,

$$x^2 + \frac{2dx}{e-1} - \frac{y^2}{e^2-1} = 0,$$

or
$$\left(x + \frac{d}{e-1}\right)^2 - \frac{y^2}{e^2-1} = \frac{d^2}{(e-1)^2}.$$

Transferring to the point $x = -\frac{d}{e-1}$, $y = 0$, we obtain

$$x^2 - \frac{y^2}{e^2-1} = \frac{d^2}{(e-1)^2},$$

which represents an hyperbola with the centre for origin, unless $d = 0$, or the given point is on the given line, in which case the locus is the two straight lines

$$\sqrt{e^2-1}\,x + y = 0, \quad \sqrt{e^2-1}\,x - y = 0.$$

If we replace the known quantity $\frac{d}{e-1}$ by a, and put

$$a^2(e^2-1)=b^2,$$

we obtain the equation

$$\frac{x^2}{a^2}-\frac{y^2}{b^2}=1,$$

which has been already discussed.

III. Suppose e to be equal to unity, or the locus to belong to Class III. The equation in this case becomes

$$y^2=4dx.$$

This represents the curve called the Parabola, which we have not yet discussed, unless $d=0$, in which case it represents two straight lines coinciding with the axis of x.

Cor. Hence we see that the simplest form, by which the three conic sections may be represented, is

$$y^2=mx+nx^2,$$

where the curve is an ellipse, hyperbola, or parabola, according as n is negative, positive, or zero.

The point O will plainly be one of the vertices of the curve, S the focus, and KX the directrix.

208. If we take the given point for origin, with any rectangular axes, and if the equation to the given line be

$$x\cos\alpha+y\sin\alpha-p=0 \text{ .... (1)},$$

then the distances of any point (xy) on the locus from the origin and (1), respectively, are

$$(x^2+y^2)^{\frac{1}{2}} \text{ and } \pm(x\cos\alpha+y\sin\alpha-p).$$

Hence the equation to the locus is

$$x^2+y^2=e^2(x\cos\alpha+y\sin\alpha-p)^2.$$

Or, still more generally, if the given point be $(x'y')$, and the given line

$$Ax + By + C = 0,$$

the same distances are

$$\{(x-x')^2 + (y-y')^2\}^{\frac{1}{2}} \text{ and } \frac{Ax+By+C}{\sqrt{A^2+B^2}},$$

and the equation to the locus is

$$(x-x')^2 + (y-y')^2 = \frac{e^2}{A^2+B^2}(Ax+By+C)^2.$$

These equations of the second degree may be interpreted by Chap. VIII.

209. *To find the polar equation to the ellipse or hyperbola, the focus being pole.*

Here we have to find the polar equation to the locus of Art. 206, the given point being a focus, and the given line the corresponding directrix. In the figure of Art. 206 let

$$SP = \rho, \quad \text{angle } PSx = \theta, \quad SX = c;$$

then
$$SP = e.PK = e(SX + SM),$$

or
$$\rho = e(c + \rho\cos\theta) \text{ .... (1)}.$$

Let $\theta = 90°$; then $\rho = ec$, or the whole chord through S, i.e. the latus rectum, is equal to $2ec$; hence, if we denote the latus rectum by $2l$, (1) may be written

$$\rho = \frac{ec}{1 - e\cos\theta}, \quad \text{or } \frac{l}{\rho} = 1 - e\cos\theta \text{.........(2)},$$

the equation required.

By the figure used it will be seen, that this is the equation, when the left-hand focus is taken in the ellipse, and the right-hand focus in the hyperbola, with their corresponding directrices, and the point P has been taken in that branch of the hyperbola, in which the pole lies. If we take the other foci and directrices in each, and the point P in the left-hand branch of the hyperbola, the equation will be

$$\frac{l}{\rho} = 1 + e \cos \theta \text{ .... (3).}$$

210. The student is recommended to trace the curves from equations (2) and (3) of Art. 209, substituting for l its value, $a(1-e^2)$ in the ellipse, and $a(e^2-1)$ in the hyperbola. He will find, in the latter, that each equation represents both branches, the positive values of ρ tracing the branch in which the pole lies, and the negative values of ρ (Art. 12) tracing the other branch. For example, if he traces from equation (2), he will observe, that, when $\theta = \cos^{-1}\frac{1}{e}$, the value of ρ becomes infinite, and the radius vector is (Art. 174) parallel to the asymptote CR (fig. Art. 166). Let this angle $= \alpha$; then as θ varies from

$$0 \text{ to } \alpha, \quad \alpha \text{ to } \pi, \quad \pi \text{ to } 2\pi - \alpha, \quad 2\pi - \alpha \text{ to } 2\pi,$$

ρ varies from

$$-(ae+a) \text{ to } -\infty, \ \infty \text{ to } ae-a, \ ae-a \text{ to } \infty, \ -\infty \text{ to } -(ae+a),$$

thus tracing the branches, $A'R'$, RA, AL, $L'A'$. Thus it appears that R', R, and L, L' are consecutive points on the hyperbola, a result which has been noticed in Art. 185.

211 (i). *To find the locus of a point P, the sum of whose distances from two given points S and H is a constant quantity.*

Let $SP = \rho$, $HP = \rho'$, $SP + HP = 2a$, $SH = 2c$, angle $PSH = \theta$; then

$$\rho + \rho' = 2a,$$

whence $$\rho'^2 = \rho^2 - 4a\rho + 4a^2,$$

but $$\rho'^2 = \rho^2 + 4c^2 - 4c\rho \cos\theta;$$

therefore $$a^2 - a\rho = c^2 - c\rho\cos\theta,$$

or $$\rho = \frac{c^2 - a^2}{c\cos\theta - a};$$

now, if we write ae for c, where e is less than unity, since $SP + HP > SH$, and therefore $a > c$, the equation becomes

$$\rho = \frac{a(1-e^2)}{1-e\cos\theta},$$

which is, evidently (Art. 209), the equation to an ellipse, whose foci are S and H, and whose eccentricity $= e$.

211 (ii). *To find the locus of a point P, the difference of whose distances from two given points S and H is a constant quantity.*

Let $HP = \rho$, $SP = \rho'$, $\pi - PHS = \theta$, then

$$\rho' - \rho = 2a,$$

whence $$\rho'^2 = \rho^2 + 4a\rho + 4a^2,$$

also $$\rho'^2 = \rho^2 + 4c^2 + 4c\rho\cos\theta$$

hence, as before, we have

$$\rho = \frac{a(e^2-1)}{1-e\cos\theta},$$

where e is greater than unity, since SH is greater than $SP - HP$, or $c > a$.

This equation (Art. 209) represents an hyperbola, whose foci are S and H, and whose eccentricity is e.

212. *Confocal Conics.*

If the equation to a central conic referred to its axes is

$$\frac{x^2}{a^2} + \frac{y^2}{b^2} = 1 \quad\text{.....................}(1);$$

then, the distance between the foci of the conic

$$\frac{x^2}{a^2+k} + \frac{y^2}{b^2+k} = 1 \quad\text{.....................} (2)$$

is $2\{(a^2+k) - (b^2+k)\}^{\frac{1}{2}}$ or $2(a^2-b^2)^{\frac{1}{2}}$; hence, by giving proper values to k, (2) may be made to represent any conic confocal with (1). The reader should observe the effect of varying k, in making the confocal take the form of ellipse, hyperbola, &c.

Ex. 1. *To find how many conics, confocal to* (1), *can be drawn through a given point* ($x'y'$).

Suppose these confocals to be represented by (2); then, if ($x'y'$) lies upon (2), we have

$$\frac{x'^2}{a^2+k}+\frac{y'^2}{b^2+k}=1,$$

or $$(a^2+k)(b^2+k)-(a^2+k)y'^2-(b^2+k)x'^2=0 \qquad (3);$$

also $$a^2+k-(b^2+k)=a^2-b^2=a^2e^2 \qquad (4),$$

two equations to determine a^2+k and b^2+k. Eliminating a^2+k, we have

$$(b^2+k)^2-(x'^2+y'^2-a^2e^2)(b^2+k)-a^2e^2y'^2=0 \qquad (5).$$

The values of b^2+k obtained from this equation, are both real by the theory of quadratic equations; and real corresponding values of a^2+k can be obtained from (4). Hence two conics confocal to (1) can be drawn through the point ($x'y'$). If the point ($x'y'$) be *on* the given conic (1), one of these is, of course, the conic itself.

Ex. 2. *Two confocal conics cut each other at right angles at all their common points; and, of the two which can be drawn through any given point, one is an ellipse and the other an hyperbola.*

Let (1) and (2) be the two confocal conics, and let ($x'y'$) be a common point; then, by subtraction,

$$\frac{x'^2}{(a^2+k)a^2}+\frac{y'^2}{(b^2+k)b^2}=0 \qquad (6);$$

but this is the condition that the tangents

$$\frac{xx'}{a^2+k}+\frac{yy'}{b^2+k}=1, \qquad \frac{xx'}{a^2}+\frac{yy'}{b^2}=1,$$

should be at right angles, which proves the first proposition. Also, in order that ($x'y'$) should be real, we see from (6) that $(a^2+k)a^2$ and $(b^2+k)b^2$ must be of different signs, which proves the second.

EXAMPLES IX.

THE following problems are enunciated, some for the ellipse, and some for the hyperbola, though many of them are equally applicable to both curves.

1. If the tangent to an hyperbola, at a point whose abscissa CM is positive, meet the transverse axis in T; $A'M$, $A'A$, $A'T$ will be in harmonical progression.

2. The distance of the centre of an ellipse from a tangent inclined to the major axis at an angle ϕ, is $= a(1 - e^2 \cos^2 \phi)^{\frac{1}{2}}$.

3. The distance of the focus of an ellipse from a tangent inclined to the major axis at an angle ϕ, is

$$a\{e \sin \phi + (1 - e^2 \cos^2 \phi)^{\frac{1}{2}}\}.$$

4. Find the angle (θ) at which the focal distance SP is inclined to the major axis, when SP is a mean proportional between the semi-axes of an ellipse, when $a = 50$, $b = 30$.

5. If in any hyperbola, three abscissæ be taken in arithmetical progression, the focal distances of the extremities of the ordinates of these points will also be in arithmetical progression.

6. Shew that the equations to the tangents to an ellipse $(3x^2 + y^2 = 3)$, inclined at an angle of 45° to the axis of x, are $y = x + 2$, $y = x - 2$.

7. If the semi-axes of an ellipse are 5 and 4, find the angle at which CP is inclined to the major axis, when an arithmetic mean between CA and CB.

*8. Find the eccentric angle (i) at the extremity of the latus rectum of an ellipse, and (ii) at the point where $x = y$.

9. If any number of hyperbolas be described, having the same transverse axis, the tangents drawn at the extremities of their latera recta will all pass through one point.

10. If the tangent at any point P in an hyperbola intersect the axis in T, and CP meet the tangent at A in E, ET is parallel to AP.

11. Shew $\tan\frac{PSH}{2}\tan\frac{PHS}{2}=\frac{1-e}{1+e}$, where P is any point in an ellipse.

12. Find the points of intersection of the ellipse and hyperbola whose equations are

$$x^2+2y^2=1,\quad 3x^2-6y^2=1,$$

and shew that at each of these points the tangent to the ellipse is the normal to the hyperbola.

13. If CA, CB be the semi-axes of an ellipse, shew that, when SBH is a right angle, $CA^2 : CB^2 = 2 : 1$.

14. Find the condition that the line $\left(\frac{x}{m}+\frac{y}{n}=1\right)$ should touch the hyperbola $\left(\frac{x^2}{a^2}-\frac{y^2}{b^2}=1\right)$.

15. The tangent to an ellipse is inclined to the major axis at an angle ϕ; shew that the area included by this tangent and the axes is $=\frac{1}{2}(a^2\tan\phi+b^2\cot\phi)$.

16. The circle described on any radius vector SP of an ellipse as diameter, will touch the circle on the axis major.

17. Find where the tangents from the foot of the directrix will meet the hyperbola, and what angle they will make with the transverse axis.

18. Find the equation to the tangent at the extremity of the latus rectum of an ellipse whose equation is $\frac{x^2}{9a^2}+\frac{y^2}{4a^2}=1$.

19. A tangent at the extremity of the latus rectum of an hyperbola meets any ordinate PM produced in R; shew that $SP = MR$, where S is the focus through which the latus rectum passes.

*20. If the sum of the eccentric angles of two points on an ellipse is constant and equal to 2α, the chord joining those points is always parallel to the tangent at the point whose eccentric angle is α.

21. Find the radius of a circle inscribed in a semi-ellipse, touching the axis minor.

22. From the point where the circle on the major axis is intersected by the minor axis produced, a tangent is drawn to the ellipse; find the point of contact.

23. If from the extremities of the minor axis two straight lines be drawn through any point in the ellipse, and intersect the axis major in Q and R, then $CQ \cdot CR = CA^2$.

24. If a rod slide between a vertical wall and a horizontal plane, any point on it traces out an ellipse.

25. If a tangent be drawn to the interior of two concentric ellipses, the axes of which are in the same straight line, meeting the exterior one in P, Q, and at P, Q tangents be drawn to the latter, intersecting in R, prove that the locus of R is an ellipse.

26. Shew that the locus of one end of a given straight line, whose other end and a given point in it move in straight lines at right angles to one another, is an ellipse.

27. If with the co-ordinates of any point in an elliptic quadrant as semi-axes, a concentric ellipse be described, the chord of the quadrant of the one will be a tangent to the other.

28. The locus of the centre of a circle touching two circles externally is an hyperbola.

29. The locus of the centre of a circle touched by one circle externally and one internally is an hyperbola.

30. Find the locus of the extremity of the perpendicular from the centre on the tangent to the hyperbola.

*31. Find the co-ordinates of tho pole of the chord of an ellipse, which passes through two points whose eccentric angles are θ and ϕ.

*32. Prove Arts. 203, 204 by means of the equation to the tangent obtained in Art. 164, Ex. 1.

33. If $3AC = 2CS$ in an hyperbola, find the inclination of the asymptotes to the transverse axis.

34. If from a point P in an hyperbola, PK be drawn parallel to the transverse axis, cutting the asymptotes in I and K, then $PK \cdot PI = a^2$, or, if parallel to the conjugate, $PK \cdot PI = b^2$.

35. Is the point (2, 3) *without* or *within* the hyperbola $2x^2 - 3y^2 = 7$? Shew that the straight line joining this point with the point (6, 4) cuts the curve.

36. If A, A' be the extremities of the major axis of an ellipse, T the point where the tangent at P meets AA', QTR a line perpendicular to AA', and meeting AP, $A'P$ in Q and R respectively, then $QT = TR$.

*37. Find the eccentricity and latus rectum of the conic

$$x^2 + 2y^2 - 2x + 4y - 6 = 0,$$

the axes being rectangular.

38. Find the equations to the asymptotes of the curve

$$3x^2 - 10xy + 3y^2 - 8 = 0;$$

and find the angle between the asymptotes of

$$x^2 - 10xy + y^2 + x + y + 1 = 0,$$

the axes in the latter case being rectangular.

39. How many normals can be drawn to an ellipse from a point on the major axis, and how many from a point on the minor axis?

40. In the equilateral hyperbola, the eccentricity is the ratio of the diagonal of a square to its side.

41. A tangent at any point P of an ellipse meets the axis major produced in T, and the axis minor produced in t; to find the locus of a point Q in Tt, such that, $QT : Qt = m : n$.

42. To find the locus of the intersection of the ordinate of any point in an ellipse produced, with the perpendicular from the centre upon the tangent at that point.

43. If the normal at P meet the axis major of an ellipse in G, and GK be drawn perpendicular to SP, $GK = e \,.\, PM$, where PM is the ordinate of P.

44. If SQ be drawn, always bisecting the angle PSC, in an ellipse, and equal to a mean proportional between SC and SP, find the eccentricity of the curve which is the locus of Q.

45. Two straight lines, such that the product of the tangents of their inclinations to the axis of x is constant, touch an ellipse; shew that the locus of their intersection is an ellipse, or hyperbola, according as the product is negative or positive.

46. Shew that the locus of the summit of a moveable right angle, one side of which touches one, and the other side the other of two confocal ellipses, is a concentric circle.

47. If P be any point in the hyperbola, S and H the foci, find the locus of the centre of the circle which is inscribed in SPH.

48. If a tangent at any point of an hyperbola be intersected by the tangents at the vertices in H and K, the circle on HK as diameter passes through the foci.

CHAPTER X.

Central Conic Sections. Conjugate Diameters. The Hyperbola referred to its Asymptotes.

213. We saw (Art. 145), that there is only one system of rectangular axes, about which the ellipse and hyperbola are so situated, that each axis should bisect all chords parallel to the other. We shall now shew, that there are an infinite number of oblique axes, which possess the above property with regard to the two curves.

214. *To find the locus of the middle points of any system of parallel chords.*

Let QQ' be one of the chords, M its middle point $(x'y')$,

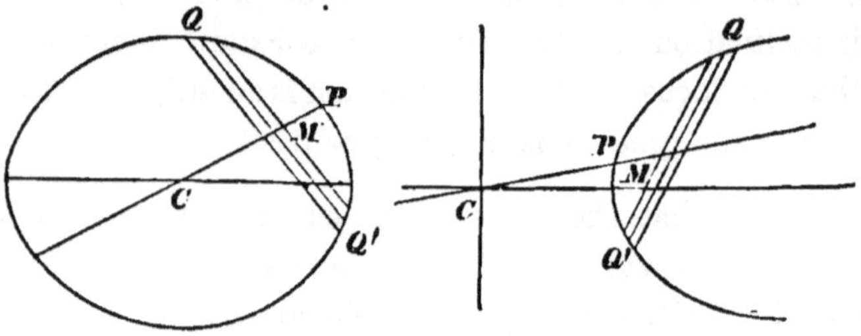

and let the equation to QQ' be

$$\frac{x-x'}{c}=\frac{y-y'}{s}=l \text{ ....} (1).$$

Now, if we substitute for x and y from this equation in the equation

$$\frac{x^2}{a^2}+\frac{y^2}{b^2}=1 \text{ ....} (2),$$

we obtain equation (2) of Art. 176, to determine the distances (l) of the point ($x'y'$) from the intersection of (1) and (2). But, since these distances (MQ, MQ') are equal in magnitude, the two values of l are equal and of opposite signs (Art. 34, Cor.); hence the coefficient of $l = 0$, or

$$\frac{cx'}{a^2} + \frac{sy'}{b^2} = 0,$$

which is a relation between the co-ordinates of the middle point of the chord QQ'. But, since the chords are parallel, s and c are the same for them all, and the same relation holds for the middle points of all; hence the equation to the locus required is

$$\frac{cx}{a^2} + \frac{sy}{b^2} = 0,$$

which represents a straight line (CM) through the centre.

If m be the tangent of the angle which the chords make with the axis of x, $m = \frac{s}{c}$, and the equation is

$$y = -\frac{b^2}{a^2m}x.$$

The equation for the hyperbola is, of course,

$$y = \frac{b^2}{a^2m}x.$$

COR. If ($x'y'$) be the point P, where CM meets the curve, we have

$$m = -\frac{b^2x'}{a^2y'},$$

but the right-hand member (Art. 181) represents the tangent of the angle, which the tangent at ($x'y'$) makes with the axis of x. Hence the tangent at the extremity of CP is parallel to the chords bisected by that line, as it evidently should be, since it may be considered as the limiting position of any chord QQ', as it moves parallel to itself up to P.

215. The straight line, which bisects any system of parallel chords, is called a *diameter* of the curve, and the chords are called the *ordinates* of that diameter. We have seen above (Art. 214), that, if the equation to one of the chords be

$$y = mx + c,$$

the equation to the diameter, of which those chords are ordinates, is

$$y = -\frac{b^2}{a^2 m} x.$$

We see then, that all diameters pass through the centre, and, conversely, since m may have any value, all lines passing through the centre are diameters. We shall see hereafter, that the same is true in the case of the parabola; but, the centre of the parabola being infinitely distant, all its diameters will consequently be parallel.

216. *If two diameters be such, that one of them bisects all chords parallel to the other, then the second will bisect all chords parallel to the first.*

Let the diameter CP $(y = mx)$ bisect all chords parallel

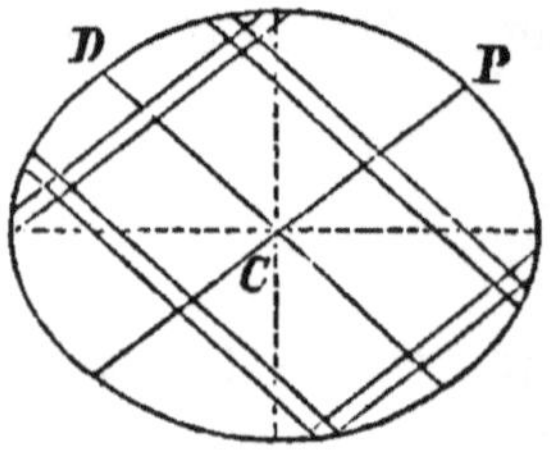

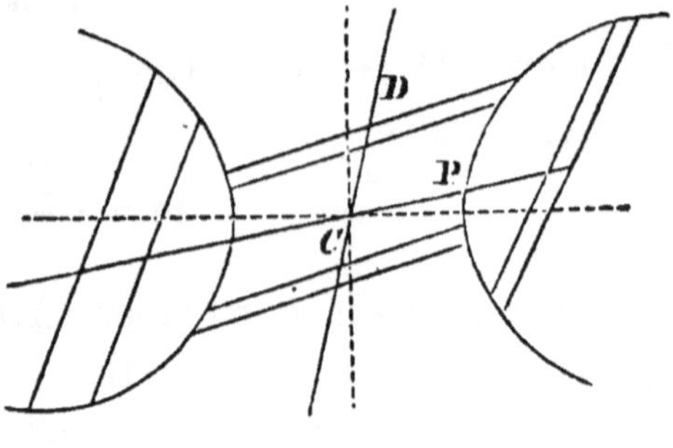

to the diameter CD $(y = m'x)$. Then, by Art. 215,

$$m = -\frac{b^2}{a^2 m'}, \quad \text{or } mm' = -\frac{b^2}{a^2};$$

and this is the only condition that must hold, in order that $(y=m'x)$ should bisect all chords parallel to $(y=mx)$.

217. Diameters so related, that each bisects every chord parallel to the other, or, more commonly, such portions of these diameters, as are intercepted by the curve, are called *Conjugate Diameters;* and the condition that the diameter $(y=mx)$ should be conjugate to the diameter $(y=m'x)$ is, by the last article,

$$mm'=-\frac{b^2}{a^2}.$$

We shall see hereafter that only central curves can have conjugate diameters.

If θ, θ' be the angles which the conjugate diameters make with the major axis, we have

in the ellipse, $\tan\theta\tan\theta'=-\dfrac{b^2}{a^2}$,

in the hyperbola, $\tan\theta\tan\theta'=\dfrac{b^2}{a^2}$;

hence, *in the ellipse the conjugate diameters fall on different sides of the axis minor;* for, if one of the angles θ, θ' be acute, and its tangent positive, the other must be obtuse, and its tangent negative; and we see by similar reasoning, that *in the hyperbola the conjugate diameters lie on the same side of the conjugate axis.*

218. Also, *of any two conjugate diameters, only one can meet the hyperbola;* for if one of the tangents $(\tan\theta)$ be less than $\dfrac{b}{a}$, the other $(\tan\theta')$ must be greater than $\dfrac{b}{a}$, and it is evident

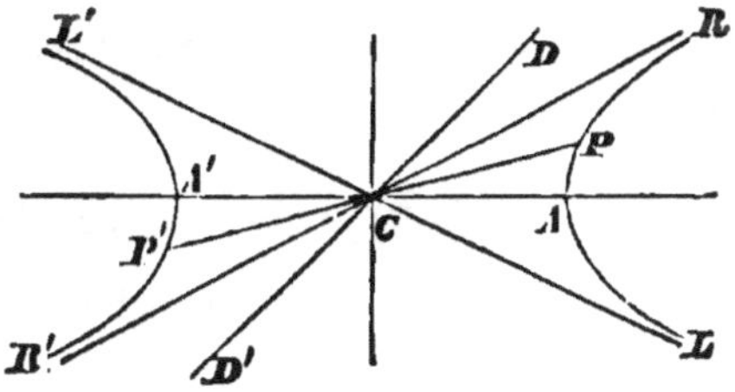

from Art. 169 that the diameter, which makes the angle θ with the axis, will fall within the angle RCL, made by the asymptotes, and meet the curve, while the diameter, which makes the angle θ' with the axis, will fall within the angle RCL', and will not meet the curve, for it has been shewn that $\tan RCA = \frac{b}{a}$.

If, in the hyperbola, $\tan\theta = \tan\theta' = \pm\frac{b}{a}$, the two diameters coincide with the asymptote RR' or with LL', according as we take the upper or lower sign. An asymptote is thus a self-conjugate diameter, and therefore should meet and *not* meet the curve, which agrees with its meeting the curve at infinity.

If $\tan\theta = -\tan\theta' = \frac{b}{a}$, in the ellipse, the angles θ and θ' are supplementary, and, by the symmetry of the figure, the conjugate diameters are equal; hence the equal conjugate diameters in an ellipse are parallel to chords joining the extremities of the major and minor axes; and if an ellipse and hyperbola have the same centre and axes, the equal conjugate diameters of the ellipse coincide with the asymptotes of the hyperbola. Their equation as one locus is

$$\frac{x^2}{a^2} - \frac{y^2}{b^2} = 0.$$

219. When the ellipse becomes a circle, $b^2 = a^2$, and

$$\tan\theta \tan\theta' = -1,$$

or all conjugate diameters of the circle are at right angles to one another.

When the hyperbola is rectangular,

$$\tan\theta \tan\theta' = 1,$$

therefore $$\theta + \theta' = 90° \text{ or } = 270°.$$

Since the asymptotes make angles of 45° and 135° with the transverse axis, the conjugate diameters of the rectangular hyperbola are equally inclined to the asymptotes.

220. We saw (Art. 195), that, when the curve is referred to its axes, as axes of co-ordinates, the polar of any point in the axis of x is parallel to the axis of y, and *vice versâ*. This may be now seen to be a property of all conjugate diameters, including the axes as a particular case. For suppose the equations to CP and CD to be

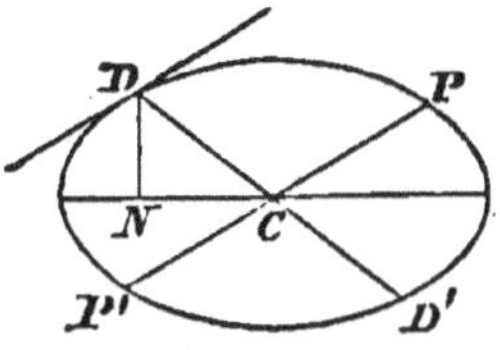

$$y = mx \text{..........(1)}, \quad y = -\frac{b^2}{a^2m}x \text{..........(2)}.$$

Now, if $(x'y')$ be any point on CD, the equation to the polar of $(x'y')$ is

$$\frac{xx'}{a^2} + \frac{yy'}{b^2} = 1 \text{ .... (3)},$$

which is the equation to a line making with the axis of x an angle whose tangent is $-\dfrac{b^2x'}{a^2y'}$; but since $(x'y')$ is on (2),

$$y' = -\frac{b^2}{a^2m}x', \text{ or } m = -\frac{b^2x'}{a^2y'},$$

and therefore (3) is parallel to (1); hence *the polar of any point on CD is parallel to CP, and vice versâ.*

Cor. If $(x'y')$ be the point D, where the diameter meets the curve, the equation to the polar

$$\frac{xx'}{a^2} + \frac{yy'}{b^2} = 1$$

will represent the tangent at D; hence, *the tangent at the extremity of any diameter is parallel to its conjugate*, as we saw in Art. 214, Cor.

221. *The co-ordinates of the extremity of any diameter being given, to find those of the extremity of the diameter conjugate to it.*

Let CP, CD (fig. Art. 220) be a pair of conjugate diameters in the ellipse, and let x', y' be the co-ordinates of P; then the equations to CP and CD (Art. 216) are

$$y = \frac{y'}{x'}x \text{........(1)}, \quad y = -\frac{b^2x'}{a^2y'}x \text{...........(2)}.$$

To find the co-ordinates of D (xy), we have from (2) the relation

$$\frac{ay}{bx} = -\frac{bx'}{ay'} \text{ .... (3)},$$

also $$a^2y^2 + b^2x^2 = a^2b^2 = a^2y'^2 + b^2x'^2 \text{........... (4)},$$

since (xy), $(x'y')$ are points on the curve;

hence from (3) $$\frac{a^2y^2 + b^2x^2}{b^2x^2} = \frac{a^2y'^2 + b^2x'^2}{a^2y'^2},$$

or from (4) $$b^2x^2 = a^2y'^2,$$

$$x = \pm \frac{a}{b}y';$$

hence $$CN = -\frac{a}{b}y', \text{ or } \frac{x}{a} = -\frac{y'}{b},$$

and from (2) $$DN = \frac{b}{a}x', \text{ or } \frac{y}{b} = \frac{x'}{a}.$$

The other pair of values $\left(x = \frac{a}{b}y', y = -\frac{b}{a}x'\right)$ have reference to the extremity D'.

222. In the hyperbola, if the diameter CP meet the curve, CD will not (Art. 218) meet it; and, indeed, it is evident that the method of the last article will give us imaginary values of the co-ordinates of D; for, as in equation (3) of the last article, we have for the co-ordinates of D (xy),

$$\frac{ay}{bx} = \frac{bx'}{ay'} \quad\text{.......................}\quad (1),$$

hence $$\frac{a^2y^2 - b^2x^2}{b^2x^2} = \frac{-(a^2y'^2 - b^2x'^2)}{a^2y'^2} \quad\text{...........}\quad (2),$$

which would give $b^2x^2 = -a^2y'^2$. If, however, we take $D\ (xy)$ as the point where CD meets the conjugate hyperbola, whose equation is

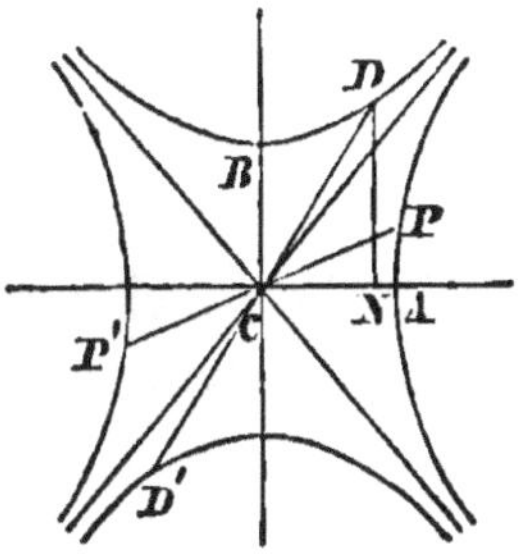

$$\frac{y^2}{b^2} - \frac{x^2}{a^2} = 1,$$

we have

$$a^2y^2 - b^2x^2 = a^2b^2 = -(a^2y'^2 - b^2x'^2),$$

whence, from (2) and (1)

$$x = CN = \frac{a}{b}y',\ y = DN = \frac{b}{a}x'.$$

The other pair of values have reference to D'.

We shall therefore *define* the extremity of the diameter conjugate to CP, as '*the point where it meets the conjugate hyperbola.*'

It is evident, by exactly the same reasoning as we have used in the case of the hyperbola itself, that, if we consider CD as a diameter of the conjugate hyperbola, CP will be the diameter conjugate to CD; and whatever is proved of the point P, as a point in the hyperbola itself, is true of the point D, as a point in the conjugate hyperbola; for instance, (Art. 220), the tangent at D is parallel to CP, and so forth.

*223. If ϕ is the eccentric angle of the point P, we shall have (Arts. 164, 221) in the ellipse,

for P, $x = a\cos\phi$, $y = b\sin\phi$; for D, $x = -a\sin\phi$, $y = b\cos\phi$;

and (Arts. 172, 222) in the hyperbola,

for P, $x = a\sec\phi$, $y = b\tan\phi$; for D, $x = a\tan\phi$, $y = b\sec\phi$.

These results will be found very useful in the solution of problems which relate to conjugate diameters.

*224. If ϕ, ϕ' be the eccentric angles of P and D in the ellipse, we have for D

$$x = a\cos\phi' = -a\sin\phi,\quad y = b\sin\phi' = b\cos\phi;$$

therefore $$\cos\phi' = -\sin\phi,\quad \sin\phi' = \cos\phi,$$

whence $$\phi' = 90^\circ + \phi.$$

If the eccentric angle ϕ' of any point in the conjugate hyperbola, be constructed as in the given hyperbola, the coordinates of the point may be written,

$$x = a\tan\phi',\quad y = b\sec\phi',$$

since these values satisfy the equation $\frac{y^2}{b^2} - \frac{x^2}{a^2} = 1$; hence, if ϕ, ϕ' be the eccentric angles of P and D in the given and conjugate hyperbola respectively, we have for D

$$x = a\tan\phi' = a\tan\phi,\quad y = b\sec\phi' = b\sec\phi;$$

therefore $$\tan\phi' = \tan\phi,\quad \sec\phi' = \sec\phi,$$

whence $$\phi = \phi'.$$

Cor. The above properties afford a simple method of constructing geometrically the diameter CD, conjugate to a given one CP. For, if we construct the eccentric angle of P by Arts. 164, 172, we may construct the eccentric angle of D as above, and thus find the point D.

225. *In the ellipse, the sum of the squares of any two semi-conjugate diameters is equal to the sum of the squares of the semi-axes; and, in the hyperbola, the same is true of their difference.*

Let $P(x'y')$ and $D(x''y'')$ be the extremities of any two semi-conjugate diameters $CP(a')$, and $CD(b')$; then

$$a'^2 = CP^2 = x'^2 + y'^2,$$

and
$$b'^2 = CD^2 = x''^2 + y''^2$$
$$= \frac{a^2 y'^2}{b^2} + \frac{b^2 x'^2}{a^2}. \quad \text{(Art. 221.)}$$

Therefore
$$a'^2 + b'^2 = (a^2 + b^2)\left(\frac{x'^2}{a^2} + \frac{y'^2}{b^2}\right)$$
$$= a^2 + b^2, \text{ in the ellipse;}$$

since in that case
$$\frac{x'^2}{a^2} + \frac{y'^2}{b^2} = 1.$$

Also
$$a'^2 - b'^2 = (a^2 - b^2)\left(\frac{x'^2}{a^2} - \frac{y'^2}{b^2}\right)$$
$$= a^2 - b^2 \text{ in the hyperbola;}$$

since in that case
$$\frac{x'^2}{a^2} - \frac{y'^2}{b^2} = 1.$$

Cor. If $a^2 = b^2$, we have in the hyperbola,
$$a'^2 - b'^2 = 0,$$
or every diameter in the equilateral hyperbola equals its conjugate; hence, by symmetry, they are equally inclined to the asymptotes, as in Art. 219.

226. *The rectangle contained by the focal distances of any point, is equal to the square of the corresponding semi-conjugate diameter.*

In the ellipse $CD^2 = a^2 + b^2 - CP^2$, (Art. 225)

but
$$CP^2 = x'^2 + y'^2 = x'^2 + b^2\left(1 - \frac{x'^2}{a^2}\right)$$
$$= b^2 + e^2 x'^2;$$

therefore
$$CD^2 = a^2 - e^2 x'^2 = (a - ex')(a + ex')$$
$$= HP \,.\, SP. \quad \text{(Art. 196.)}$$

A similar investigation for the hyperbola would lead us to the same result.

227. *To find the length of the perpendicular from the centre on the tangent at any point P (x'y'), in terms of the semi-diameter, conjugate to CP.*

If p be the length of a perpendicular from the origin on the line $\frac{xx'}{a^2}+\frac{yy'}{b^2}=1$, we have (Art. 52, Cor. 2)

$$p^2=\frac{1}{\left(\frac{x'^2}{a^4}+\frac{y'^2}{b^4}\right)}=\frac{a^2b^2}{\left(\frac{b^2x'^2}{a^2}+\frac{a^2y'^2}{b^2}\right)};$$

therefore $p=\frac{ab}{b'}$; for $b'^2=\frac{b^2x'^2}{a^2}+\frac{a^2y'^2}{b^2}$, as in Art. 225.

228. *All parallelograms, whose sides are formed by straight lines passing through the extremity of one diameter and parallel to its conjugate, are equal in area.*

By Art. 220, these lines are tangents to the curve, for it is there proved that the tangent at the extremity of any diameter is parallel to its conjugate; and, by Art. 181, the tangents at the extremities of any diameter are parallel to one another.

Let PP', DD' be two conjugate diameters, and let the

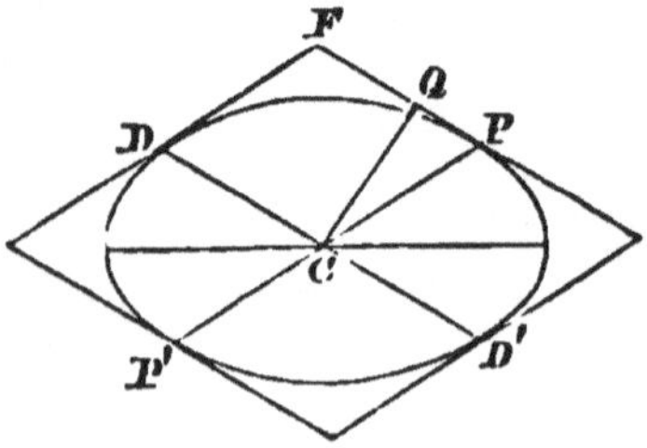

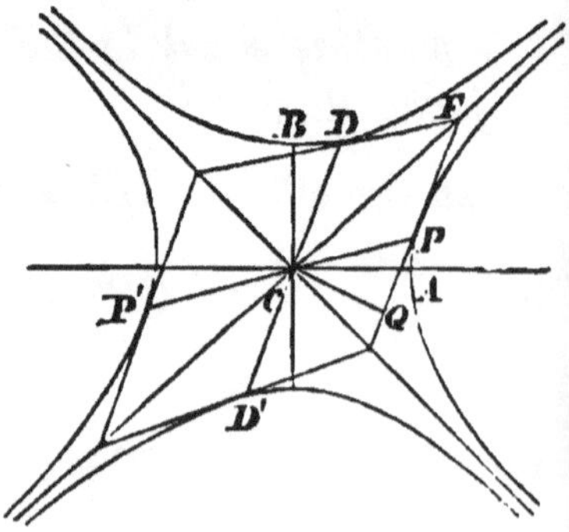

sides of the parallelogram be tangents at P, P', D, D'. Then the area of the whole parallelogram

$$= 4 \text{ times the parallelogram } CPFD$$
$$= 4 \,.\, CQ \,.\, CD,$$

where CQ is the perpendicular from the centre on the tangent at P,

$$= 4 \frac{ab}{b'} \,.\, b' \text{ (by Art. 227)}$$
$$= 4ab.$$

229. From the last article may be found the angle between any two conjugate diameters which are given; for, if this angle be $= \gamma$, and the given semi-conjugate diameters be a', b', we have

$$a'b' \sin \gamma = \text{parallelogram } CPFD$$
$$= ab;$$

hence
$$\sin \gamma = \frac{ab}{a'b'}.$$

This equation together with the relations

$$a'^2 + b'^2 = a^2 + b^2, \text{ in the ellipse,}$$
$$a'^2 - b'^2 = a^2 - b^2, \text{ in the hyperbola,}$$

determine also the magnitude of two conjugate diameters that contain an angle γ. Their position is known from the equation

$$\tan \theta \tan (\theta + \gamma) = -\frac{b^2}{a^2},$$

where θ is the angle that CP makes with the axis of x.

230. The angle γ, PCD, in the ellipse, is always, except in the case of the axes, greater than a right angle, P being supposed in the first and D in the second quadrant; for if m and m' be the tangents of the angles, that CP and CD make with the major axis,

$$\tan \gamma = \frac{m' - m}{1 + mm'} \quad \text{.......................} \quad (1),$$

and since $$m' = -\frac{b^2}{a^2 m},$$

we have $$\tan\gamma = -\frac{(a^2m^2 + b^2)}{m(a^2 - b^2)} \quad\text{.................}\quad (2),$$

and therefore, since m is always positive, and a greater than b, γ is $> 90^\circ$, unless $m = 0$, in which case $\gamma = 90^\circ$.

When $a = b$, $\tan\gamma = \infty$, or the conjugate diameters of the circle are all at right angles to each other, as we found in Arts. 132, 219.

Also, since (Art. 229) $a'b' \sin\gamma = ab$, $\sin\gamma$ will be the least when $a'b'$ is the greatest; but

$$2a'b' = a'^2 + b'^2 - (a' - b')^2 = a^2 + b^2 - (a' - b')^2,$$

and $a'b'$ is the greatest, when $a' = b'$. Hence $\sin\gamma$ has its least value, when $a' = b'$; or the equi-conjugate diameters contain the largest obtuse angle, which is determined by the equation

$$\sin\gamma = \frac{2ab}{a^2 + b^2}.$$

231. *To find the equation to the ellipse or hyperbola, when referred to any two conjugate diameters* ($2a'$, $2b'$) *as axes.*

We saw (Art. 145), that there is only one system of *rectangular* axes, to which, when a central curve is referred, its equation is of the form

$$Px^2 + Qy^2 = R,$$

an equation which asserts that all chords parallel to one axis are bisected by the other. But every diameter (Art. 216) bisects the chords parallel to its conjugate; hence there are an infinite number of *oblique* axes, which will give the equation in the above form, the only limitation being, that they should pass through the centre, and that the angles (θ, θ'), which they make with the major or transverse axis, should be subject to the condition

$$\tan\theta\tan\theta' = -\frac{b^2}{a^2}, \text{ in the ellipse,}$$

$$\tan\theta\tan\theta' = \frac{b^2}{a^2}, \text{ in the hyperbola.}$$

Hence, the reasoning of Arts. 150, 158 applies to conjugate diameters, in precisely the same manner as to the axes of the curve; and the equations to the ellipse and hyperbola, referred to any pair of conjugate diameters, the parts of which intercepted by the curve are $2a'$ and $2b'$, are

$$\frac{x^2}{a'^2} + \frac{y^2}{b'^2} = 1, \text{ and } \frac{x^2}{a'^2} - \frac{y^2}{b'^2} = 1.$$

232. Since this equation is of precisely the same form as the equation to the curve referred to its axes, it follows that every property, that has been deduced from the latter, may be deduced from the former, so long as those properties do not depend upon the inclination of the axes; and, with this limitation, everything which has been proved of the axes, is true of any pair of conjugate diameters.

For example, the equation to the tangent at any point ($x'y'$), when the curve is referred to any conjugate diameters ($2a'$, $2b'$), is

$$\frac{xx'}{a'^2} + \frac{yy'}{b'^2} = 1,$$

and the intercept on the axis of x, and the subtangent, are, as before, $\frac{a'^2}{x'}$ and $\frac{a'^2 - x'^2}{x'}$: Also, if we wish to draw a tangent to the curve from an external point $(x'y')$, we have, as in Article 123,

$$\frac{x^2}{a'^2}+\frac{y^2}{b'^2}=1, \quad \frac{xx'}{a'^2}+\frac{yy'}{b'^2}=1,$$

as equations to determine the points of contact.

233. The equation to the asymptotes, when the hyperbola is referred to any pair of conjugate diameters, is (Art. 186)

$$\frac{x^2}{a'^2}-\frac{y^2}{b'^2}=0;$$

also, the equations to the conjugate hyperbola and its asymptotes, when referred to the same (Art. 222) conjugate diameters, are

$$\frac{y^2}{b'^2}-\frac{x^2}{a'^2}=1, \quad \frac{y^2}{b'^2}-\frac{x^2}{a'^2}=0.$$

234. We may obtain a simple geometrical method for constructing a pair of conjugate diameters, containing a given angle, as follows.

The reasoning of Art. 216 is equally applicable, when the curve is referred to any pair of conjugate diameters[1]; hence we see that, when the curve is referred to a pair whose semi-lengths are a', b', the condition that the two lines

$$y=mx, \quad y=m'x$$

should represent two conjugate diameters is

$$mm'=-\frac{b'^2}{a'^2}.$$

[1] In the proof of Art. 214, if ω is the angle between the conjugate diameters, we have (Art. 34),

$$s=\frac{\sin\theta}{\sin\omega}, \quad c=\frac{\sin(\omega-\theta)}{\sin\omega}, \text{ and } \therefore m=\frac{\sin\theta}{\sin(\omega-\theta)}, \quad m'=\frac{\sin\theta'}{\sin(\omega-\theta')}.$$

Now, if the curve be referred to these two diameters, the equation may be written

$$y^2 = -\frac{b'^2}{a'^2}(x^2 - a'^2),$$

which equation may be split into the two

$$y = -k\frac{b'^2}{a'^2}(x - a'), \quad y = \frac{1}{k}(x + a'),$$

where k is a perfectly arbitrary quantity; now these two straight lines (i) pass through the extremities of the diameter which is the axis of x; (ii) intersect in the curve, since by eliminating k between them we have the equation to the curve; (iii) fulfil the same conditions, with regard to their inclinations to the axis of x as a pair of conjugate diameters, for

$$\left(\frac{1}{k}\right)\left(-k\frac{b'^2}{a'^2}\right) = -\frac{b'^2}{a'^2}.$$

Straight lines drawn in this manner from the extremities of any diameter to a point in the curve, are called *Supplemental Chords;* hence diameters parallel to any pair of supplemental chords are conjugate.

Hence, to draw a pair of conjugate diameters containing any given angle, describe on any diameter a segment of a circle containing that angle, and join the point where it meets the curve with the extremities of the assumed diameter. We thus obtain a pair of supplemental chords inclined at the given angle. The straight lines drawn through the centre parallel to these, will be the required conjugate diameters.

The property of supplemental chords was demonstrated for the circle (Art. 111), when it was shewn that all supplemental chords in the circle are at right angles, as are all conjugate diameters.

235. The above property is evident geometrically thus. Let PP' be any diameter, and Q any point in the curve; draw CR, CR' to the middle points of PQ, $P'Q$ respectively; then CR and CR' are (Euc. VI. 2) parallel to $P'Q$ and PQ respectively; but CR bisects all chords parallel to PQ, or CR'; and CR' bisects all chords parallel to $P'Q$, or CR. Hence CR, CR' are conjugate, and PQ, $P'Q$ are parallel to conjugate diameters.

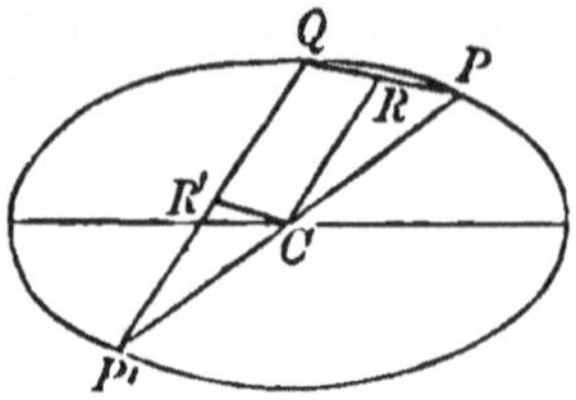

236. *Tangents at the extremities of any chord intersect in the diameter of which the chord is an ordinate.*

If we take that diameter and its conjugate as the axes of x and y, the equation to the tangent will be

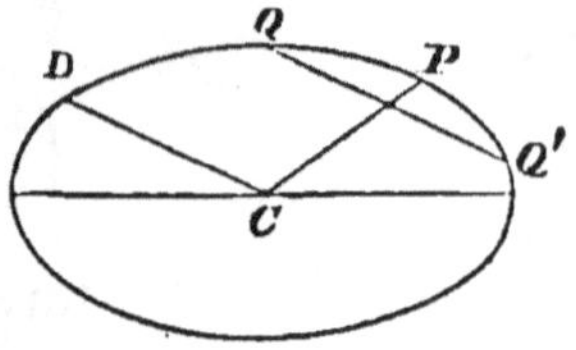

$$\frac{xx'}{a'^2} \pm \frac{yy'}{b'^2} = 1,$$

according as we take $Q(x'y')$ or $Q'(x', -y')$ as the point of contact.

In each case, when $y = 0$, x has the same value $= \frac{a'^2}{x'}$, or the tangents meet CP produced in the same point.

This may also be proved as follows. Let the equation to the chord be

$$Ax + By + C = 0 \ldots\ldots\ldots\ldots\ldots\ldots\ldots (1);$$

then, by Art. 215, the equation to the diameter that bisects it may be written

$$Bb^2x - Aa^2y = 0 \ldots\ldots\ldots\ldots\ldots\ldots\ldots (2).$$

If $(x'y')$ be the point of intersection of tangents at the extremities of (1), then, by Art. 123, (1) is equivalent to

$$\frac{xx'}{a^2} + \frac{yy'}{b^2} = 1 \ldots\ldots\ldots\ldots\ldots\ldots\ldots (3),$$

whence $$\frac{Aa^2}{x'} = \frac{Bb^2}{y'} = -\frac{C}{1} \qquad (4);$$

and the values of x' and y' in (4) will be seen to satisfy (2). Hence the point of intersection is on the diameter.

Cor. Conjugate diameters are (Art. 129) conjugate lines through the centre, the poles of which are infinitely distant. Cf. Art. 220.

237. *The Asymptotes.*

The following articles relate exclusively to the hyperbola, since the asymptotes of the ellipse have been shewn to be imaginary.

The diagonals of all parallelograms, described as in Art. 228, are parallel to the asymptotes.

Take $CP\,(=a')$ and $CD\,(=b')$ as the axes of x and y; then F (fig. Art. 228) is the point $(a'b')$, and the equations to CF, PD are

$$\frac{x}{a'} - \frac{y}{b'} = 0, \quad \frac{x}{a'} + \frac{y}{b'} = 1.$$

They are therefore parallel to the asymptotes, whose equations are (Art. 233)

$$\frac{x}{a'} - \frac{y}{b'} = 0, \quad \frac{x}{a'} + \frac{y}{b'} = 0.$$

238. Hence, if any two conjugate diameters CP, CD be given in position, we can find the asymptotes, by completing the parallelogram $CPFD$, the diagonals of which will shew the direction of the asymptotes; or, if the asymptotes be given, we can find the position of the diameter conjugate to any diameter CP, whose position is given; for, if we draw PO parallel to one asymptote to meet the other in O, and produce PO to D, making $OD = PO$, CD will be the diameter conjugate to CP.

239. *If any line (RR′) cut an hyperbola and its asymptotes, the portions (RQ), (R′Q′) intercepted between the curve and the asymptotes are equal.*

The equation to the two asymptotes, considered as one locus, is (Art. 169)

$$\frac{x^2}{a^2} - \frac{y^2}{b^2} = 0 \text{ ....} (1).$$

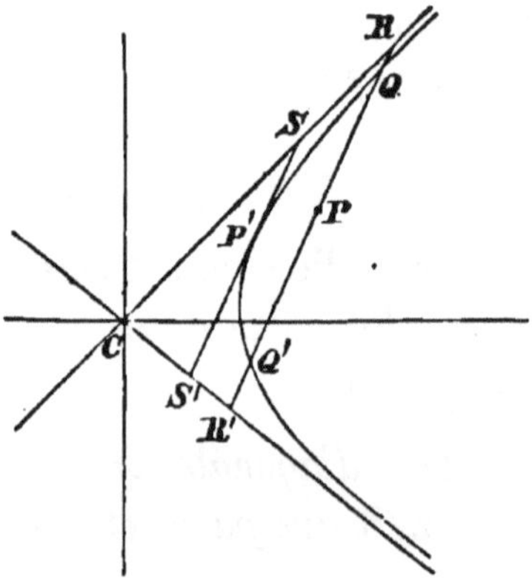

Now let the equation to RR', passing through a point $P(x'y')$, be

$$\frac{x-x'}{c} = \frac{y-y'}{s} = l \text{......} (2);$$

then, for the distances of P from R and R', we have from (1) and (2)

$$\frac{(cl+x')^2}{a^2} - \frac{(sl+y')^2}{b^2} = 0 \text{.................} (3);$$

and the condition that $P(x'y')$ should be the middle point of RR' is, exactly as in Art. 214,

$$\frac{cx'}{a^2} - \frac{sy'}{b^2} = 0 \text{.......................} (4);$$

but this is the condition (Art. 214) that $(x'y')$ should be on the diameter bisecting QQ'. Hence, if P be the middle point of RR', it is also the middle point of QQ', and we have

$$PQ = PQ', \; PR = PR', \text{ and } \therefore \; RQ = R'Q'.$$

Cor. Equation (4), which is the condition that $(x'y')$ should be the middle point of the chord RR', is also (Art. 180) the condition that the line should be a tangent to the curve at the point $(x'y')$; hence, if the middle point $(x'y')$ of RR' be *on* the curve, RR' is a tangent, or *the portion of the tangent intercepted by the asymptotes is bisected at the point of contact.* This follows directly from the consideration, that the tangent $SP'S'$ is the limiting position of RR', when it is moved parallel to itself up to P'.

240. *From the equation to the hyperbola referred to its axes, to derive the equation when referred to its asymptotes as axes.*

Let the lower asymptote be taken for axis of x, and let CM, PM be the original co-ordinates, CN, PN the new ones; draw NQ perpendicular to CM, and NV parallel to it, meeting PM produced in V, and let the angles RCM, LCM each $=\alpha$; then $\tan\alpha = \frac{b}{a}$, and, if $a^2+b^2=m^2$, we have

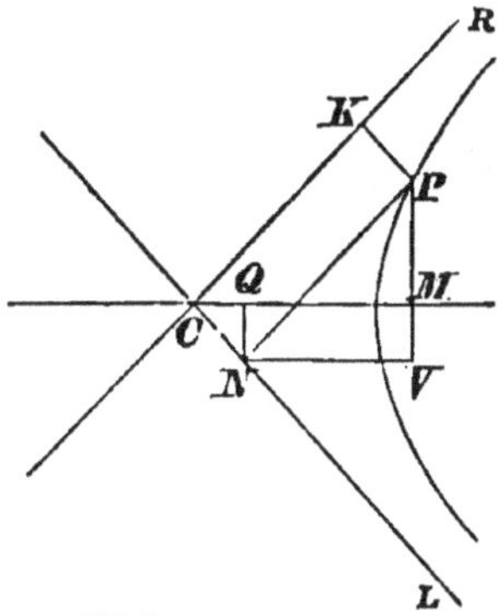

$$\cos\alpha = \frac{a}{m}, \quad \sin\alpha = \frac{b}{m}.$$

Now $$CM = NV + CQ = PN\cos\alpha + CN\cos\alpha,$$
$$PM = PV - QN = PN\sin\alpha - CN\sin\alpha;$$

hence we must write in the equation to the hyperbola,

for x, $\qquad (y+x)\cos\alpha$, or $\dfrac{a(y+x)}{m}$,

for y, $\qquad (y-x)\sin\alpha$, or $\dfrac{b(y-x)}{m}$,

which gives $$(y+x)^2-(y-x)^2=m^2,$$

or $$xy=\frac{a^2+b^2}{4},$$

which is therefore the required equation. The equation to the conjugate hyperbola is (Art. 175)

$$xy=-\frac{a^2+b^2}{4}.$$

241. The result of the last article may be obtained independently as follows. Let the equation to the hyperbola referred to its asymptotes be

$$Ax^2+2Hxy+By^2+2Gx+2Fy+C=0 \text{.........(1).}$$

Putting $y = 0$, we have for the points where the curve meets the axis of x,

$$Ax^2 + 2Gx + C = 0 \quad\ldots\ldots\ldots\ldots\ldots\ldots(2);$$

but, since the two roots of this equation are infinite, we have (Appendix) $A = 0$, $G = 0$. Similarly $B = 0$, $F = 0$, and equation (1) becomes

$$2Hxy + C = 0, \text{ or } xy = c^2 \ldots\ldots\ldots\ldots\ldots\ldots (3).$$

If now h and k be the co-ordinates of the vertex, and ω the angle between the asymptotes, we have

$$hk = c^2 \ldots\ldots\ldots(4), \quad CA^2 = h^2 + k^2 + 2hk \cos\omega \ldots\ldots\ldots(5).$$

But since CA bisects the angle between the axes, $h = k$, and therefore from (4), $= c$; hence from (5)

$$2c^2 (1 + \cos\omega) = a^2 \ldots\ldots\ldots\ldots\ldots\ldots (6);$$

but $\quad \tan\frac{\omega}{2} = \frac{b}{a}, \quad \therefore 1 + \cos\omega = 2\cos^2\frac{\omega}{2} = \frac{2a^2}{a^2 + b^2};$

therefore $\quad c^2 = \dfrac{a^2 + b^2}{4}.$

242. The equation then to an hyperbola referred to its asymptotes is $xy = k$, where k is a constant, and is positive or negative, according as the curve lies in the angles where x and y are both positive or both negative, or in those where x and y are of different signs; also the equations to the asymptotes are $x = 0$, $y = 0$. Hence, if any transformation whatsoever of axes be made, so that (Art. 74) we have to write

$$\text{for } x,\ ax + by + c; \quad \text{for } y,\ a'x + b'y + c',$$

the equation to the hyperbola will become

$$(ax + by + c)(a'x + b'y + c') = k \ldots\ldots\ldots (1),$$

where $\quad ax + by + c = 0 \ldots\ldots(2), \quad a'x + b'y + c' = 0 \ldots\ldots(3),$ are the asymptotes, since $x = 0$, $y = 0$ are transformed to these equations. Hence the equation to any hyperbola, whatever be the origin or axes, may be written in the form (1).

Conversely, if any equation of the second degree can be written as (1), it represents an hyperbola with (2) and (3) for asymptotes; for suppose the lengths of straight lines drawn from (xy), parallel to (3) to meet (2), and parallel to (2) to meet (3), to be (Art. 52, Cor. 1) respectively

$$p\,(ax+by+c),\ \ q\,(a'x+b'y+c');$$

then, since equation (1) may be written

$$p\,(ax+by+c)\times q\,(a'x+b'y+c')=kpq,$$

it asserts that the product of two such lines, drawn from any point (xy) on the locus, is constant; hence, if (2) and (3) are taken as axes, we shall have $xy =$ a constant, which equation (Art. 150) must represent an hyperbola; for it cannot be broken into two linear equations, unless the constant $= 0$, in which case it represents the axes.

Since the equation $xy=k$ represents the conjugate hyperbola if the sign of k is changed, so also does equation (1), which is obtained from it; and all the curves obtained from (1) by varying k only, have the same centre and asymptotes. Also, if we draw the lines (2) and (3), and mark their positive and negative sides (Arts. 55, 56), it is easy to see in which angles the hyperbolas lie. If the positive sign is used with k, the curves must lie in the ++ and −− compartments formed by the lines; for then the expressions

$$ax+by+c,\ \ a'x+b'y+c'$$

are both positive or both negative. If the negative sign is used, the curves lie in the +− and −+ compartment.

Ex. The equation

$$(y-x)(x+2y-3)=7,$$

represents an hyperbola, of which the asymptotes are the lines

$$y-x=0, \qquad x+2y-3=0.$$

The curve lies in the ++ and in the −− compartments formed by the lines, as in the figure.

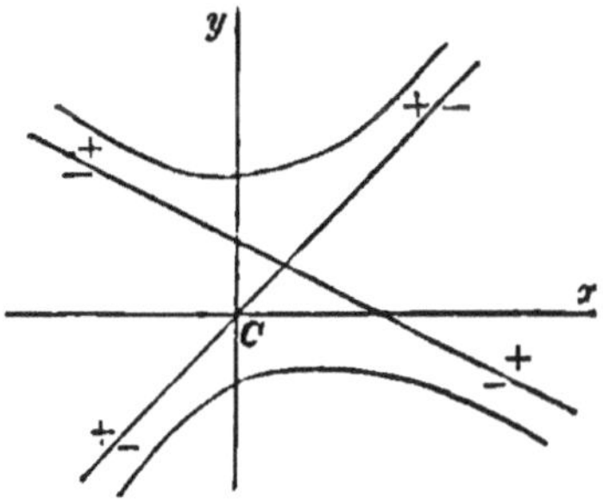

243. *To find the equation to the tangent at any point $(x'y')$, when the asymptotes are the co-ordinate axes.*

Let ω be the angle between the asymptotes, and let the equation to a line cutting the hyperbola in $(x'y')$ be (Art. 34)

$$\frac{x-x'}{c}=\frac{y-y'}{s}=l \text{ .... } (1),$$

where
$$s=\frac{\sin\alpha}{\sin\omega},\quad c=\frac{\sin(\omega-\alpha)}{\sin\omega};$$

then, substituting from this equation in the equation to the hyperbola

$$xy=\frac{a^2+b^2}{4} \text{ .... } (2),$$

we have
$$(sl+y')(cl+x')=\frac{a^2+b^2}{4},$$

or
$$scl^2+(sx'+cy')\,l=0 \text{ .... } (3),$$

since $(x'y')$ is a point on the curve, and therefore

$$x'y'=\frac{a^2+b^2}{4}.$$

Equation (3) will give us, by reasoning exactly similar to that used in Art. 117, as a condition that equation (1) should represent a tangent at $(x'y')$,

$$sx'+cy'=0 \text{ .... } (4),$$

and from (1)
$$\frac{y-y'}{x-x'}=\frac{s}{c},$$

$$=-\frac{y'}{x'}, \text{ from } (4);$$

hence
$$xy'+yx'=2x'y', \text{ or } xy'+yx'=\frac{a^2+b^2}{2},$$

which is the required equation.

244. Since the equation

$$xy' + yx' = \frac{a^2 + b^2}{2}$$

is not altered, when x changes places with x', and y with y', it follows (Art. 131) that this is the equation to the polar of $(x'y')$, when the hyperbola is referred to its asymptotes, and that the properties of Arts. 123—127, 129, 130 may be proved by it.

245. We may make use of the equation to the chord of contact in the above form, to shew on which branch of the hyperbola the tangents fall, that are drawn from any external point $(x'y')$. The points of contact are determined (Art. 123, Cor.) by the equations

$$xy' + yx' = \frac{a^2 + b^2}{2} \ldots\ldots (1), \quad xy = \frac{a^2 + b^2}{4} \ldots\ldots\ldots (2).$$

Eliminating y between (1) and (2), we have

$$4y'x^2 - 2(a^2 + b^2)\,x + (a^2 + b^2)\,x' = 0 \ldots\ldots\ldots\ldots (3).$$

If x_1, x_2 are the roots of (3), we have

$$x_1 + x_2 = \frac{a^2 + b^2}{2y'} \ldots\ldots (4), \quad x_1x_2 = (a^2 + b^2)\,\frac{x'}{4y'} \ldots\ldots\ldots (5).$$

Now, if x' and y' are both positive or both negative, the abscissæ of the points of contact are so also; for from (5) they must be of the same sign, and from (4) that sign must be the sign of y'; hence, if $(x'y')$ lies in the same angle of the asymptotes as the curve, the points of contact lie in that angle. If the point $(x'y')$ lies in either of the two angles which do not contain the curve, so that x' and y' have different signs, the product of the roots of (3) is negative; hence, in this case, the abscissæ of the points of contact have opposite signs, and the tangents fall upon both branches.

EXAMPLES X.

*1. PROVE Arts. 221, 225—227 by means of the eccentric angle.

2. If CP, CQ be semi-diameters, at right angles to each other,

$$\frac{1}{CP^2}+\frac{1}{CQ^2}=\frac{1}{a^2}+\frac{1}{b^2}.$$

3. If, from the focus S of an ellipse, perpendiculars be drawn on CP, CD, conjugate diameters, these perpendiculars produced backwards will intersect CD and CP in the directrix.

4. If ρ, r and ρ', r' be respectively the focal distances of two points, P, D, the extremities of a pair of conjugate diameters of an ellipse, then

$$\rho r+\rho' r'=a^2+b^2.$$

5. If a tangent to an hyperbola at P cut off CT, Ct from the axes, then, $PT \, . \, Pt = CD^2$, CD being the semi-conjugate diameter.

6. In the rectangular hyperbola all diameters at right angles to one another are equal.

7. From the extremities P, D of two conjugate diameters, normals are drawn to the major axis of an ellipse; the sum of the squares of these two $=\frac{b^2}{a^2}(a^2+b^2)$.

8. If the tangent at the vertex A cut any two conjugate diameters of an ellipse produced in T and t, then, $AT \, . \, At = b^2$.

9. The lengths of the equal conjugate diameters of an ellipse are $\sqrt{2(a^2+b^2)}$, and the eccentric angles of their extremities are 45° and 135°.

10. The locus of the middle points of chords of an ellipse, which pass through a fixed point, is an ellipse with the same eccentricity, and if the fixed point be the focus, the major axis of the ellipse is SC.

11. The tangent at any point of an hyperbola is produced to meet the asymptotes; shew that the triangle cut off is of constant area.

12. If the asymptotes of the hyperbola are axes, shew that the equation to one directrix is $x + y - a = 0$.

13. If any two tangents be drawn to an hyperbola, and their intersections with the asymptotes be joined, the joining lines will be parallel.

14. Shew that the locus of the points of quadrisection of all parallel chords in a circle is a concentric ellipse.

15. If the angle between the equal conjugate diameters of an ellipse is 60°, find the eccentricity.

16. If a be the angle between two conjugate diameters which make angles θ, θ' with the axis major,

$$\cos a = e^2 \cos\theta \cos\theta'.$$

*17. CP, CD are semi-conjugate diameters of an ellipse, and PF is a perpendicular let fall from P on CD or CD produced; determine the locus of F.

*18. The chords joining the extremities of the conjugate diameters of an ellipse will all touch in their middle points a concentric ellipse with axes $a\sqrt{2}$, $b\sqrt{2}$, coincident with those of the original curve.

19. If a circle be described from the focus of an hyperbola, with radius equal to half the conjugate axis, it will touch the asymptotes in the points where they are cut by the directrix.

20. Trace the curve, referred to rectangular axes,

$$\frac{(4x-3y)^2}{4}+\frac{(3x+4y+6)^2}{9}=25.$$

21. The radius of a circle, which touches an hyperbola and its asymptotes, is equal to that part of the latus rectum produced, which is intercepted between the curve and the asymptote.

22. The equation to the diameter conjugate to

$$\frac{x}{c}-\frac{y}{s}=0 \quad \text{is} \quad \frac{x}{c}+\frac{y}{s}=0,$$

the hyperbola being referred to its asymptotes.

23. An ellipse being traced upon a plane, draw the axes and the directrix, and find the focus.

24. Find the angle between the asymptotes of the hyperbola $xy=bx^2+c$, the axes being rectangular; and write the equation to the conjugate hyperbola.

25. Tangents are drawn to an hyperbola, and the portions intercepted by the asymptotes are divided in a given ratio; shew that the locus of the point of division is an hyperbola.

26. Draw the asymptotes of the hyperbolas

$$xy-2x-3y-2=0, \quad xy+2x^2+3=0,$$

and place the curves in the proper angles.

*27. Find the locus of the intersection of tangents to an ellipse, which are parallel to conjugate diameters.

28. Find the equation to the locus of the middle points of all chords of a given length, in an ellipse.

29. If two concentric equilateral hyperbolas be described, the axes of the one being the asymptotes of the other, they will intersect at right angles.

30. If P be the middle point of a straight line AB, which is so drawn as to cut off a constant area from the corner of a square, its locus is an equilateral hyperbola.

31. If S and H be the foci of an equilateral hyperbola, and a circle be described upon SH, then the quadrantal chord of this circle will be a tangent to that described upon the transverse axis.

32. If a be the acute angle between the axes of co-ordinates of the ellipse $(x^2 + y^2 = c^2)$, find the lengths of the axes and the eccentricity.

33. If AA' be any diameter of a circle, PQ any ordinate to it, then the locus of the intersections of AP, $A'Q$ is an equilateral hyperbola.

34. In an equilateral hyperbola, focal chords parallel to conjugate diameters are equal.

35. If a series of straight lines have their extremities in two straight lines at right angles to one another, and all pass through a given point, the locus of their middle points is an equilateral hyperbola.

36. PQ is an ordinate to the axis major AA' of an ellipse, meeting the curve in P and Q; draw AP, $A'Q$ intersecting in R; the locus of R is an hyperbola with the same centre and axes.

37. If tangents be drawn, making a given angle with the axes of all ellipses having the same foci, the locus of the point of contact is an equilateral hyperbola.

38. If normals be drawn to an ellipse from a given point within it, the points where they meet the curve will all lie in an equilateral hyperbola which passes through the given point, and has its asymptotes parallel to the axes of the ellipse.

39. Find the locus of the middle points of chords in a circle, which touch a concentric ellipse.

*40. If normals be drawn from the extremities of conjugate diameters to an hyperbola, and the point of their intersection be joined to the centre, this line produced will be perpendicular to the straight line passing through the extremities of the conjugate diameters.

41. Given in position, a straight line AB and a point P outside it; a straight line PM is drawn, intersecting AB in C, from the extremity M of which a perpendicular MD on AB intercepts CD of a given magnitude; find the locus of M.

42. The locus of the centres of all circles, which cut off from the directions of two sides of a triangle chords equal to two given straight lines, is an equilateral hyperbola, having two conjugate diameters in the directions of these sides.

43. A straight line passes through a given point and is terminated in the sides of a given angle; find the locus of the point which divides it in a given ratio.

44. From a point P perpendiculars are dropped upon the sides of a given angle, so as to contain a quadrilateral of given area; shew that the locus of P is an hyperbola, whose centre is the vertex of the given angle.

45. Given the base of a triangle and the difference of the tangents of the base angles; shew that the locus of the vertex is an hyperbola, of which the perpendicular through the centre of the base is an asymptote.

46. If about the exterior focus of an hyperbola, a circle be described with radius equal to the semi-conjugate axis, and tangents be drawn to it from any point in the hyperbola, the straight line joining the points of contact will touch the circle described on the transverse axis as diameter.

47. If, from the centre of an equilateral hyperbola, a straight line be drawn through any point P, and if ϕ and ϕ' be the angles which this line and the polar of P respectively make with the transverse axis, then

$$\tan\phi \tan\phi' = 1.$$

48. Prove that the circle which passes through any three of the four points, in which the equilateral hyperbola

$$x^2 + 2hxy - y^2 + 2gx + 2fy + c = 0$$

cuts the rectangular co-ordinate axes, is equal to the circle

$$x^2 + y^2 + 2gx + 2fy = 0.$$

49. Find the locus of the middle points of a system of parallel chords, drawn between an hyperbola and the conjugate hyperbola.

50. If, in two concentric hyperbolas, whose axes are coincident, two points be taken, whose abscissæ are as the transverse axes of the hyperbolas, the locus of the middle point of the straight line joining them is an hyperbola, whose axes are arithmetic means between those of the given hyperbolas.

51. If tangents be drawn from different points of an ellipse, of lengths equal to n times the semi-conjugate diameter at the point, the locus of their extremities will be a concentric ellipse with semi-axes equal to $a\sqrt{n^2+1}$, $b\sqrt{n^2+1}$.

52. If a length $PQ = CD$ be taken in the normal to an ellipse, the locus of the point Q is a circle, whose radius $= a - b$ or $a + b$, according as Q is taken within or without the ellipse.

CHAPTER XI.

The Parabola.

246. We saw (Art. 153) that there is one pair of rectangular axes, to which when the parabola is referred, its equation may be written in the simple form

$$y^2 = Lx.$$

We shall now proceed to determine its form and principal properties from this equation, and shall suppose L to represent a positive quantity.

Since $y = \pm\sqrt{Lx}$, when $x = 0$, $y = \pm 0$; hence the origin is a point in the curve, and the line $(x = 0)$ meets the curve in two coincident points; that is to say, the axis of y is a tangent to the curve. No part of the curve can lie on the left side of the origin, for negative values of x would render y imaginary. It must be symmetrical with regard to the axis of x, since every value of x gives two equal values of y with opposite signs; also, as x increases indefinitely in a positive direction, y increases indefinitely in both positive and negative directions; hence the form of the curve is that of the figure: the point A is the vertex, and Ax the axis of the parabola.

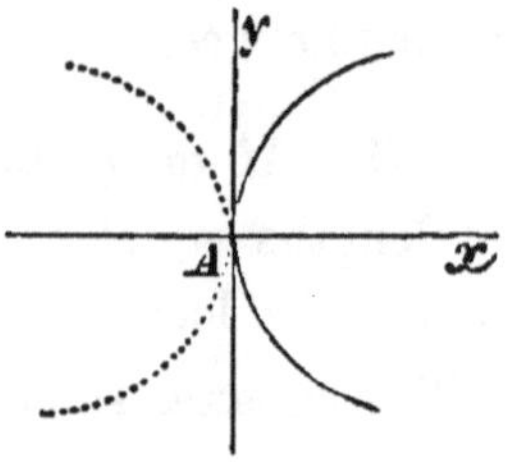

If the equation be

$$y^2 = -Lx,$$

no positive values of x will give real values for y, but x may have any negative value; hence, in this case, no part of the curve lies on the right of the origin, and, by exactly the same reasoning as before, the curve may be seen to correspond to the dotted line in the figure. It is proved to be concave to the axis of x in Art. 258, and this is also evident from Art. 249.

247. The parabola, like the hyperbola, has infinite branches, with this important difference in their nature. The tangent to the hyperbola, and consequently the direction of the branch, tends ultimately to coincide with a straight line making a finite angle with the axis of x, viz. the asymptote; while the tangent to the parabola, as will be shewn hereafter (Art. 254, Cor.), tends ultimately to become parallel to that axis and infinitely distant from it.

248. We saw (Art. 153), that the parabola might be considered as a central curve, with its centre removed to an infinite distance. We may therefore regard the parabola as an elongated ellipse; and, as this analogy is very useful in enabling us to foresee the properties of the curve, we shall prove the following proposition.

249. *If we suppose the distance between one vertex and focus of an ellipse to be given, while the axis major increases without limit, the curve will ultimately become a parabola.*

The equation to the ellipse, when the vertex A' is origin, is (Art. 165)

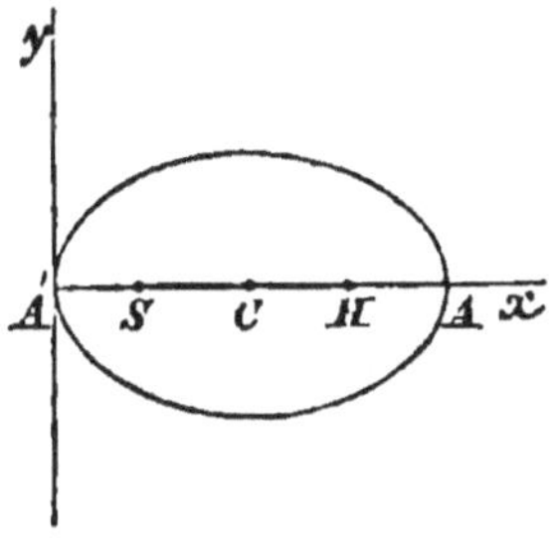

$$y^2 = \frac{b^2}{a^2}(2ax - x^2),$$

and, in order to find out what this equation becomes under the proposed circumstances, it will be necessary to express b in terms of a and the distance (d) $A'S$ between the vertex and focus, which is supposed to remain finite. Now,

$$A'S + SC = a,$$

or
$$d + \sqrt{a^2 - b^2} = a,$$

whence
$$b^2 = 2ad - d^2;$$

and the equation becomes

$$y^2 = \frac{2ad - d^2}{a^2}(2ax - x^2),$$

or
$$y^2 = \left(2d - \frac{d^2}{a}\right)\left(2x - \frac{x^2}{a}\right),$$

or, when $a = \infty$,
$$y^2 = 4dx,$$

which is the equation to a parabola.

Since $\frac{b^2}{a^2} = \frac{2ad - d^2}{a^2}$, and therefore vanishes, when a is infinite, we have $e^2 = 1 - \frac{b^2}{a^2} = 1$, when $a = \infty$; hence the parabola may be considered as an ellipse whose eccentricity $= 1$.

The same property may be proved in the same manner for the hyperbola: in that case, $b^2 = 2ad + d^2$, which value must be substituted in the equation

$$y^2 = \frac{b^2}{a^2}(2ax + x^2).$$

We shall, for the future, use the equation to the parabola in the form

$$y^2 = 4dx,$$

derived from its analogy with the ellipse; and we shall call that point on the axis of x, at a distance $= d$ from the vertex, which was the focus of the ellipse, the *focus* of the parabola.

*250. The co-ordinates of any point in the parabola, as in the ellipse and hyperbola, may be expressed by a single variable. For the values

$$x = d\mu^2,\ \ y = 2d\mu,$$

evidently satisfy the equation $y^2 = 4dx$. The value of μ will be seen (Art. 254, Cor.) to be $\cot\alpha$, where α is the angle which the tangent at the point makes with the axis of x.

Ex. 1. *To find the equation to the chord joining two points defined by the single variable,* μ.

Let the points be

$$d\mu^2,\ 2d\mu;\qquad d\mu'^2,\ 2d\mu';$$

then the equation to the chord is

$$\frac{y - 2d\mu}{x - d\mu^2} = \frac{2d\mu' - 2d\mu}{d\mu'^2 - d\mu^2},$$

$$= \frac{2}{\mu + \mu'} \text{..........................(1)};$$

whence $$(\mu + \mu')\,y = 2x + 2d\mu\mu'.$$

If $\mu = \mu'$, the chord becomes a tangent, and the equation to the tangent is

$$\mu y = x + d\mu^2 \text{.........................(2)}.$$

If $\tan^{-1} m$ is the angle which the tangent makes with the axis of x, $m = \frac{1}{\mu}$, and (2) becomes

$$y = mx + \frac{d}{m} \text{.........................(3)},$$

as in Art. 256.

Ex. 2. *To find the equation to the normal at the point* (μ).

Since the normal passes through the point ($d\mu^2$, $2d\mu$), and is perpendicular to the tangent (2), its equation is

$$y - 2d\mu = -\mu(x - d\mu^2),$$

or $$y + \mu x = 2d\mu + d\mu^3 \text{......................(4)}.$$

If $\tan^{-1} m$ is the angle which the normal makes with the axis of x, $m = -\mu$, and equation (4) becomes

$$y = mx - 2dm - dm^3,$$

as in Art. 261.

251. *To find the length of a straight line drawn from a point $(x'y')$ to meet the parabola.*

As in Art. 114, let the equation to the line be

$$\frac{x-x'}{c}=\frac{y-y'}{s}=l \quad\text{..................}\ (1);$$

then, for the distances of $(x'y')$ from the points of section of the line and the parabola ($y^2=4dx$), we have

$$(sl+y')^2=4d\,(cl+x');$$

therefore $$s^2l^2+2\,(sy'-2dc)\,l+y'^2-4dx'=0\ \text{.........}(2),$$

or $$Pl^2+Ql+R=0.$$

Now this equation will always give two values for l; hence every straight line meets the parabola in two real, coincident, or imaginary points, according as the roots of (2) are real and unequal, real and equal, or imaginary.

We shall hereafter have occasion to consider the following particular forms that this equation may assume.

If $P=0$, one value of l (Appendix) becomes infinite.

If $P=0$ and $Q=0$, both values of l become infinite.

If $R=0$, the point $(x'y')$ is on the curve, and one value of l becomes $=0$.

If $R=0$ and $Q=0$, both values of l become $=0$, and the line passes through two coincident points of the curve, and is a tangent.

If $Q=0$, the roots of the equation are equal and of opposite signs, and $(x'y')$ is therefore the middle point of the chord.

252. If $P=0$, that is, if $s=0$ in equation (2), the line (1) is parallel to the axis of the parabola, since s and c are the sine and cosine of the angle which (1) makes with the

axis. In this case then, one value of l becomes infinite; hence *a straight line drawn parallel to the axis of a parabola meets the curve in one point only at a finite distance from the origin, and in one point at an infinite distance.*

If $P=0$, $Q=0$, both values of l become infinite; but this gives

$$s=0, \quad y'=\frac{2dc}{s}=\frac{2dc}{0}=\infty;$$

so that the line is parallel to the axis of the parabola by the first condition, and infinitely distant from it by the second. Hence no straight line at a finite distance from the origin meets the parabola, as the asymptotes meet the hyperbola, altogether at an infinite distance. In other words, the parabola has no asymptotes.

253. We have said that a straight line, drawn parallel to the axis of a parabola, meets the curve in one finite point, and in one point at infinity. We adopt this language, because it is in accordance with the algebraic result; it is a short way of enunciating what may be stated more clearly as follows. Let K be the point $(x'y')$, and KPD a straight line cutting the parabola in P and D, and let KL be drawn parallel to the axis; then, if the point D moves along the curve to an infinite distance, and the line DK turns about K, KD will tend to coincide with KL as its limiting position. Without this explanation, the statement is not intelligible, as the line KL does not actually meet the curve at infinity; indeed, as shewn in Art. 246, the curve becomes indefinitely distant from it. As in the case of the hyperbola, our statement must be held to assert, that the curve tends

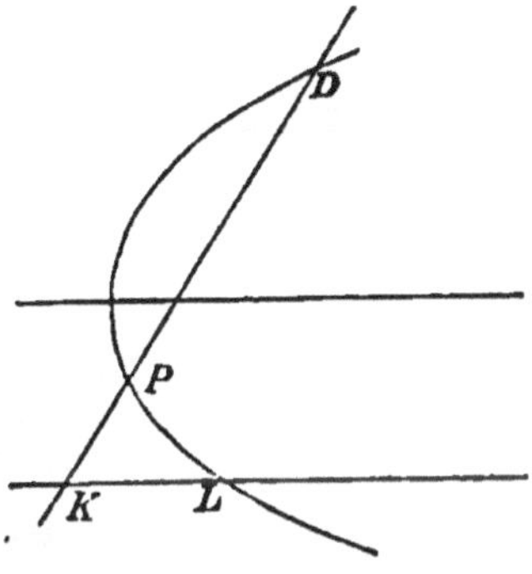

to become a straight line parallel to KL. We have explained in Art. 42 the meaning of the assertion, that parallel straight lines meet at infinity.

254. *To find the equation to a straight line touching the parabola in the point* $(x'y')$.

We shall proceed exactly as in the case of the circle and ellipse, and shall simply point out the steps in the proof.

Let the equation to a line cutting the curve in $(x'y')$ be

$$\frac{x-x'}{c}=\frac{y-y'}{s}=l\ldots\ldots\ldots\ldots(1);$$

then, for the distances of $(x'y')$ from the points of section of the line and the parabola $(y^2=4dx)$, we have, as in Art. 251,

$$s^2l^2+2(sy'-2dc)\,l+y'^2-4dx'=0,$$

or $$s^2l^2+2(sy'-2dc)\,l=0\ldots\ldots\ldots\ldots(2),$$

since $y'^2=4dx'$. Equation (2) gives us $l=0$, as it should, and

$$s^2l+2(sy'-2dc)=0.$$

If the line (1) be a tangent at $(x'y')$, l vanishes, and we have

$$sy'-2dc=0\ldots\ldots\ldots\ldots(3).$$

Eliminating s and c by equations (1) and (3), we have

$$(y-y')\,y'=2d\,(x-x');$$

or $$yy'=2dx-2dx'+y'^2,$$

or, since $y'^2=4dx'$, we have

$$yy'=2d\,(x+x'),$$

for the equation to the tangent at $(x'y')$.

COR. If α be the angle which the tangent makes with the axis of x, we have

$$\tan\alpha=\frac{2d}{y'}.$$

When $y'=0$, $\tan\alpha=\infty$, or the tangent at the vertex is perpendicular to the axis, as we saw in Art. 246. Also as y' increases from 0 to ∞, $\tan\alpha$ decreases from ∞ to 0, or the tangent, and therefore the direction of the curve, tends continually to become parallel to the axis. This agrees with Art. 253.

255. If we make $y=0$ in the equation to the tangent, we have $x=-x'$ or $AT=AM$. Hence the subtangent $MT=2x'$, and is bisected at the vertex.

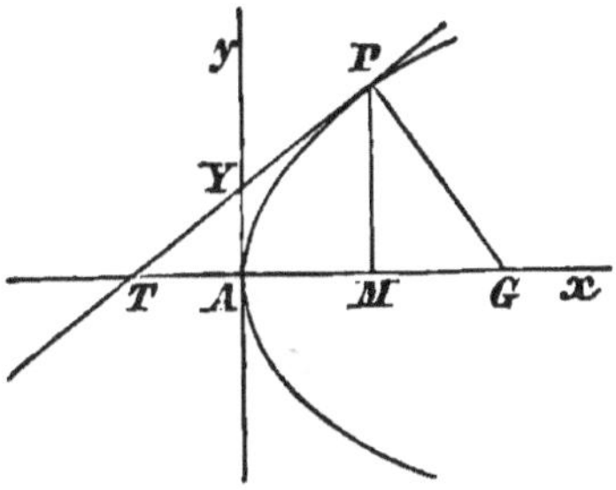

Also, writing $x=0$ in the equation to the tangent, we have for the intercept AY,

$$y=\frac{2dx'}{y'}=\frac{y'}{2};$$

hence, when x' and y' become infinite, the intercepts of the tangent at $(x'y')$ on the axes become infinite, or the tangent has no limiting position at a finite distance from the origin, as it has in the case of the hyperbola, where it becomes the asymptote.

256. *To find the equation to the tangent in terms of its inclination to the axis.*

This equation can be found independently, or deduced from the equation of Art. 254 by the methods of Arts. 119, 120. The resulting equation to the tangent is

$$y=mx+\frac{d}{m}.$$

Cor. Hence if a straight line and a parabola be represented by the equations

$$y=mx+b,\quad y^2=4dx,$$

the condition of tangency is $b=\dfrac{d}{m}$.

257. *To determine the tangents to a parabola which pass through a given point $(x'y')$.*

The equation to the tangent is

$$y = mx + \frac{d}{m} \quad \text{.......................} (1),$$

and, since it passes through $(x'y')$, we have

$$y' = mx' + \frac{d}{m};$$

or

$$m^2 - \frac{y'}{x'}m + \frac{d}{x'} = 0 \quad \text{.................} (2);$$

which equation gives two values of m, and therefore shews that, in general, two tangents may be drawn to the curve through a given point.

If μ, μ' be the two roots of the equation, the equations to the two tangents will be

$$y - y' = \mu(x - x'), \quad y - y' = \mu'(x - x').$$

258. The roots of equation (2) (Art. 257) are real and different, real and equal, or imaginary, according as

$$y'^2 - 4dx' > = < 0;$$

and it is easy to see, that this inequality gives the condition that $(x'y')$ should be without, on, or within the curve. Hence no real tangent can be drawn to the parabola from within the curve, &c., as in Art. 188.

By reference to Arts. 55, 121 Cor., 188, it will be seen in every case, if $\phi(xy) = 0$ is the equation to the locus, and $(x'y')$ any point in the plane, that $\phi(x'y')$ changes sign when $(x'y')$ crosses the locus.

259. *To determine the locus of the intersection of two tangents at right angles to one another.*

If in (Art. 257) the two tangents are at right angles,

$$\mu\mu = -1.$$

Hence, from (2), since $\mu\mu'$ is the product of the roots,

$$\frac{d}{x'} = \mu\mu' = -1,$$

or $x = -d$ is the equation required, which represents a straight line perpendicular to the axis of x, at a distance $= d$ on the negative side of the origin. It will be seen hereafter (Art. 263) that this is the *directrix* of the parabola.

260. It will be remembered that the equation to the same locus in the case of the ellipse (Art. 189), was found to be a circle whose equation is

$$x^2 + y^2 = a^2 + b^2 \ldots\ldots\ldots\ldots\ldots\ldots\ldots(1),$$

and we shall now shew that, when the ellipse passes into a parabola, this circle becomes the directrix. For, transferring the origin in (1) to A' by writing $x - a$ for x, we have

$$x^2 - 2ax + y^2 - b^2 = 0 \ldots\ldots\ldots\ldots\ldots\ldots(2);$$

but if $A'S = d$, $b^2 = 2ad - d^2$ (Art. 249), and (2) becomes

$$x^2 - 2ax + y^2 - 2ad + d^2 = 0 \ldots\ldots\ldots\ldots\ldots(3).$$

Dividing (3) by a and then making a infinite, we obtain the equation

$$x = -d,$$

the equation (Art. 263) to the directrix of the parabola. From this analogy, (1) is called the *director circle* of the ellipse.

261. *To find the equation to the normal to a parabola at any point* $(x'y')$.

Since the normal passes through $(x'y')$, its equation is

$$y - y' = m(x - x'),$$

and, since it is perpendicular to the line

$$yy' = 2d(x + x') \text{.....................} (1),$$

we have $m = -\frac{y'}{2d}$, and the equation is therefore

$$y - y' = -\frac{y'}{2d}(x - x') \text{................} (2).$$

Putting $y = 0$ in this equation, we have (fig. Art. 255), for the intercept of the normal on the axis of x,

$$x = 2d + x', \text{ or } AG = 2d + AM.$$

Also $MG = AG - AM$, or the *subnormal* $= 2d$.

COR. If m is the tangent of the angle that the normal makes with the axis of x, we have

$$m = -\frac{y'}{2d}; \text{ also } y'^2 = 4dx';$$

therefore $\qquad y' = -2dm, \quad x' = dm^2;$

hence, substituting these values in equation (2), we obtain

$$y = mx - 2dm - dm^3.$$

262. If in the equation to the tangent

$$yy' = 2d(x + x'),$$

we write x', x for x, x', and y', y for y, y', respectively, the equation remains unchanged. Hence (Art. 131) all the theories of poles and polars proved for the circle in Arts. 123—127, 129, 130, are equally true for the parabola; and the proofs will require no alteration, except that we must write the equations to the parabola and its tangent, in the place of the equations to the circle and its tangent. The student should convince himself of the truth of the above assertion, by writing out the articles with the requisite changes.

263. If $y'=0$, the equation to the polar becomes

$$x+x'=0,$$

which shews us that the polar of any point on the axis of x is parallel to the axis of y.

The polar of the focus, whose co-ordinates are (Art. 249) $x=d$, $y=0$, will have for its equation

$$x+d=0,$$

a straight line perpendicular to the axis of the parabola, lying to the left of the vertex, and at a distance from it $=d$. The polar of the focus is called the *directrix* of the parabola.

264. *To find the distance of any point in the parabola from the focus.*

Since the co-ordinates of the focus (S) are $x=d$, $y=0$, the square of the distance of any point $P(x'y')$ from it (Art. 7)

$$=(x'-d)^2+y'^2;$$

but, if $(x'y')$ be a point on the curve,

$$y'^2=4dx';$$

hence $$SP^2=(x'-d)^2+4dx'=(d+x')^2,$$

or $$SP=d+x'.$$

265. *The distance from the directrix of any point in the parabola, is equal to its distance from the focus.*

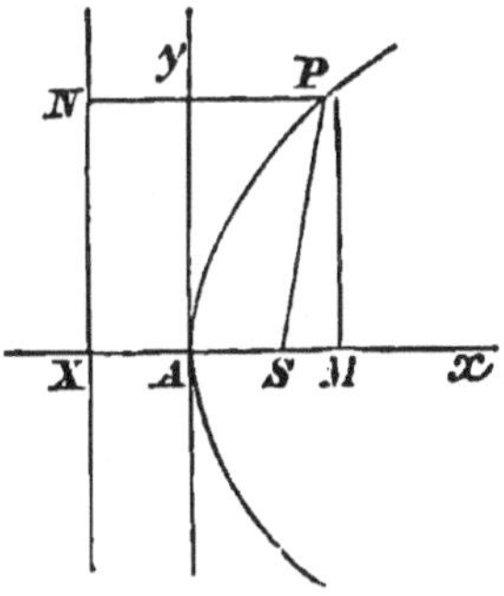

The equation to the directrix is (Art. 263)

$$x+d=0 \ldots\ldots\ldots\ldots\ldots(1);$$

and, if a perpendicular be dropped from any point $P(x'y')$ in the curve upon (1), we have (Art. 54)

$$PN=x'+d=SP.$$

This property is analogous to that proved for central curves (Art. 199), since we shewed (Art. 249) that the parabola might be regarded as an ellipse whose eccentricity $(e) = 1$.

266. The property proved in the last article will enable us to describe a parabola mechanically, by means of a ruler and cord. For let a ruler RNK, right-angled at N, slide along a line LX, and let a cord whose length is $= NR$ be fastened at R and at a point S; and while NK slides along LX, let a pencil P be moved, so as to keep a portion of the string stretched against RN. Then P will trace out a parabola; for the distance PN will be always equal to SP. LX will be the directrix, and S the focus.

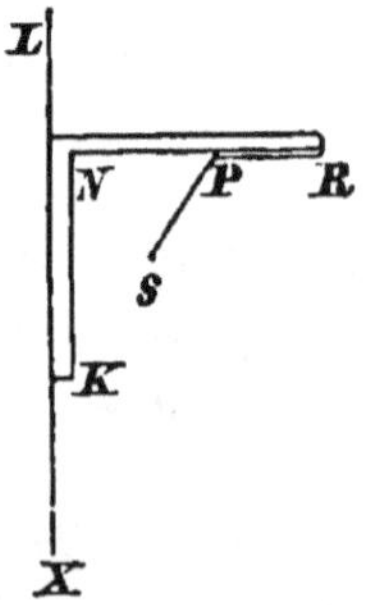

267. The double ordinate through the focus may be found, by putting $x = d$ in the equation $y^2 = 4dx$; then

$$y^2 = 4d^2, \text{ or } y = \pm 2d;$$

hence the double ordinate $= 4d$. This quantity, as in the case of the ellipse, is called the *Latus Rectum* of the curve.

The latus rectum of the ellipse or hyperbola

$$= \frac{2b^2}{a} \text{ (Art. 200)} = \frac{2(2ad \mp d^2)}{a} \text{ (Art. 249)},$$

$$= 4d,$$

when a becomes infinite, or the curve passes into a parabola.

*268. *The tangent and the normal at any point make equal angles with the focal distance of the point and the line drawn through it parallel to the axis.*

Draw PX parallel to Ax; then the equation to SP, since it passes through $(x'y')$, $(d, 0)$ is

$$\frac{y}{x-d}=\frac{y'}{x'-d}\text{..............}(1),$$

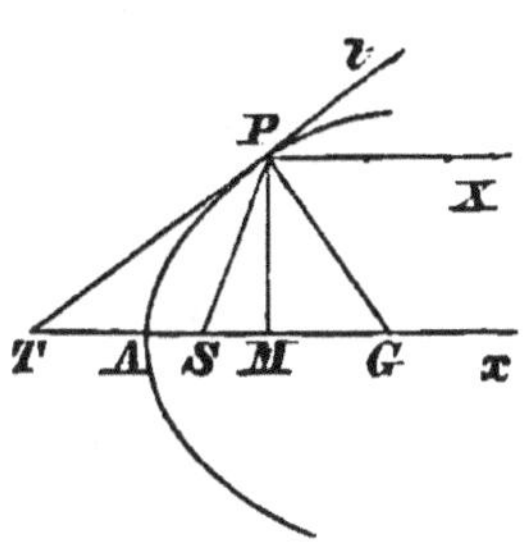

and to PX, $y-y'=0$ .... (2);

hence the equation to PG, the bisector of the angle between (1) and (2), is

$$\frac{-(x'-d)y+y'x-dy'}{\{y'^2+(x'-d)^2\}^{\frac{1}{2}}}=-(y-y')\dots(3);$$

making $y=0$ in (3), we have, since (Art. 264)

$$y'^2+(x'-d)^2=(x'+d)^2,$$

$$AG=x=2d+x';$$

hence (Art. 261) PG is the normal at P.

It will be seen that this is a modification of the property proved for the ellipse (Art. 201); for we may suppose the line PX to be in the direction of another focus H, at an infinite distance, and the angles tPX and SPT to correspond to the angles which the tangent makes with the two focal distances in the ellipse.

269. The above may be shewn geometrically thus:

$$ST=AS+AT=d+x' \text{ (Art. 255)},$$

$$SG=SM+MG=x'-d+2d=d+x';$$

therefore (Art. 264) $ST=SG=SP$;

hence the angle SPG = the angle SGP = the angle GPX,

and the angle tPX = the angle PTS = the angle SPT.

Ex. *The exterior angle between two tangents to a parabola is half the angle between the focal distances of the points of contact.*

Let PT and $P'T'$ be the two tangents, T being the nearer to A; and let them intersect in Q, and let TQ be produced to V; then

$$PSx=2PTS, \quad P'Sx=2P'T'x;$$

therefore

$$PSP'=PSx-P'Sx,$$
$$=2(PTS-P'T'x),$$
$$=2\,VQP'.$$

270. *To find the locus of the extremities of perpendiculars dropped from the focus on the tangent.*

The equation to the tangent in terms of its inclination to the axis is (Art. 256)

$$y = mx + \frac{d}{m};$$

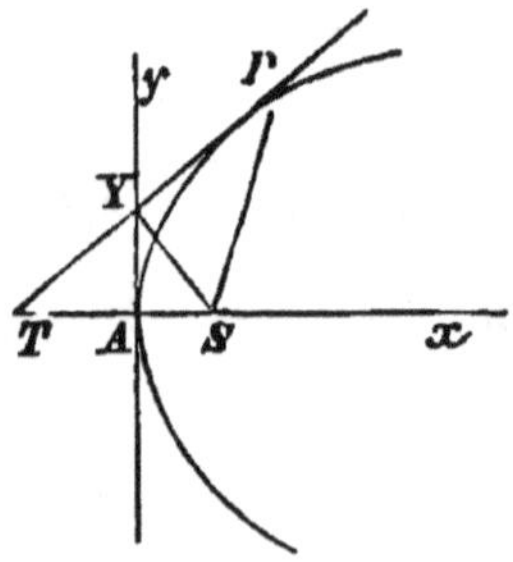

hence the equation to a straight line SY, drawn through the focus $(d, 0)$, and perpendicular to this tangent, is

$$y = -\frac{1}{m}(x-d).$$

If we combine these equations in any way, the resulting equation will be satisfied by the co-ordinates of Y. Our object is to eliminate m, which quantity particularizes the tangent; hence, subtracting, we have

$$\left(m+\frac{1}{m}\right)x = 0.$$

Therefore $x=0$ is the equation to the required locus, which evidently represents the axis of y, or the tangent at the vertex. This equation may be obtained from the corresponding equation for central conics, as in Article 249.

271. *To find the length of the perpendicular from the focus on the tangent.*

Let $SY=p$, $SP=\rho$ (fig. Art. 270); then for the perpendicular from $S(d, 0)$ on the line

$$yy' - 2d(x+x') = 0,$$

we have $$p^2 = \frac{(2d^2+2dx')^2}{y'^2+4d^2} = \frac{4d^2(d+x')^2}{4d(d+x')} = d(d+x');$$

therefore $$p^2 = d\rho.$$

272. *Any focal chord is perpendicular to the straight line joining its pole with the focus.*

By definition, the directrix is the polar of the focus, and therefore (Art. 129) the polar of any point in the directrix will pass through the focus. Hence we may assume the pole of any focal chord to be a point $(-d, y')$ in the directrix; then the equation to the chord is (Art. 123)

$$yy' = 2d(x-d) \quad\text{..................(1)},$$

and, the equation to a perpendicular to this line, through the focus $(d, 0)$, is

$$y = -\frac{y'}{2d}(x-d) \quad\text{.................... (2)},$$

and, when $x=-d$ in (2), $y=y'$, or the line represented by (2) passes through the point $(-d, y')$, which is the pole of the chord; hence the truth of the proposition.

273. *To shew that the locus of a point, whose distance from a given point is equal to its distance from a given straight line, is a parabola.*

This is the converse of what was proved in Art. 265, where it was shewn, that the distance of any point in the parabola from the focus is equal to its distance from the directrix. It has already been proved as a particular case of Art. 206.

274 (i). *To find the polar equation to the parabola, the focus being the pole.*

Here we have to find the polar equation to the locus of Art. 273, the given point being the focus, and the given line the directrix. Exactly as in Art. 209, if we make $e=1$, we obtain as the equation

$$\frac{l}{\rho} = 1 - \cos\theta,$$

where l is put for half the latus rectum, and is equal to $2d$.

274 (ii). *Confocal parabolas.*

The equation to a parabola with the focus as origin is

$$y^2 = 4d(x+d) \dots\dots\dots\dots\dots\dots (1),$$

and by varying d (1) can be made to represent any parabola, confocal to (1) and with the axis coincident.

Ex. *To find how many parabolas, confocal and with coincident axes, can be drawn through a given point* $(x'y')$.

If the point $(x'y')$ lies upon (1), we have the equation

$$4d^2 + 4dx' - y'^2 = 0 \dots\dots\dots\dots\dots\dots (2),$$

a quadratic in d, whose two roots (always real) will determine the two confocal parabolas which can be drawn through the point. Also, since (2) may be written

$$\frac{4d^2}{y'^2} + 2\frac{x'}{y'} \cdot \frac{2d}{y'} - 1 = 0 \dots\dots\dots\dots\dots\dots (3),$$

it follows that the product of the two values of $\frac{2d}{y'}$ is -1, or the two tangents (Art. 254) at $(x'y')$ are at right angles. The parabolas will be turned in opposite directions, the two values of d in (2) being of different signs.

275. *Diameters.*

We saw (Art. 153) that there is only one system of rectangular axes, which will give the equation to the parabola under the form $y^2 = Lx$. Here the axis of x is a diameter, for it bisects all chords parallel to the axis of y; and the axis of y is a tangent (Art. 246) at the extremity of that diameter. The same will be the case if the general equation be reduced to the form $y^2 = Lx$ with oblique axes, as in Art. 155; so that, if the general equation of the second degree represent a parabola, the equations

$$\sqrt{A}x + \sqrt{B}y = 0, \quad 2Gx + 2Fy + C = 0,$$

will represent the diameter through the origin and the tangent at its vertex. We shall now consider these other diameters, and shew that the form of the equation, when the axes are a diameter and the tangent at its vertex, is always $y^2 = Lx$.

276. *To find the locus of the middle points of any system of parallel chords.*

We shall proceed as in the case of central curves. Let QQ' be one of the chords, M $(x'y')$ its middle point, and let the equation to QQ' be

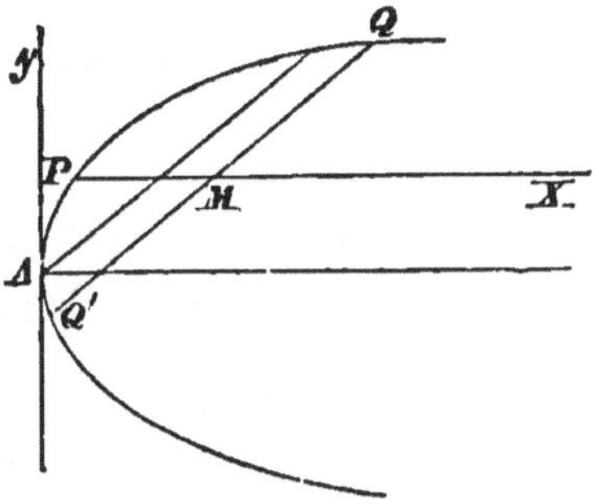

$$\frac{x-x'}{c}=\frac{y-y'}{s}=l.$$

Then for the distance (l) from $(x'y')$ of the points of section of the chord and parabola ($y^2=4dx$), we have equation (2) of Art. 251 to determine MQ and MQ'; but, since these distances are equal in magnitude, the two values of l are equal and of opposite signs; hence the coefficient of $l=0$, or $sy'-2dc=0$; and, since s and c are the same for all the chords, this relation holds for the ordinates of all the middle points; hence the equation required is

$$sy-2dc=0,$$

which represents a straight line (PX) parallel to Ax.

If m be the tangent of the angle that the chords make with the axis, the equation becomes

$$y=\frac{2d}{m}, \text{ since } m=\frac{s}{c}.$$

Cor. It is evident, as in central curves, that the tangent at P, the extremity of PX, is parallel to the chords; for the ordinate y' of $P=\frac{2d}{m}$, and the equation to the tangent is

$$yy'=2d(x+x'), \quad \text{or} \quad y=m(x+x'),$$

which is the equation to a straight line parallel to the chords.

277. As in central curves, the straight line which bisects any system of parallel chords is called a *diameter* of the

parabola, and the chords are called the *ordinates* of the diameter. We see then that the equations to a chord and the diameter of which it is an ordinate are, respectively,

$$y = mx + c \text{ and } y = \frac{2d}{m}.$$

Hence *all diameters of a parabola are parallel to the axis;* this agrees with the fact that the centre of a parabola is at an infinite distance. Conversely, all straight lines parallel to the axis may be considered as diameters; for by giving a suitable value to m in the equation $y = \frac{2d}{m}$, y may receive any value we please.

278. *The polar of any point in a diameter is parallel to the ordinates of that diameter.*

We saw (Art. 263), that this is true when the diameter is the axis of the parabola. Let $y = \frac{2d}{m}$ be the equation to any diameter; then the equation to the polar of any point $(x'y')$ in it is

$$yy' = 2d(x + x'),$$

and, since $y' = \frac{2d}{m}$, this equation becomes

$$y = m(x + x'),$$

which is the equation to a straight line parallel to the chords which the diameter in question bisects. If the point $(x'y')$ be the extremity A' of the diameter (fig. Art. 280), the equation

$$yy' = 2d(x + x')$$

represents the tangent at A'; hence, *the tangent at the extremity of any diameter is parallel to the ordinates of that diameter,* as we saw in Art. 276.

279. Since all diameters of a parabola are parallel, it cannot have *conjugate* diameters: it has however properties which correspond to the properties of conjugate diameters in central curves, and which may be foreseen by regarding the parabola as deduced from the ellipse by the method of Art. 249.

If we refer the ellipse to any diameter and the tangent at its vertex, as axes, the equation will be

$$y^2 = \frac{b'^2}{a'^2}(2a'x - x^2) \qquad (1),$$

which results from writing $x - a'$ for x in the equation to the ellipse referred to any two conjugate diameters $2a'$ and $2b'$. Now this equation is of the same form as the equation to the ellipse, when the vertex is origin, the major axis the axis of x, and a tangent at the vertex the axis of y. Hence, regarding the parabola as an elongated ellipse, we may conceive (fig. Art. 280) that any diameter $A'X$ has a conjugate at an infinite distance, parallel to the tangent $A'Y$, and we foresee that the equation to the parabola, when referred to $A'X$, $A'Y$ as axes, will be in the same form ($y^2 = Lx$), as when it is referred to the axis of the curve and the tangent at its vertex as axes.

For suppose fig. Art. 280 to represent part of an ellipse, A' the extremity of a diameter, and the origin in equation (1); and suppose that $A'X$ meets Ax in the centre C, and there is a diameter CD ($= b'$) conjugate to CA' ($= a'$); let $SA' = d'$; then (Art. 226) we have

$$SA' \,.\, HA' = CD^2, \text{ or } d'(2a - d') = b'^2;$$

hence equation (1) may be written

$$y^2 = \left(2d' - \frac{d'^2}{a}\right)\left(2x - \frac{x^2}{a'}\right)\frac{a}{a'}.$$

Now, when the centre is removed to an infinite distance, both a and a' become infinite; also

$$a : a' = \sin CA'A : \sin CAA',$$

which is a ratio of equality when $A'C$ becomes parallel to Ax, as it does, when their point of intersection (C) becomes infinitely distant; hence in this case equation (1) becomes

$$y^2 = 4d'x.$$

We shall, in the next article, prove this property independently.

280. If we transfer the origin to the extremity A' $(x'y')$ of one of these diameters, we have, writing $x + x'$ for x and $y + y'$ for y in the equation

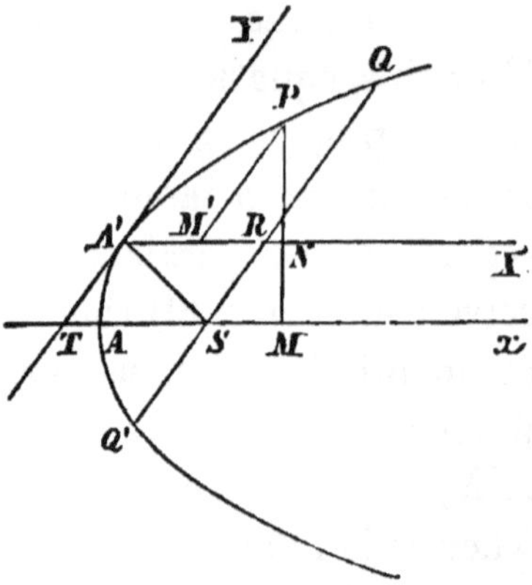

$$y^2 = 4dx,$$
$$(y + y')^2 = 4d\,(x + x'),$$

or $\quad y^2 + 2y'y = 4dx \ldots\ldots\ldots(1),$

since $\quad y'^2 = 4dx'.$

If we now preserve the axis of x, and take a new axis of y, ($A'Y$) inclined at an angle θ to the axis of x; then

$$\text{the old } y = PN = PM' \sin\theta,$$
$$\text{the old } x = A'N = A'M' + PM' \cos\theta,$$

$A'M'$ and PM' being the new co-ordinates of the point P, when $A'X$, $A'Y$ are axes. Hence, writing $y \sin\theta$ for y and $x + y\cos\theta$ for x in (1), the equation becomes

$$y^2 \sin^2\theta + 2y'y \sin\theta = 4d\,(x + y\cos\theta),$$

or $\quad y^2 \sin^2\theta + y\,(2y' \sin\theta - 4d\cos\theta) = 4dx \ldots\ldots\ldots(2).$

Now, in order that (2) may be reduced to the form $y^2 = Lx$, we must have

$$2y' \sin\theta - 4d\cos\theta = 0, \text{ or } \tan\theta = \frac{2d}{y'};$$

but this is the tangent of the angle which the tangent at A' $(x'y')$ makes with the axis of x; hence, when we use a diameter and the tangent at its vertex as co-ordinate axes, the equation to the parabola will be

$$y^2 = \frac{4d}{\sin^2\theta}x,$$

where θ is the inclination of the axes.

But $$\frac{d}{\sin^2\theta} = d\,\text{cosec}^2\,\theta = d\,(1 + \cot^2\theta)$$

$$= d\left(1 + \frac{y'^2}{4d^2}\right) = d\left(1 + \frac{x'}{d}\right)$$

$$= d + x' = SA';$$

and the equation to the parabola, referred to any diameter and the tangent at its vertex, is

$$y^2 = 4d'x,$$

where d' is the distance of the origin from the focus. This includes the case of the axis of the parabola and a tangent at its vertex.

281. The quantity $4SA'$ is called the *parameter* of the diameter which passes through A'; when the diameter is the axis of the parabola, it is sometimes called the *principal parameter*, as well as the *latus rectum*. In all cases it equals the double ordinate through the focus; for draw QQ' through the focus parallel to the tangent at A'; then (Art. 269)

$$A'R = ST = SA',$$

and $$QR^2 = 4SA' \,.\, A'R = 4SA'^2,$$

or $$QR = 2SA', \text{ and } QQ' = 4SA'.$$

282. Since the equation to a parabola, referred to *any* diameter and the tangent at its vertex, is of the same form as when the diameter is the axis, it follows that every property, which has been proved for the latter system, is true for all the others, so long as the property does not depend upon the inclination of the axes. For example, if we have for the equation to the parabola referred to such axes,

$$y^2 = 4d'x,$$

we shall find, exactly as in Art. 254, that the equation to the tangent at any point $(x'y')$ is

$$yy' = 2d'(x + x');$$

hence, as before, the subtangent is double the abscissa; also the equation to a straight line joining the points of contact of two tangents drawn from any point $(x'y')$ is

$$yy' = 2d'(x + x'),$$

and so on.

283. *Tangents at the extremities of any chord will intersect in the diameter of which the chord is an ordinate.*

Let the parameter of the diameter be $4d'$; then, if we take the diameter and the tangent at its extremity as axes, the equation to the tangent will be (fig. Art. 280),

$$\pm yy' = 2d'(x + x'),$$

according as we take $Q(x'y')$, or $Q'(x', -y')$, as the point of contact. In each case, when $y = 0$, x has the same value $= -x'$, or the tangent meets the diameter produced in the same point.

The student will find no difficulty in adapting the latter part of Art. 236 to the case of the parabola.

EXAMPLES XI.

1. FIND the length of the side of an equilateral triangle, one of whose angles is at the focus and the other two on the parabola.

2. Draw the curves $x^2 + 2y = 0$ and $y = 2x - x^2$, and find their points of intersection, the axes being rectangular.

3. The rectangle contained between two ordinates y_1, y_2 of a parabola ($y^2 = 4dx$) is equal to d^2; find the magnitudes of y_1 and y_2, the distance between them being $= d$.

4. Shew that a straight line, drawn from the point of the parabola of which the abscissa is $8d$, and cutting the axis at the point $x = 4d$, will, if produced, meet the curve again at the point $x = 2d$, and be a normal at that point, $4d$ being the latus rectum.

5. From any point there cannot be drawn more than three normals to a parabola. If the point be on the axis, and the abscissa less than $2d$, one only can be drawn.

6. The tangent at any point of a parabola will meet the directrix and latus rectum produced in two points equidistant from the focus.

7. Two equal parabolas have a common axis; a straight line, touching the interior and bounded by the exterior, will be bisected in the point of contact.

8. Find a parabola which shall touch a given circle at a given point, its axis being coincident with a given diameter.

9. To prove that the area of a triangle inscribed in a parabola is equal to

$$\frac{1}{8d}(y' \sim y'')\,(y'' \sim y''')\,(y''' \sim y'),$$

where y', y'', y''' are the ordinates of the vertices of the triangle, $y^2 = 4dx$ being the equation to the curve.

10. If a parabola intersect a circle in four points, prove that the ordinates of the points of intersection which lie on one side of the axis of the parabola are together equal to the sum of the ordinates of the points of intersection which lie on the other side of the axis.

11. Two tangents are drawn to a parabola at points whose co-ordinates are a, b, a', b'. To find the point in which they intersect.

12. Trace the curve $y = x - x^2$, and determine whether the straight line $x + y = 1$ is a tangent to it.

13. From points in the exterior of two equal parabolas having the same axis, tangents are drawn to the interior one; they will touch it at the extremities of diameters whose distance from one another is constant.

14. Three parabolas, having their axes parallel, intersect; shew that the three chords passing through their points of contact pass through one point.

15. Two tangents to a parabola make angles whose tangents are μ, μ' with the axis; find the equation to the tangent at the extremity of the diameter of which the chord of contact is an ordinate.

16. Find the locus of the middle points of chords passing through any fixed point, and adapt the proof to (i) the focus, (ii) the vertex, and (iii) the foot of the directrix of a parabola.

17. If A be the vertex, S the focus, and PSp the focal chord of a parabola, prove that the rectilinear triangle PAp varies as the square root of the distance Pp.

18. If a straight line be drawn from the foot of the directrix of a parabola, making an angle 45° with the axis, it will touch all parabolas having the same axis and directrix.

19. Find the equation to the normal at the extremity of the latus rectum of the parabola whose equation is $y^2 = 4d(x - d)$, and find its distance from the origin of co-ordinates.

20. Lp is a normal to the parabola at L, the extremity of the latus rectum, meeting the parabola again in p. Shew that the diameter in which the tangents at L and p intersect, passes through the other extremity of the latus rectum.

21. Two ordinates to a parabola meet the axis in points equidistant from the focus, and the vertex is joined with the point where one of the ordinates meets the parabola; find the equation to the locus of the point where this line intersects the other ordinate.

22. Two tangents are drawn to a parabola, making angles θ, θ' with the axis. Prove that (i) if $\sin\theta \, . \sin\theta'$ be constant, the locus of the intersection of the tangents is a circle, whose centre is in the focus; (ii) if $\tan\theta \, . \tan\theta'$ be constant, the locus is a straight line perpendicular to the axis; (iii) if $\cot\theta + \cot\theta'$ be constant, the locus is a straight line parallel to the axis; (iv) if $\cot\theta - \cot\theta'$ be constant, the locus is a parabola equal to the original parabola.

23. A series of triangles are constructed on a given base, their vertices being in a straight line parallel to the base; shew that the perpendiculars through the extremities of the base to the sides of these triangles, will intersect in a parabola, whose latus rectum is the distance between the lines.

24. To find the equation to a parabola, referred to the two tangents at the extremities of the latus rectum as axes.

25. To find the area of a triangle included between the tangents to parabolas $y^2 = 4dx$, $y^2 = 4\delta x$, at points, the common abscissa of which is a, and the portion of the ordinate intercepted between the two curves.

26. To find the magnitude of the ordinate of such a point in a parabola ($y^2 = 4dx$) that the intercepts on the axes of co-ordinates of a tangent drawn to the curve at this point may be equal to each other.

27. The three altitudes of any triangle described about a parabola all pass through a single point in the directrix.

28. To find the distance of the vertex and focus from the tangent in terms of the inclination of the tangent to the axis of x.

29. Find the locus of the centre of the circle which shall always touch a given circle and a given straight line.

30. From the vertex of a parabola a straight line is drawn inclined at an angle 45° with the tangent at any point; find the equation to the curve which is the locus of their intersection.

31. In the focal distance SP take Sp equal to the ordinate PM. Find the polar equation to the locus traced out by the point p.

32. From two points in a diameter of a parabola two pairs of tangents are drawn to the curve; the trigonometrical tangents of the inclination of one pair to the axis are μ_1, μ_2, and of the other μ_3, μ_4; to prove that

$$\frac{1}{\mu_1}+\frac{1}{\mu_2}=\frac{1}{\mu_3}+\frac{1}{\mu_4}.$$

33. The vertex of a parabola is taken for the centre of a given circle; to find the equation to a straight line touching both circle and parabola.

34. If from any point Q of the line BQ, which is perpendicular to the axis CAB of a parabola whose vertex is A, QP be drawn parallel to the axis to meet the parabola in P; shew that, if CA be taken $=AB$, the locus of the intersection of AQ and CP is the original curve.

35. A parabola being traced upon a plane, draw the axis and directrix, and find the focus.

36. Two equal tangents cannot be drawn to a parabola, except from a point on the axis.

37. Transform the equation to the tangent to the ellipse in the form $y=mx+\sqrt{a^2m^2+b^2}$, into the corresponding equation for the parabola.

38. Given the radius vector, drawn from the focus to any point of a parabola, and the angle it makes with the curve; find the latus rectum and the position of the vertex.

39. The locus of the centre of a circle which passes through a given point and touches a given straight line is a parabola.

40. If from the focus of a parabola straight lines be drawn to meet the tangents at a given angle, prove that the locus of their points of intersection will be that tangent to the parabola, the inclination of which to the axis is equal to the given angle.

41. The abscissa and double ordinate of a parabola are h and k, and the diameters of the circumscribed and inscribed circles are R and r. Prove that $R + r = h + k$.

42. If PQ be a chord of a parabola which is a normal at P, and the tangents at P and Q intersect in a point T, shew that PT is bisected in the directrix.

43. To find the equation to all parabolas which are touched by the straight lines $y = \pm \frac{x}{2}$.

44. Find the equation to the parabola, referred to the tangent and normal at the positive extremity of the latus rectum, as axes.

45. Two normals to a parabola ($y^2 = 4dx$) are always at right angles to each other; to find the locus of their intersection.

46. To find the equations to all the common chords of the two curves $y^2 = 2cx - x^2$, $y^2 = 4dx$.

47. To prove that a series of circles, of which the centres are in a parabola, and which pass through the focus, all touch the directrix.

48. The distance of a point from one given straight line varies as the square of its distance from another given straight line; shew that its locus is a parabola, having the second line as a diameter, and the first as a tangent at its vertex.

49. If the focus is origin, the equation to the tangent to the parabola, in the form of Art. 27, is

$$x \cos a + y \sin a + \frac{d}{\cos a} = 0.$$

50. From a point P, the concourse of two tangents (PQ, PQ') to a parabola, $PABC$ is drawn meeting the curve in A, C, and QQ' in B. PA, PB, PC are in harmonical progression.

51. To find the locus of the intersection of perpendiculars from the focus on the normal.

52. If two tangents be drawn to a parabola, prove that a third tangent, parallel to the chord joining the points of contact, will bisect the parts of the other tangents, which are included between their point of intersection and their points of contact.

53. The abscissæ of two points in a parabola, reckoned along the axis, are x, $3x$, and the corresponding focal distances r, $2r$; to find the position of the former of these points.

54. Any number of parabolas are described having the same vertex and axis, and any straight line is drawn at right angles to the common axis. If any points whatever in this line be taken as poles, to prove that all the polars belonging to all the parabolas will intersect in a single point.

55. The centre of an ellipse coincides with the vertex of a parabola, and the axis major of the ellipse is perpendicular to the axis of the parabola; required the proportion of the axes of the ellipse that it may cut the parabola at right angles.

56. Given a point where a parabola intersects a given diameter, and also the parameter of that diameter; shew that the locus of the vertex is an ellipse.

57. A parabola slides between rectangular axes; find the locus of (i) the focus, and (ii) the vertex.

CHAPTER XII.

General properties of Conic Sections.

In this chapter we shall shew how the chief properties of the loci of the second degree may be deduced from the general equation of the second order without reduction. We shall begin by tracing the loci. If the axes are rectangular, the method of Chap. VIII. may be used for this purpose, and is to be preferred, if we wish to determine the elements of the locus, such as the axes, position of foci, &c.; but the following method may be used with great advantage, when the axes are oblique, and in all cases where we wish simply to trace the form of the locus, without determining its elements.

284. Solving the equation

$$Ax^2 + 2Hxy + By^2 + 2Gx + 2Fy + C = 0 \text{.........(1)},$$

(which we shall call $\phi(xy) = 0$), so as to obtain y in terms of x, we have, as in Art. 62,

$$y = -\frac{Hx + F}{B} \pm \frac{1}{B}\sqrt{(H^2 - AB)\,x^2 + 2\,(HF - BG)\,x + F^2 - BC}.$$

Let us draw the right line (DT in figs. of Arts. 286, 287, 292) whose equation is

$$y = -\frac{Hx + F}{B} \text{ or } Hx + By + F = 0 \text{.........(2)};$$

then, in order to obtain the ordinates corresponding to any abscissa x' of the locus represented by (1), we have only to increase and diminish the ordinate of (2) corresponding to x' by the quantity

$$\frac{1}{B}\sqrt{(H^2 - AB)\,x'^2 + 2\,(HF - BG)\,x' + F^2 - BC} \ldots (3).$$

This line (2) then bisects every chord of the locus parallel to the axis of y, and is therefore a diameter.

Cor. Similarly, from the solution as a quadratic in x, we find that the line

$$Ax + Hy + G = 0 \ldots\ldots\ldots\ldots (4)$$

is the diameter bisecting all chords parallel to the axis of x. Equations (2) and (4) are those obtained in Art. 141, to determine the centre.

285. In order then to trace the locus, we must examine expression (3), which we shall call Y; for, as long as the values given to x render the quantity under the root positive, we can find two points of the locus corresponding to every abscissa; if they make it $= 0$, the two points coincide; and, if they make it negative, the value of Y is imaginary, or no point of the locus corresponds to the abscissa in question. For the sake of brevity we shall write

$$Y = \frac{1}{B}\sqrt{(H^2 - AB)\,(x^2 + 2Qx + R)},$$

where $$Q = \frac{HF - BG}{H^2 - AB}, \quad R = \frac{F^2 - BC}{H^2 - AB},$$

and may be either positive or negative quantities. The student will observe, that $-Q$ is the abscissa of the centre obtained in Art. 141; and this will be seen to agree with the results hereafter obtained.

Now the expression $x^2 + 2Qx + R$ may be written

$$(x + Q)^2 - (Q^2 - R).$$

We have then three cases:

(i) If $Q^2 - R$ is positive, that is, if the roots of the equation

$$x^2 + 2Qx + R = 0$$

are real and unequal, the expression can be broken up into real factors,

$$(x + Q + \sqrt{Q^2 - R})(x + Q - \sqrt{Q^2 - R}),$$

or $$(x - a)(x - b),$$

where $a + b = -2Q$, $ab = R$. This expression is positive for all real values of x, except for those which lie between a and b; and it attains its greatest negative value, $-(Q^2 - R)$, when

$$x = -Q = \frac{a + b}{2}.$$

Also, if we make $x = -Q + h$ or $= -Q - h$, the values of the expression are the same.

(ii) If $Q^2 - R = 0$, or the roots of the equation are real and equal, the expression becomes $(x + Q)^2$.

(iii) If $Q^2 - R$ is negative, that is, if the roots of the equation are imaginary, the expression cannot be broken up into real factors. We shall write it

$$(x + Q)^2 + D^2,$$

where the symbol D^2 is used to denote the positive quantity $R - Q^2$. This expression can never be negative for any real value of x, and has its smallest positive value when $x = -Q$; also, if we make $x = -Q + h$ or $= -Q - h$, the values of the expression are the same.

286. We shall now apply these results to the three cases, when $H^2 - AB$ is negative, positive, or zero.

I. When $H^2 - AB$ is negative. In this case we know, by Art. 150, that the locus is an ellipse or circle, a point, or imaginary.

(i) *Let the roots be real and unequal;* then Y, which in this case may be written

$$Y = \frac{1}{B}\sqrt{(H^2 - AB)(x-a)(x-b)},$$

is $=0$, when $x=a$ and when $x=b$, and is real for those values of x only that lie between a and b, since these are the only values that make $(x-a)(x-b)$ negative; hence the locus lies wholly between the two lines ($D'L$ and DR), parallel to the axis of y, whose equations are

$$x - a = 0, \quad x - b = 0.$$

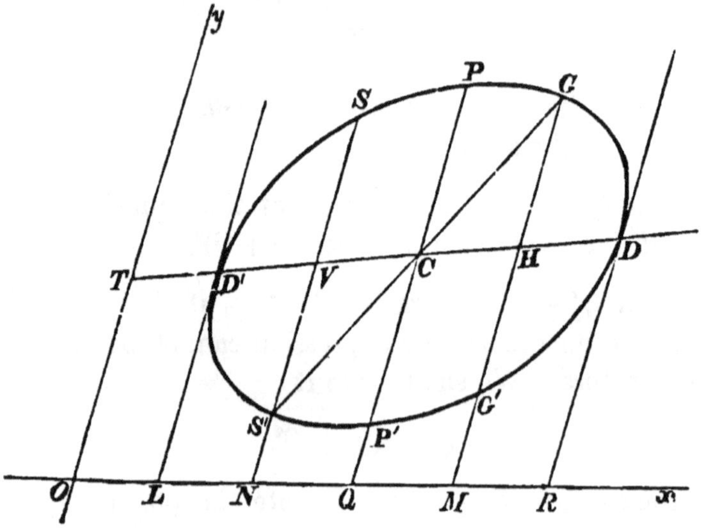

These lines are tangents to the curve at the points where the diameter (2) DT cuts it; for, if we put $x=a$ in the equation to the curve and diameter, the two ordinates to the curve will become equal, and will equal the ordinate of the diameter. The value of each is

$$y = -\frac{Ha+F}{B};$$

Hence the line $x-a=0$ passes through two coincident points of the curve, and is a tangent. Similarly $x-b=0$ is a tangent at the other extremity of the diameter. It will be observed that they are both parallel to the ordinates of the diameter.

The value of Y will be greatest, when $x=\frac{a+b}{2}$ (*i.e.* $=-Q$), the abscissa of C, the point midway between D and D'; for we have shewn that this value of x gives the expression

$$x^2+2Qx+R, \text{ or } (x-a)(x-b),$$

its largest negative value. This point is the centre of the curve; for, if we write

$$x=\frac{a+b}{2}+h, \text{ or } x=\frac{a+b}{2}-h,$$

we have shewn that the values of $(x-a)(x-b)$, and therefore of Y, are the same. From this it follows, that if we take $QM=QN=h$, we shall have $CV=CH$, and

$$HG=HG'=VS=VS';$$

hence, as in Art. 140, GCS' is a straight line, and is bisected in C, which is therefore the centre. The curve then, which is evidently an ellipse, has the form given in the figure. CP is the diameter conjugate to CD, since it is parallel to the ordinates of CD.

Ex. Let the equation be

$$2x^2-4xy+4y^2-2x-8y+9=0,$$

whence
$$2y-x-2\pm\sqrt{-(x-1)(x-5)}=0.$$

The curve has then for diameter the line $2y-x-2=0$; it lies wholly between the lines $x-1=0$, $x-5=0$; and the largest value of Y, corresponding to $x=3$, is $Y=1$. The co-ordinates of the centre are $x=3$, $y=\frac{5}{2}$.

(ii) *Let the roots be real and equal.* Then

$$Y = \frac{1}{B}\sqrt{H^2 - AB}\,(x + Q),$$

and the equation to the locus is

$$Hx + By + F \pm \sqrt{H^2 - AB}\,(x + Q) = 0;$$

hence, since $\sqrt{H^2 - AB}$ is imaginary, the equation can only be satisfied by the values of x and y which make

$$Hx + By + F = 0, \text{ and } x + Q = 0,$$

and the locus may be considered as an ellipse reduced to a point, or as two imaginary straight lines which intersect in C.

Ex. Let the equation be

$$5x^2 + 4xy + y^2 - 10x - 2y + 10 = 0,$$

whence $$y + 2x - 1 \pm \sqrt{-1}\,(x - 3) = 0,$$

an equation which is satisfied by one pair of real values only, namely,

$$x = 3, \quad y = -5.$$

(iii) *Let the roots be imaginary.* Then

$$Y = \frac{1}{B}\sqrt{(H^2 - AB)\{(x + Q)^2 + D^2\}},$$

and, since $(x + Q)^2 + D^2$ can never be negative, Y is never real; and the locus therefore may be called an imaginary ellipse.

Ex. Let the equation be

$$5x^2 - 8xy + 4y^2 - 3x + 4y + 2 = 0;$$

solving for y, we have for the quantity under the root

$$-(x^2 + x + 1), \text{ or } -\left\{\left(x + \frac{1}{2}\right)^2 + \frac{3}{4}\right\},$$

a quantity which can never be positive. The locus is therefore imaginary.

287. II. When $H^2 - AB$ is positive. In this case we know, by Art. 150, that the locus is either an hyperbola or two intersecting straight lines.

(i) *Let the roots be real and unequal.* Then

$$Y = \frac{1}{B}\sqrt{(H^2 - AB)(x-a)(x-b)}.$$

As in the case of the ellipse, the diameter DT, whose equation is

$$Hx + By + F = 0,$$

meets the curve in the points D', D, and the lines $D'L$, DR, whose equations are $x - a = 0$, $x - b = 0$, are tangents at those

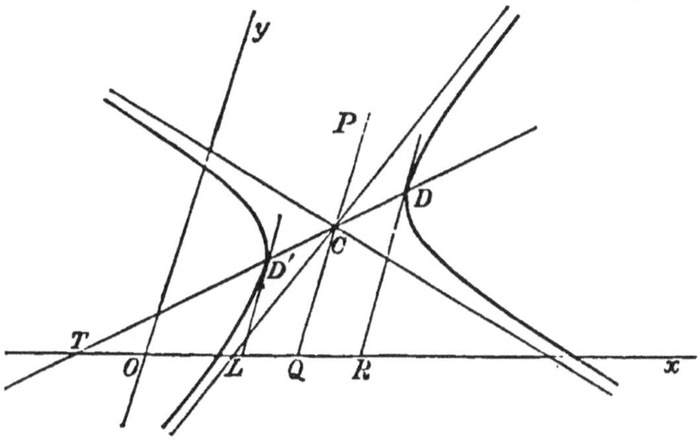

points. Also, since $(x-a)(x-b)$ is negative for values of x between a and b, Y is imaginary for those values; hence no part of the curve lies between the parallels $D'L$, DR. Beyond those limits any number of points of the locus may be found, equidistant from the diameter DT; and, since Y is real when x has any real value, positive or negative, except those which lie between a and b, we shall obtain four infinite branches, and the curve, which is evidently an hyperbola, has the form given in the figure. As in the ellipse, C, the middle point of DD', may be shewn to be the centre, and CP the diameter conjugate to CD.

Ex. Let the equation be

$$3x^2-8xy+4y^2-x+4y+5=0,$$

whence $$2y-2x+1\pm\sqrt{x^2-3x-4}=0,$$

or $$2y-2x+1\pm\sqrt{(x+1)(x-4)}=0.$$

The curve then has $2y-2x+1=0$ for a diameter; and $x+1=0$, $x-4=0$ are the tangents at the points where it meets the curve. It has four infinite branches, but no part of it lies between $x=-1$ and $x=4$.

(ii) *Let the roots be real and equal.* Then the equation to the locus is

$$Hx+By+F\pm\sqrt{H^2-AB}\,(x+Q)=0,$$

which represents two straight lines which intersect in C, for which point

$$Hx+By+F=0, \text{ and } x+Q=0.$$

The line DT is still a diameter of the locus, which may be called a Rectilinear Hyperbola.

(iii) *Let the roots be imaginary.* Then

$$Y=\frac{1}{B}\sqrt{(H^2-AB)\{(x+Q)^2+D^2\}};$$

and, since $(x+Q)^2+D^2$ is positive for every real value of x, the value of Y is real for every such value. Since Y does not vanish for any real value of x, the diameter DT

$$Hx+By+F=0,$$

which bisects all chords parallel to the axis of y, does not meet the curve; for the ordinate of the curve cannot be made equal to that of the diameter. As before, any number of points may be found equidistant from the diameter DT, by taking values of Y corresponding to different values of x. The least values of Y will be when $x=-Q$, represented by CP, CP' in the figure; and on each side of this line the values of Y increase indefinitely, forming an hyperbola with four infinite branches, as in the figure. As in the ellipse, it may be shewn that C is the centre; also CP is

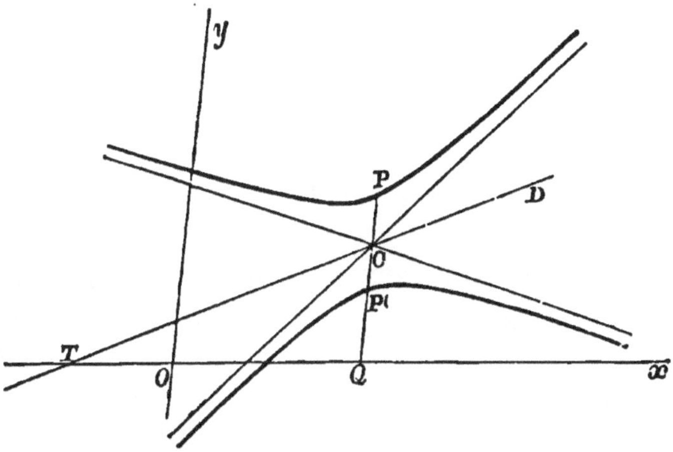

parallel to the ordinates of the diameter CD, and is therefore the diameter conjugate to it.

Ex. Let the equation be

$$3x^2 - 8xy + 4y^2 + 6x - 4y - 4 = 0,$$

whence $$2y - 2x - 1 \pm \sqrt{x^2 - 2x + 5} = 0,$$

or $$2y - 2x - 1 \pm \sqrt{(x-1)^2 + 4} = 0.$$

Since the quantity under the root can never $=0$, the curve does not meet the diameter $2y - 2x - 1 = 0$; and, since it can never be negative, there are points on the curve corresponding to every abscissa. The smallest value of Y is 1, corresponding to $x = 1$. The co-ordinates of the centre are $x = 1$, $y = \frac{3}{2}$.

288. It is often useful in tracing the figure of a curve, to find the points where it cuts the axis of x, by putting $y = 0$ in the equation, and *vice versâ*, and to see, in each case, whether the points so found are possible and different, possible and coincident, or imaginary.

Ex. Let the equations be

$$4x^2 + 7xy + 9y^2 + 4x + 6y + 1 = 0 \text{.........................(1)},$$

$$2x^2 + 5xy + y^2 + x - 2y + 1 = 0 \text{.........................(2)}.$$

It will be found that (1) touches the axis of x where $x=-\frac{1}{2}$, and the axis of y where $y=-\frac{1}{3}$; and that (2) does not meet the axis of x, and touches the axis of y where $y=1$.

We shall be aided in tracing the figure of the hyperbola, if we find the asymptotes, which we now proceed to do.

289. *To find the equation to the asymptotes of a conic section from the general equation.*

We have seen (Art. 186) that the equations to a conic and its asymptotes differ only in the constant term. Hence, if $\phi(xy)=0$ is the equation to a conic, the equation to its asymptotes is $\phi(xy)+\mu=0$, where μ is determined by the condition that the equation should split into factors, and so represent two straight lines. Let then the equation to the asymptotes be

$$Ax^2+2Hxy+By^2+2Gx+2Fy+C+\mu=0.$$

Proceeding exactly as in Art. 62, but writing $C+\mu$ for C, we obtain the condition

$$AB(C+\mu)+2FGH-AF^2-BG^2-(C+\mu)H^2=0,$$

whence $$\mu=\frac{ABC+2FGH-AF^2-BG^2-CH^2}{H^2-AB}.$$

Cf. Arts. 186, 242.

Ex. 1. If the equation to the conic is

$$3x^2-8xy+4y^2+6x-4y-4=0,$$

the equation to the asymptotes will be

$$3x^2-8xy+4y^2+6x-4y-4+\mu=0.$$

Solving for y, we have

$$2y-2x-1\pm\sqrt{x^2-2x+5-\mu}=0;$$

hence $\mu=4$, and the equation to the asymptotes is

$$3x^2-8xy+4y^2+6x-4y=0,$$

or $$(x-2y+2)(3x-2y)=0.$$

Ex. 2. If a conic and its asymptotes are represented by

$$\phi(xy) = 0, \qquad \phi(xy) + \mu = 0,$$

and the co-ordinates of the centre are $x'y'$; then, since the asymptotes pass through the centre, we have $\phi(x'y') + \mu = 0$, and their equation becomes

$$\phi(xy) - \phi(x'y') = 0.$$

291. The foregoing results have been obtained on the supposition that the equation can be solved as a quadratic in y; that is, we have supposed that B does not $= 0$. This must always be the case with the loci of Class I., for, if A or B were to vanish, $H^2 - AB$ could not be negative. But if $H^2 - AB$ is positive, or the locus belongs to Class II., either A or B may vanish. If B vanishes, we have, solving for x,

$$Ax + Hy + G \pm \sqrt{(H^2 - AB)y^2 + 2(HG - AF)y + G^2 - AC} = 0,$$

which will give results similar to those obtained above, and the locus may be traced in the same manner.

If both A and B vanish, the equation is of the form

$$2Hxy + 2Gx + 2Fy + C = 0;$$

hence we cannot solve the equation as a quadratic, and trace the locus by the method of the preceding articles. If we multiply by $\frac{H}{2}$, the equation may be written

$$(Hx + F)(Hy + G) = FG - \frac{CH}{2},$$

a form of equation which has been fully explained in Art. 242.

Ex. Let the equation be

$$2xy + 6x - y - 8 = 0,$$

whence

$$(2x - 1)(y + 3) = 5,$$

which represents an hyperbola, whose asymptotes are $2x - 1 = 0$, $y + 3 = 0$.

292. III. When $H^2 - AB = 0$, the locus is (Art. 153) a parabola, two parallel straight lines, two coincident straight lines, or imaginary. In this case if we draw the diameter DT whose equation is

$$Hx + By + F = 0,$$

we shall have, for the quantity to be added and subtracted from the ordinates of this line, in order to obtain the ordinates of the curve,

$$Y = \frac{1}{B}\sqrt{2(HF - BG)x + F^2 - BC},$$

or
$$= \frac{1}{B}\sqrt{2(HF - BG)(x - a)},$$

where a may be either positive or negative.

We shall then have three cases:

(i) $HF - BG$ positive. Then, as in the ellipse, if we draw the line DR whose equation is $x - a = 0$, it will be a

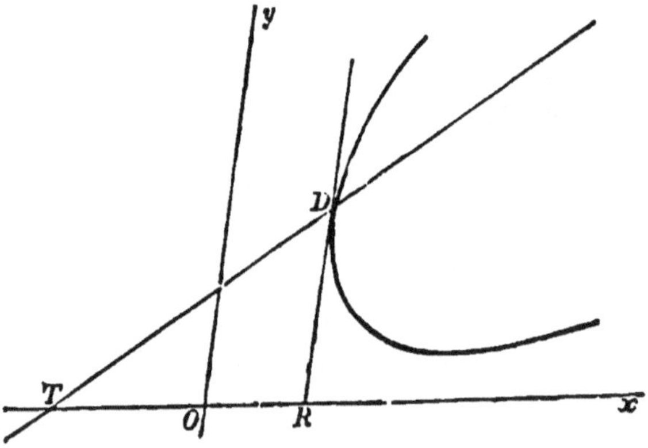

tangent to the curve at the point D. Also, if x is less than OR (or a), the values of Y become imaginary, or no part of the curve lies to the left of DR. For all values of x from OR to infinity, Y is real, and the curve, which we see to be a parabola, may be traced as before, and will have two infinite branches, as in the figure.

(ii) $HF-BG$ negative. In this case Y is imaginary, if $x-a$ is >0, that is, if $x>OR$; hence the curve has no part to the right of D. It will be a parabola, with the infinite branches turned in the direction opposite to those of the figure.

Ex. 1. Let the equation be

$$x^2-2xy+y^2+3x-2y-1=0,$$

whence
$$y-x-1\pm\sqrt{-(x-2)}=0.$$

The curve cuts the diameter $y-x-1=0$ at a point where $x=2$, $y=3$, and has no part lying to the right of the tangent $x-2=0$. Its infinite branches will be turned in the direction opposite to those of the figure.

Ex. 2. Let the equation be

$$x^2-2xy+y^2+x-2y+3=0,$$

whence
$$y-x-1\pm\sqrt{x-2}=0.$$

The curve cuts the diameter $y-x-1=0$ in the points $x=2$, $y=3$. No part of the curve lies to the left of the line, $x-2=0$, parallel to the axis of y, and this line is a tangent. The form is that of the figure.

(iii) $HF-BG=0$. In this case

$$Y=\frac{1}{B}\sqrt{F^2-BC},$$

and is real, zero, or imaginary, according as the quantity under the root is positive, zero, or negative. In the first case, the locus is two straight lines, parallel to the diameter DT whose equation is

$$Hx+By+F=0,$$

and equidistant from it; in the second, these two lines coincide with the diameter; in the third, the equation has no real geometrical signification, but may be said to represent two imaginary straight lines parallel to DT. The locus may be called a Rectilinear Parabola.

293. An equation belonging to Class III. is not always capable of being solved as a quadratic in both x and y. It can, however, always be solved as a quadratic in one of them; for, if both A and B were to vanish, since $H^2 - AB = 0$, H would also vanish, and the equation would not be of the second degree. Hence the locus may always be traced by the method above.

*294. The two following articles will shew how certain results can be obtained by means of Art. 76 (ii), without reducing the equations.

(I) *To find the eccentricity of the conic $\phi(xy) = 0$, the axes being rectangular.*

Suppose (Art. 150) the equation to be reduced to the form

$$A'x'^2 + B'y'^2 + \phi(x'y') = 0 \dots\dots\dots\dots\dots (1);$$

then it is easily seen, from Arts. 158, 174, that

$$e^2 = \frac{B' - A'}{B'}, \quad \text{or} \quad e^2 = \frac{A' - B'}{A'},$$

according as the foci of the conic are on the axis of x or of y. Now the sum and product of these two values of e^2, are each equal to

$$\frac{(A' - B')^2}{-A'B'} = \frac{(A' + B')^2 - 4A'B'}{-A'B'} = \frac{(A - B)^2 + 4H^2}{H^2 - AB}.$$

If we denote this quantity by Q, e^2 will be determined by the equation

$$e^4 - Qe^2 + Q = 0 \dots\dots\dots\dots\dots\dots\dots (2).$$

If the conic is an ellipse, $H^2 - AB$, and therefore Q, is negative; and one value of e^2 must from (2) be positive, and the other negative. The real value of e is the eccentricity required. If the conic is an hyperbola, $H^2 - AB$ is positive, and both values of e^2 are positive, since both the sum and the product of the roots of (2) are positive. In this case equation (1) represents hyperbolas lying (Arts. 186, 242) between the same asymptotes, but in different angles of the asymptotes, according to the sign of $\phi(x'y')$; and these have not the same eccentricity. To select the proper value of e^2, we must use the methods of Chap. VIII.

If 2θ is the angle between the asymptotes of an hyperbola, in which lies the transverse axis of the curve, we have (Arts. 67, 174, 186)

$$\frac{4(H^2 - AB)}{(A+B)^2} = \tan^2 2\theta = \frac{4(\sec^2\theta - 1)}{(2-\sec^2\theta)^2} = \frac{4(e^2-1)}{(2-e^2)^2},$$

from which equation (2) may be obtained.

Ex. *To find the eccentricities of the conics*

$$3x^2 + 2xy + 3y^2 - 16y + 23 = 0 \ldots\ldots\ldots\ldots (1),$$

$$x^2 - 10xy + y^2 + x + y + 1 = 0 \ldots\ldots\ldots\ldots (2).$$

In (1) $A = 3$, $H = 1$, $B = 3$, and therefore $Q = -\frac{1}{2}$; hence the equation becomes $2e^4 + e^2 - 1 = 0$, from which $e^2 = -1$ or $\frac{1}{2}$. Since $H^2 - AB$ is negative, the conic is an ellipse, and its eccentricity $= \frac{1}{\sqrt{2}}$.

In (2) $A = 1$, $H = -5$, $B = 1$, and therefore $Q = \frac{25}{6}$; hence the equation becomes $6e^4 - 25e^2 + 25 = 0$, from which $e^2 = \frac{5}{2}$ or $\frac{5}{3}$. Since $H^2 - AB$ is positive, the conic is an hyperbola. The co-ordinates of the centre are $x = y = \frac{1}{8}$, and the equation may (Art. 151, Ex.) be reduced successively to

$$x^2 - 10xy + y^2 + \frac{9}{8} = 0, \text{ and } 4x^2 - 6y^2 - \frac{9}{8} = 0.$$

This gives $\frac{5}{3}$ as the proper value of e^2.

(II) *To find the length of the equi-conjugate diameters, and the equation to them, in the ellipse whose equation is*

$$Ax^2 + 2Hxy + By^2 + C = 0 \ldots\ldots\ldots\ldots\ldots(1),$$

the axes being inclined at an angle ω.

Equation (1), reduced to the axes of the curve by the transformation of Art. 72, becomes

$$A'x^2 + B'y^2 + C = 0 \ldots\ldots\ldots\ldots\ldots\ldots(2),$$

where $-A'B' = \dfrac{H^2 - AB}{\sin^2\omega}$, $A' + B' = \dfrac{A + B - 2H\cos\omega}{\sin^2\omega}$.

Also the equation to the equi-conjugate diameters (Art. 218) is

$$A'x^2 - B'y^2 = 0 \ldots\ldots\ldots\ldots\ldots\ldots\ldots\ldots(3).$$

At the intersection of (2) and (3),

$$x^2 = -\frac{C}{2A'}, \quad y^2 = -\frac{C}{2B'};$$

then, if r is the length of the semi-diameter,

$$r^2 = x^2 + y^2 = -\frac{C(A' + B')}{2A'B'} = \frac{C(A + B - 2H\cos\omega)}{2(H^2 - AB)}.$$

If a circle with this radius is described, concentric with the ellipse, its equation with the original axes is

$$x^2 + 2xy\cos\omega + y^2 = \frac{C(A + B - 2H\cos\omega)}{2(H^2 - AB)} \ldots\ldots(4).$$

By the reasoning of Art. 63 (ii), we have, for the equation to two straight lines joining the origin to the intersections of (1) and (4),

$$Ax^2 + 2Hxy + By^2 + \frac{2(H^2 - AB)}{A + B - 2H\cos\omega}(x^2 + 2xy\cos\omega + y^2) = 0 \ldots\ldots(5).$$

But these intersections are the extremities of the equi-conjugate diameters; hence (5) is the equation required.

295. *To find the length of a straight line drawn from any point (x′y′) to meet a conic section.*

Let the equations to the line and conic, referred to any axes whatsoever, be

$$\frac{x-x'}{c}=\frac{y-y'}{s}=l \text{..................(1)},$$

$$Ax^2+2Hxy+By^2+2Gx+2Fy+C=0 \text{.........(2)}.$$

Then, as in Art. 114, we have, to determine the distances of $(x'y')$ from the points of intersection of (1) and (2), the equation

$$\left.\begin{matrix} Ac^2 \\ +2Hcs \\ +\ Bs^2 \end{matrix}\right| l^2+2\left.\begin{matrix}(Ax'+Hy'+G)c \\ +2(Hx'+By'+F)s \\ \end{matrix}\right| l+\phi(x'y')=0 \text{......(3)},$$

where $\phi(x'y')=Ax'^2+2Hx'y'+By'^2+2Gx'+2Fy'+C.$

The remarks on the corresponding equation, obtained in Art. 176, may be applied without alteration to (3). We leave the student to verify results already obtained about the asymptotes, by equating to zero the coefficients of l^2 and l, as in Arts. 177, 178.

296. The rectangle on the segments of the chord will be equal to the product of the roots of equation (3), and therefore

$$=\frac{\phi(x'y')}{Ac^2+2Hcs+Bs^2};$$

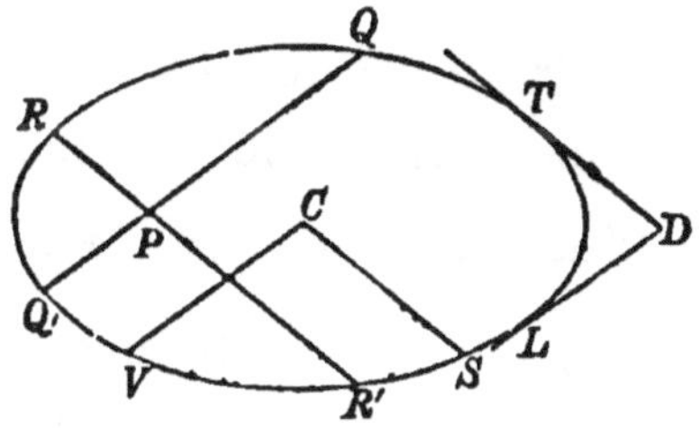

and, if another chord be drawn through the same point $(x'y')$, and its direction be determined by s' and c', quantities corresponding to s and c above, the rectangle on its segments

$$= \frac{\phi(x'y')}{Ac'^2 + 2Hc's' + Bs'^2};$$

hence the ratio of the rectangles is

$$Ac'^2 + 2Hc's' + Bs'^2 : Ac^2 + 2Hcs + Bs^2,$$

a ratio which does not depend upon the point $(x'y')$, and which remains the same, as long as the chords make, respectively, the same angles with the axes; hence, *if QQ', RR' be two chords of a conic, and P their point of intersection, the ratio $PQ \cdot PQ' : PR \cdot PR'$ is not altered by moving each chord parallel to itself, and so shifting the position of P in any manner.* The reader will observe that this conclusion might have been deduced from the equations of Arts. 114, 176, 251.

Cor. 1. Let CV, CS be semi-diameters parallel to QQ', RR' respectively; then

$$\frac{PQ \cdot PQ'}{PR \cdot PR'} = \frac{CV^2}{CS^2}.$$

Cor. 2. Let QQ', RR' move parallel to themselves, till they become tangents at L and T'; then $PQ \cdot PQ'$ becomes DL^2, and $PR \cdot PR'$ becomes DT'^2; hence

$$\frac{DL^2}{DT'^2} = \frac{CV^2}{CS^2}, \text{ or } \frac{DL}{DT'} = \frac{CV}{CS}.$$

Cor. 3. Let Q, Q', R, R' be the four points where a circle intersects the conic; then (Euc. III. 35, 36)

$$PQ \cdot PQ' = PR \cdot PR'; \quad \therefore CV = CS;$$

hence the diameters parallel to QQ' and RR' are equal, and therefore equally inclined to an axis of the conic; i.e. *if a conic and a circle intersect in four points, any two chords passing through the four points are equally inclined to an axis of the conic.*

Ex. *If α, β, γ, δ be the eccentric angles of the four points of intersection of a circle and an ellipse, then will $\alpha+\beta+\gamma+\delta=2n\pi$.*

The chords joining two of the points must make the same angle with one side of the axis, as the chord joining the other two does with the other; and the equations to the chords being (Art. 164, Ex. 1)

$$\frac{x}{a}\cos\frac{\alpha+\beta}{2}+\frac{y}{b}\sin\frac{\alpha+\beta}{2}=\cos\frac{\alpha-\beta}{2},$$

$$\frac{x}{a}\cos\frac{\gamma+\delta}{2}+\frac{y}{b}\sin\frac{\gamma+\delta}{2}=\cos\frac{\gamma-\delta}{2};$$

we have
$$\tan\frac{\alpha+\beta}{2}=-\tan\frac{\gamma+\delta}{2},$$

or
$$\frac{\alpha+\beta}{2}=n\pi-\frac{\gamma+\delta}{2}; \text{ hence } \alpha+\beta+\gamma+\delta=2n\pi.$$

COR. 4. A particular case of Cor. 3 is when three of the points of intersection approach indefinitely near to one another. The circle is then called the *Circle of Curvature* at the point to which the three points have approached; and it will be observed, that one circle and one only can be drawn through these points, since three points determine a circle. If then QQ' and RR' move parallel to themselves, till the points Q, R, Q' approach indefinitely near to one another, the result of Cor. 3 may be thus stated: *The common chord of a conic and the circle of curvature at any point, and the tangent to the conic at that point, are equally inclined to an axis of the conic.*

*297. *To find the equation to the tangent of a conic section at the point $(x'y')$.*

Exactly as in Art. 117, if $(x'y')$ is on the curve,

$$\phi(x'y')=0,$$

and the condition that equation (1) (Art. 295) should be a tangent is

$$(Ax'+Hy'+G)c+(Hx'+By'+F)s=0 \ldots\ldots (4).$$

From (1) and (4) we obtain the equation to the tangent

$$\frac{y-y'}{x-x'}=-\frac{Ax'+Hy'+G}{Hx'+By'+F}\ldots\ldots\ldots\ldots\ldots\ldots(5)^1.$$

Multiplying up, and remembering that $(x'y')$ is a point on (2), we get equation (5) in the form

$$(Ax'+Hy'+G)x+(Hx'+By'+F)y+Gx'+Fy'+C=0\ldots(6).$$

COR. If the curve passes through the origin, $C=0$; and the equation to the tangent at the origin (0, 0) is

$$Gx+Fy=0;$$

hence, *if a conic passes through the origin, the equation to the tangent at the origin is obtained by equating to zero the terms of lowest dimensions in the equation to the conic.*

*298. Equation (6) of Art. 297 is not altered, if x changes place with x' and y with y'; hence (Art. 131) the equation to the polar is of the same form as the equation to the tangent, and all the theories of poles and polars proved for the circle in Arts. 123—127, 129, 130, can be proved for any conic by means of this equation.

The equation to the polar of the origin, found by putting $x'=y'=0$ in (6), is

$$Gx+Fy+C=0.$$

*299. *If any quadrilateral ABFD* (fig. Art. 104) *be inscribed in a conic, then each of the points E, C, G is the pole of the straight line joining the other two.*

Take EB, EF as the axes of x and y, and let

$$EA=a,\quad EB=a',\quad ED=b,\quad EF=b';$$

[1] The reader of the Differential Calculus will observe that this equation is

$$(x-x')\frac{d\phi}{dx'}+(y-y')\frac{d\phi}{dy'}=0.$$

then the equations to CA and CB are

$$\frac{x}{a}+\frac{y}{b}=1\ \ldots\ldots\ldots(1), \qquad \frac{x}{a'}+\frac{y}{b'}=1\ \ldots\ldots\ldots(2);$$

and the equations to AF and BD are

$$\frac{x}{a}+\frac{y}{b'}=1\ \ldots\ldots\ldots(3), \qquad \frac{x}{a'}+\frac{y}{b}=1\ \ldots\ldots\ldots(4).$$

Now by adding (1) and (2), we obtain the equation

$$x\left(\frac{1}{a}+\frac{1}{a'}\right)+y\left(\frac{1}{b}+\frac{1}{b'}\right)=2\ \ldots\ldots\ldots\ldots\ldots(5),$$

which is therefore (Art. 43) the equation to a straight line passing through C. But we obtain the same equation by adding (3) and (4); therefore also (5) passes through G. Hence (5) is the equation to CG.

Now, if the equation to the conic is

$$Ax^2+2Hxy+By^2+2Gx+2Fy+C=0\ldots\ldots\ldots(6),$$

a and a' are found by putting $y=0$ in (6), and are therefore the roots of the equation $Ax^2+2Gx+C=0$; hence (Appendix)

$$\frac{1}{a}+\frac{1}{a'}=-\frac{2G}{C}; \text{ similarly } \frac{1}{b}+\frac{1}{b'}=-\frac{2F}{C},$$

and equation (5) becomes

$$Gx+Fy+C=0,$$

which is (Art. 298) the polar of the origin; hence CG is the polar of E. Similarly GE is the polar of C; and hence, by Art. 130, EC is the polar of G. The points E, C, G are called *a conjugate triad* with respect to any conic passing through A, B, F, D. Cf. Art. 340.

*300. *To find the equation to the normal to a conic section at a point $(x'y')$, the axes being rectangular.*

The equation to the normal will be, from equation (5) of Art. 297,

$$\frac{y-y'}{x-x'}=\frac{Hx'+By'+F}{Ax'+Hy'+G}.$$

COR. If the curve passes through the origin, the equation to the normal at the origin becomes

$$Fx-Gy=0.$$

*301. *To find the equation to the diameter of a conic section, bisecting chords drawn parallel to a given line.*

In Art. 295 let (1) represent one of the chords, and let the given line be $y=mx$, so that $\frac{s}{c}=m$. Then, if $(x'y')$ be the middle point of the chord, we have, as in Art. 132, for the equation to the diameter,

$$Ax+Hy+G+(Hx+By+F)\,m=0,$$

or $$x(A+mH)+y(H+mB)+G+mF=0 \quad\text{......(1)}.$$

If this be written in the form $y=m'x+b$, we have

$$m'=-\frac{A+mH}{H+mB},$$

or $$A+H(m+m')+Bmm'=0 \quad\text{...........(2)}.$$

The symmetry of this equation shews that chords parallel to $y=m'x$ are also bisected by a diameter parallel to $y=mx$. It is therefore the condition that the lines $y=mx$ and $y=m'x$ should be parallel to conjugate diameters of the conic $\phi(xy)=0$.

COR. If $y=mx$ and $y=m'x$ are a pair of conjugate diameters of the conic

$$Ax^2+2Hxy+By^2+C=0 \quad\text{................(3)},$$

and are represented by the equation

$$ax^2+2hxy+by^2=0 \quad\text{....................(4)},$$

then (Art. 63) m and m' are the roots of the equation

$$bM^2 + 2hM + a = 0;$$

therefore $$m + m' = -\frac{2h}{b}, \qquad mm' = \frac{a}{b};$$

and from (2) we have the relation

$$Ab + Ba = 2Hh \ldots\ldots\ldots\ldots\ldots\ldots\ldots(5),$$

as the condition that the straight lines (4) should be conjugate diameters of the conic (3). See Examples XII. 53.

Ex. *To find the length of the equi-conjugate diameters of* (3) *and the equation to them, from condition* (5).

Write the equation to a concentric circle, radius r, and get the equation to straight lines joining the origin with the intersections of this circle and (3), as in Art. 294 (II); then apply condition (5), to determine r.

*302. *Tangents at the extremities of any chord intersect in the diameter of which the chord is an ordinate.*

*303. *The polar of any point on a diameter and the tangents at the extremities of a diameter are parallel to the ordinates.*

We leave the student to prove these two propositions for any conic. As we have found the equation to the diameter bisecting any system of parallel chords, and shewn that the equation to the polar is of the same form as the equation to the tangent, he will find no difficulty in imitating the proofs of Art. 220 and the latter part of Art. 236.

Ex. 1. *To find the equation to the axes of the conic $\phi(xy) = 0$, the axes of co-ordinates being rectangular.*

Since a conic is symmetrical with respect to an axis, tangents from any point $(x'y')$ on an axis are equal. Let the directions of two tangents from $(x'y')$ be determined, as in Art. 295, by c, s and c', s', and let their lengths be l and l'; then, since the roots of the equation (3) are equal in this case,

$$l^2 = \frac{\phi(x'y')}{Ac^2 + 2Hcs + Bs^2} = \frac{(Pc + Qs)^2}{(Ac^2 + 2Hcs + Bs^2)^2},$$

where $P = Ax' + Hy' + G$, $Q = Hx' + By' + F$; and we have similar equations for l', c', s'. But, if $l^2 = l'^2$,

$$Ac^2 + 2Hcs + Bs^2 = Ac'^2 + 2Hc's' + Bs'^2,$$

therefore $$P^2c^2+2PQcs+Q^2s^2=P^2c'^2+2PQc's'+Q^2s'^2;$$

whence $$A\,(c^2-c'^2)-B\,(s'^2-s^2)=2H\,(c's'-cs)\text{.......................(1)},$$

and $$P^2\,(c^2-c'^2)-Q^2\,(s'^2-s^2)=2PQ\,(c's'-cs)\text{.....................(2)}.$$

But $c^2+s^2=1=c'^2+s'^2$, or $c^2-c'^2=s'^2-s^2$; hence dividing (2) by (1), and writing for P, Q their values, we find that the point $(x'y')$ must lie on the conic

$$\frac{(Ax+Hy+G)^2-(Hx+By+F)^2}{A-B}=\frac{(Ax+Hy+G)\,(Hx+By+F)}{H}\text{...(3)},$$

which is therefore the equation to the axes. Cf. Arts. 146, 341 (ii), Ex. 1.

Ex. 2. *To find the equation to a locus of a point P, such that the line joining P to the centre of the conic is perpendicular to the polar of P.*

Let $(x'y')$ be a point on the locus, which is (Art. 195) the axis; then the polar of $(x'y')$ is

$$(Ax'+Hy'+G)\,x+(Hx'+By'+F)\,y+Gx'+Fy'+C=0\text{.........(1)}.$$

Also the equation to a straight line joining P with the centre may (Art. 141) be written

$$Ax+Hy+G+k\,(Hx+By+F)=0\text{.....................(2)};$$

If (2) is at right angles to (1), we have (Art. 47, Cor. 2),

$$(A+kH)\,(Ax'+Hy'+G)+(H+kB)\,(Hx'+By'+F)=0\text{.........(3)};$$

and, since $(x'y')$ is a point on (2),

therefore $$Ax'+Hy'+G+k\,(Hx'+By'+F)=0\text{.....................(4)}.$$

Eliminating k between (3) and (4), we find that $(x'y')$ must lie on the conic (3), Ex. 1.

Ex. 3. *To find the equation to the axis of the parabola whose equation, with rectangular axes, is*

$$(ax+by)^2+2Gx+2Fy+C=0\text{.......................(1)}.$$

This may be deduced from Ex. 1, by writing a^2, ab, b^2 for A, H, B; but the following is easier.

The polar of any point $(x'y')$ on the axis, with regard to (1), is

$$(a^2x'+aby'+G)\,x+(abx'+b^2y'+F)\,y+Gx'+Fy'+C=0\text{.........(2)};$$

and the *direction* of the axis is given (Art. 156, Cor.) by the equation

$$ax+by=0\text{.......................................(3)}.$$

Since (2) is (Art. 263) perpendicular to (3), we have

$$a\,(a^2x'+aby'+G)+b\,(abx'+b^2y'+F)=0;$$

whence $$a^2\,(ax'+by')+Ga+b^2\,(ax'+by')+Fb=0.$$

The equation to the axis is therefore

$$(a^2+b^2)\,(ax+by)+Ga+Fb=0.$$

The same result is obtained from Art. 156, by writing the value of κ in equation (7).

304. *Polar equation, the focus being pole.*

It is proved in Arts. 209, 274 (i), that the focal polar equation to any conic may be written

$$\frac{l}{\rho} = 1 - e\cos\theta,$$

where $l =$ half the latus rectum, and $e = 1$ for the parabola, the left-hand focus being the pole in the ellipse, and the right-hand in the hyperbola.

Ex. *In any conic section the semi-latus rectum is an harmonic mean between the segments, made by the focus, of any focal chord.*

Let PSp be the focal chord; then

$$\frac{l}{SP} = 1 - e\cos\theta, \quad \frac{l}{Sp} = 1 - e\cos(\pi + \theta);$$

therefore $$\frac{1}{SP} + \frac{1}{Sp} = \frac{1 - e\cos\theta}{l} + \frac{1 + e\cos\theta}{l} = \frac{2}{l},$$

which proves the proposition.

305. *To find the polar equation to the chord of a conic section, the focus being the pole, and thence to deduce the polar equation to the tangent.*

Let the equation to the conic and the chord be

$$\frac{l}{\rho} = 1 - e\cos\theta \ldots (1), \quad \frac{l}{\rho} = A\cos\theta + B\cos(\theta - \alpha) \ldots (2)[1];$$

and let the angular co-ordinates of two points on the conic be $\alpha - \beta$, $\alpha + \beta$; then, if (2) passes through $\alpha - \beta$, we have for this point, from (1) and (2),

$$1 - e\cos(\alpha - \beta) = \frac{l}{\rho} = A\cos(\alpha - \beta) + B\cos\beta;$$

therefore $$(A + e)\cos(\alpha - \beta) + B\cos\beta - 1 = 0 \ldots\ldots\ldots (3).$$

[1] The student should satisfy himself that *any* straight line can be represented by equation (2).

Similarly, if (2) passes through $\alpha+\beta$,

$$(A+e)\cos(\alpha+\beta)+B\cos\beta-1=0\ldots\ldots\ldots(4).$$

Subtracting (4) from (3), we have

$$A+e=0, \text{ and therefore } B\cos\beta-1=0;$$

hence, if (2) is the chord joining the two points, it becomes

$$\frac{l}{\rho}=\sec\beta\cos(\theta-\alpha)-e\cos\theta\ldots\ldots\ldots\ldots\ldots(5).$$

If the chord becomes a tangent at the point α, $\beta=0$, and (5) becomes

$$\frac{l}{\rho}=\cos(\theta-\alpha)-e\cos\theta.$$

306. *In any conic section, if SP, SQ be two radii vectores, and PT, QT tangents at P and Q, then ST bisects the angle PSQ, unless PT, QT be drawn to different branches of the hyperbola, in which case ST bisects the angle supplementary to PSQ.*

Take the left-hand focus in the ellipse and the right-hand focus in the hyperbola, and let the vectorial angles of P and

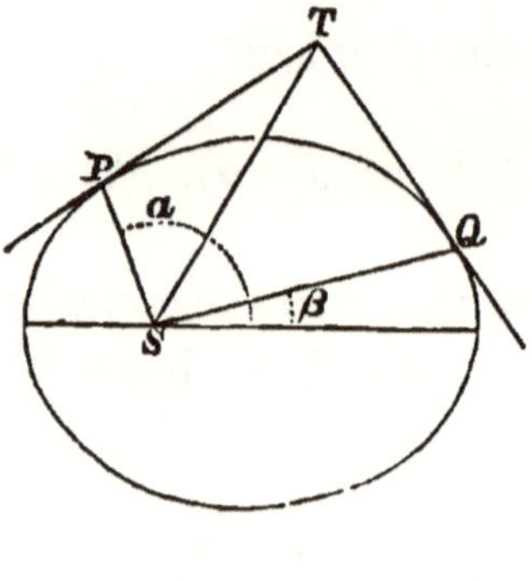

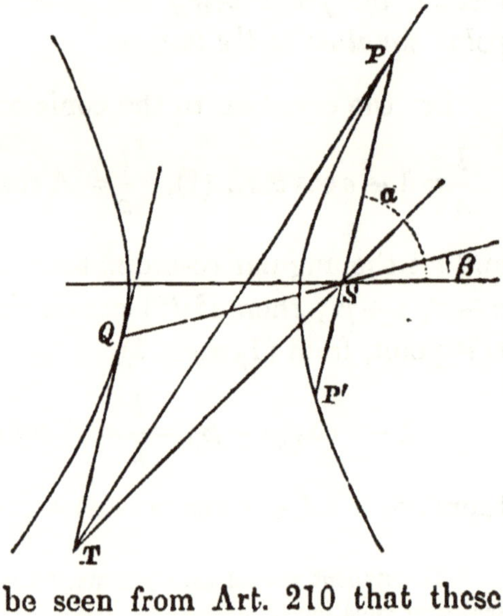

Q be α and β; it will be seen from Art. 210 that these

angles are as in the figure; for in the polar equation used in Art. 304, the outer branch of the hyperbola corresponds to the *negative* values of the radii vectores, but the vectorial angle is formed by the positive direction of the radius vector. Then the equations to PT, QT are

$$\frac{l}{\rho} = \cos(\theta - \alpha) - e\cos\theta, \quad \frac{l}{\rho} = \cos(\theta - \beta) - e\cos\theta;$$

and therefore at the point T we have from these two equations

$$\cos(\theta - \alpha) = \cos(\theta - \beta).$$

Now we cannot have $\theta - \beta = \theta - \alpha$, since α and β are by hypothesis not equal; we therefore take $\theta - \beta = \alpha - \theta$;

hence $$\theta = \frac{\alpha + \beta}{2}.$$

In the figure drawn for the hyperbola, this value of θ is the angle which TS produced makes with Sx, and ST bisects the supplemental angle QSP'. It will be seen that in this case PT, QT subtend supplemental angles at the focus; for

$$QST = TSP' = 180^\circ - TSP.$$

The figure for the parabola is similar to that drawn for the ellipse.

Ex. 1. *If the chord of a conic subtends at the focus a constant angle, it will always touch a fixed conic.*

Let the angle be 2β; then the equation to the chord is

$$\frac{l}{\rho} = \sec\beta\cos(\theta - \alpha) - e\cos\theta,$$

or $$\frac{l\cos\beta}{\rho} = \cos(\theta - \alpha) - e\cos\beta\cos\theta,$$

which is a tangent to the conic

$$\frac{l\cos\beta}{\rho} = 1 - e\cos\beta\cos\theta,$$

a conic with the same focus as the first, and whose eccentricity is $e\cos\beta$, and latus rectum $2l\cos\beta$.

Ex. 2. *To find the locus of the pole of the chord in Ex. 1.*

The co-ordinates of the pole $(\rho'\theta')$ will satisfy the equations to the two tangents at the extremities of the chord; hence

$$\frac{l}{\rho'}=\cos(\theta'-\alpha+\beta)-e\cos\theta', \quad \frac{l}{\rho'}=\cos(\theta'-\alpha-\beta)-e\cos\theta';$$

hence, as in Art. 306, $\theta'=\alpha$. Substituting for α in either equation, we get

$$\frac{l}{\rho'}=\cos\beta-e\cos\theta';$$

hence the pole lies upon a conic whose equation is

$$\frac{l\sec\beta}{\rho}=1-e\sec\beta\cos\theta.$$

Ex. 3. *To find the equation to the chord of contact of tangents from the point $(\rho'\theta')$.*

Let $\alpha-\beta$, $\alpha+\beta$ be the angular co-ordinates of the points, the tangents at which pass through $(\rho'\theta')$; then the equation to the chord through these points is

$$\frac{l}{\rho}=\sec\beta\cos(\theta-\alpha)-e\cos\theta;$$

and the equations to the tangents are the same as in Ex. 2. Hence, as in Ex. 2, we have

$$\theta'=\alpha, \qquad \frac{l}{\rho'}=\cos\beta-e\cos\theta'.$$

Substituting for α and β in the equation to the chord, we have for the equation required

$$\left(\frac{l}{\rho}+e\cos\theta\right)\left(\frac{l}{\rho'}+e\cos\theta'\right)=\cos(\theta-\theta').$$

Ex. 4. *The circle, which passes through the points of intersection of three tangents to a parabola, will pass through the focus.*

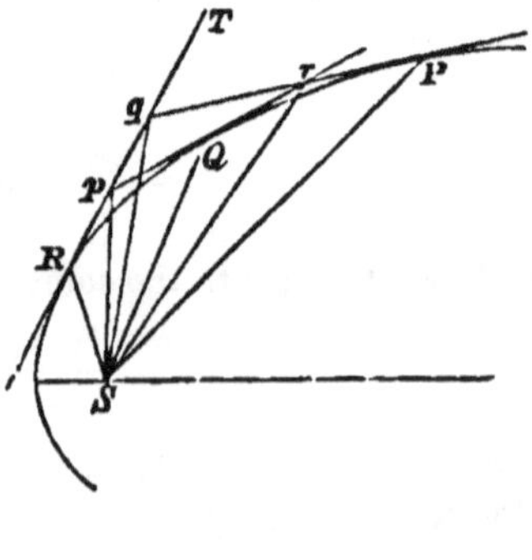

Let P, Q, R be the points of tangency, and let the tangents at P, Q intersect in r; at Q, R, in p; at R, P, in q; then by Art. 269, Ex.

$$Tqr=\frac{RSP}{2}=\frac{RSQ+QSP}{2},$$
$$=pSQ+QSr,$$
$$=pSr;$$

that is, pSr is the supplement of pqr, which (Euc. III. 22) proves the proposition.

307. *To find the equation to a conic, when the tangent and normal at any point are the axes of y and x.*

Since the axis of y is a tangent at the origin, when $x=0$ in the general equation

$$Ax^2+2Hxy+By^2+2Gx+2Fy+C=0,$$

the values of y become each $=0$; hence $F=0$, $C=0$, and the equation is reduced to

$$Ax^2+2Hxy+By^2+2Gx=0,$$

where the axis of x is any straight line drawn through the point of contact. If the axes are rectangular, they are the tangent and normal at the origin.

Ex. *If, through a given point on a conic, any two straight lines at right angles to each other be drawn to meet the curve, the straight line joining their extremities will pass through a fixed point on the normal of the given point.*

The equation to the conic, referred to the tangent and normal at the point, is

$$Ax^2+2Hxy+By^2+2Gx=0 \text{........................} (1).$$

Let the equations to the two lines be

$$y=mx, \quad y=-\frac{1}{m}x,$$

or as one locus

$$(y-mx)\left(y+\frac{1}{m}x\right)=0,$$

whence

$$y^2-\left(m-\frac{1}{m}\right)xy-x^2=0 \text{........................} (2).$$

Multiplying (2) by B and subtracting it from (1), we have

$$(A+B)x^2+\left\{2H+B\left(m-\frac{1}{m}\right)\right\}xy+2Gx=0,$$

as the equation to a locus which passes through (Art. 43) the intersections of (1) and (2). But this equation represents two straight lines, namely, the tangent at the given point ($x=0$), and

$$(A+B)x+\left\{2H+B\left(m-\frac{1}{m}\right)\right\}y+2G=0,$$

which must represent the chord joining the extremities of the two lines.

The point, where this line cuts the normal, which is the axis of x, is found by making $y=0$ in the equation; thus we obtain

$$x = -\frac{2G}{A+B};$$

hence the chord cuts the normal at an invariable distance from the origin. In the circle the point of section is the centre, by Euc. III. 31.

308. *To find the equation to a conic section, when the axes of co-ordinates are tangents to the curve.*

Suppose the axes of x and y to touch the conic at distances a and b from the origin. Putting $y=0$ and $x=0$ successively in the equation $\phi(xy)=0$, we obtain

$$Ax^2 + 2Gx + C = 0 \ \ldots\ldots (1), \qquad By^2 + 2Fy + C = 0 \ \ldots\ldots (2),$$

where the roots of (1) are each $=a$, and the roots of (2) are each $=b$; hence (Appendix)

$$\frac{1}{a^2} = \frac{A}{C}, \quad \frac{1}{a} = -\frac{G}{C}; \quad \frac{1}{b^2} = \frac{B}{C}, \quad \frac{1}{b} = -\frac{F}{C};$$

and the equation $\phi(xy)=0$ becomes, after dividing by C and substituting these values,

$$\frac{x^2}{a^2} + \frac{2H}{C}xy + \frac{y^2}{b^2} - \frac{2x}{a} - \frac{2y}{b} + 1 = 0,$$

or $$\left(\frac{x}{a} + \frac{y}{b} - 1\right)^2 + kxy = 0, \text{ where } k = \frac{2H}{C} - \frac{2}{ab}.$$

309. If the conic belongs to the parabola class,

$$\left(k + \frac{2}{ab}\right)^2 = \frac{4}{a^2b^2}, \text{ or } k^2 + \frac{4k}{ab} = 0;$$

therefore $$k = 0, \text{ or } k = -\frac{4}{ab}.$$

If $k=0$, the equation represents two coincident straight lines. If $k=-\frac{4}{ab}$, it becomes

$$\left(\frac{x}{a}+\frac{y}{b}-1\right)^2=\frac{4xy}{ab}, \quad \text{or } \frac{x}{a}+\frac{y}{b}\pm 2\sqrt{\frac{xy}{ab}}=1;$$

therefore $$\sqrt{\frac{x}{a}}\pm\sqrt{\frac{y}{b}}=\pm 1, \quad \text{or } \sqrt{\frac{x}{a}}+\sqrt{\frac{y}{b}}=1;$$

for the latter form is equivalent to the former, if we remember that the radicals may be positive or negative.

*310. *Any straight line drawn through the intersection of two tangents to a conic section, is harmonically divided by the curve and the chord of contact.*

Take the two tangents as axes; then the equation to the conic is

$$\left(\frac{x}{a}+\frac{y}{b}-1\right)^2+kxy=0 \qquad (1),$$

and the equation to the chord of contact is

$$\frac{x}{a}+\frac{y}{b}=1 \qquad (2).$$

Let the equation to any straight line through the origin be

$$\frac{x}{c}=\frac{y}{s}=l \qquad (3),$$

where l is the distance of any point (xy) from the origin. Hence, to find the distance from the origin of the points of intersection of (1) and (3), we have

$$\left(\frac{cl}{a}+\frac{sl}{b}-1\right)^2+kcsl^2=0, \quad \text{or } \left(\frac{c}{a}+\frac{s}{b}-\frac{1}{l}\right)^2+kcs=0 \ldots (4).$$

Now, if $\frac{1}{l'}, \frac{1}{l''}$ be the roots of this quadratic in $\frac{1}{l}$, that is, if l', l'' are the values of l at the points where (1) and (3) intersect, we have

$$\frac{1}{l'} + \frac{1}{l''} = 2\left(\frac{c}{a} + \frac{s}{b}\right) \quad \text{..................} (5).$$

Also, for the point of intersection of (2) and (3), we have

$$\frac{cl}{a} + \frac{sl}{b} = 1, \text{ or } \frac{1}{l} = \frac{c}{a} + \frac{s}{b} \quad \text{..............} (6);$$

hence, denoting this value by l_1, we have from (5) and (6)

$$\frac{2}{l_1} = \frac{1}{l'} + \frac{1}{l''},$$

or l_1 is an harmonic mean between l' and l''.

Cor. *Any straight line drawn through the pole is harmonically divided by the curve and the polar.*

This has been proved for the case when the pole is *without* the conic. If the pole is *within* the conic, let A be the pole, BC the polar (fig. Art. 340), and let AB be any straight line drawn through A, meeting the polar in B; then, if two tangents are drawn from B, the chord of contact will (Art. 129) pass through A, and therefore the straight line AB is harmonically divided, as above.

311. *Conditions necessary to determine a locus of the second degree.*

We observed (Art. 24), that the general equation of the first degree has in reality only two independent constants, though apparently containing three, and that, consequently, a straight line could be subjected to two independent conditions only, since these would give two relations between the constants, which would suffice to determine them. The general equation of the second degree contains six coefficients; but, as we may divide all the terms by one of these coefficients, we see that the equation contains five independent constants only. To determine a locus of the second degree, we must give the values of these five constants, or give five independent equations between them, by which they may be determined; but, in this case, it is necessary to examine, whether the equations admit of a system of real solutions, or of more than one, and whether the resulting equation of the second degree represents a curve, or one of those varieties, which have been explained above. The simplest condition would be, that the locus should pass through five points; for the co-ordinates of these points, substituted successively in the general equation, will give five equations to determine the five constants. These equations, being of the first degree, will admit of one system of real solutions, and one only, if they are *consistent* and *independent;* but we have not shewn that this condition will always be satisfied. We shall therefore prove the following theorem.

312. *Through five real points, no four of which are in the same straight line, one conic section and one only can be drawn.*

We will first consider the case, where no three are in the same straight line. Let the axes be so chosen that two of the points are on the axis of x, and two upon the axis of y, and suppose the points on the axis of x and y, respectively, to be at distances a, a' and b, b' from the origin. Let the equation to the conic be

$$Ax^2 + 2Hxy + By^2 + 2Gx + 2Fy + C = 0 \ldots\ldots (1);$$

then the values of a, a' and b, b' will be found by putting y and x successively $= 0$ in (1), and will therefore be the roots of the equations

$$Ax^2 + 2Gx + C = 0 \ldots\ldots\ldots\ldots\ldots\ldots\ldots (2),$$

$$By^2 + 2Fy + C = 0 \ldots\ldots\ldots\ldots\ldots\ldots\ldots (3);$$

therefore $$\frac{1}{aa'}=\frac{A}{C},\quad \frac{1}{a}+\frac{1}{a'}=-\frac{2G}{C};$$

$$\frac{1}{bb'}=\frac{B}{C},\quad \frac{1}{b}+\frac{1}{b'}=-\frac{2F}{C}.$$

Hence the general equation becomes, after dividing by C and substituting,

$$\frac{x^2}{aa'}+2\frac{H}{C}xy+\frac{y^2}{bb'}-\left(\frac{1}{a}+\frac{1}{a'}\right)x-\left(\frac{1}{b}+\frac{1}{b'}\right)y+1=0\ldots\ldots(4).$$

Let $(x'y')$ be the fifth point; then, by substituting x', y' in (4), we shall obtain a simple equation to determine $\frac{H}{C}$. Hence one conic and one only can be drawn through the five points.

If three of the five points be in one straight line, we shall have $(x'y')$ on one of the axes, and therefore $x'=0$ or $y'=0$; in either case the value of $\frac{H}{C}$ from (4) would be infinite, and therefore equation (4) would, by dividing by $\frac{H}{C}$, become $xy=0$, and would represent the two axes of co-ordinates, which form one of the system of conics which can be drawn through the first four points. We might have foreseen, that the locus in this case could not be an ellipse, hyperbola, or parabola, since these curves cannot be cut by a straight line in more than two points.

If more than three of the points are in one straight line, the coefficients of (1) cannot all be determined by the method of this article; and it is obvious that this ought to be the case, since more than one pair of straight lines can then be made to pass through the five points.

Thus we have proved, that we can always find a real equation of the second degree, and one only, which is satisfied by the co-ordinates of the five points. This will always represent a real geometrical locus, since the imaginary loci of the second degree (Arts. 150, 153) could not be satisfied by the co-ordinates of these points.

Ex. *To shew that equation* (4) *of the conic passing through four points can be put in the form*

$$\left(\frac{x}{a}+\frac{y}{b}-1\right)\left(\frac{x}{a'}+\frac{y}{b'}-1\right)+kxy=0,$$

where k is arbitrary, and to interpret this equation.

The equation is of the form $S+kS'=0$, and represents (Art. 43, Cor.) a system of conics, passing through the intersections of the conics $S=0$, $S'=0$, which are two pairs of straight lines drawn through the four points.

313. We may say, generally, that a conic can be made to fulfil five independent conditions, where each condition enables us to eliminate one constant. Thus the position of the centre must be counted as two conditions, for it gave us (Art. 141) two relations between the coefficients, and enabled us to eliminate two. Again, the position of the focus must count as two; for the general equation to a conic with a focus at a given point $(x'y')$ may be written (Art. 208)

$$(x-x')^2+(y-y')^2=(Px+Qy+R)^2 \dots\dots\dots(1),$$

involving three undetermined constants only. It must not be assumed that five conditions such as the above will always give us one conic and one only, as in Art. 312.

Ex. 1. *How many parabolas can be drawn through four points?*

We have here five conditions, since $H^2-AB=0$ is one. Suppose no three of the points to be in one straight line; then we obtain equation (4) of Art. 312 for a conic passing through the four points. If the conic is a parabola, we have

$$\left(\frac{H}{C}\right)^2=\frac{1}{aa'bb'}.$$

If the product $aa'bb'$ is positive, there will be two parabolas passing through the four points; if the product is negative, no real parabola can be drawn through them. If more than two points are on the same straight line, the parabola is rectilinear, or parallel straight lines.

Ex. 2. *How many conics with a given focus can be described about a triangle?*

Let the focus be $(x'y')$, and the angular points be (x_1y_1), (x_2y_2), (x_3y_3); then, if we write δ_1^2 for the known quantity $(x_1-x')^2+(y_1-y')^2$, so that δ_1 is the distance between (x_1y_1) and the focus, we have, from equation (1) (Art. 313),

$$\delta_1 = \pm(Px_1+Qy_1+R),\quad \delta_2 = \pm(Px_2+Qy_2+R),\quad \delta_3 = \pm(Px_3+Qy_3+R).$$

Every combination of signs gives a system of equations, to determine P, Q, R, and there are eight combinations; but evidently, if we change the signs in the three equations, it is equivalent to changing the signs of P, Q, R, which would leave equation (1) the same. Hence there are four different solutions, and four different conics can be described about the triangle.

314. When a conic has been subjected to four conditions only, there will remain one arbitrary constant in the equation, and there will be an infinity of curves fulfilling the given conditions. We may then find the locus of any of the remarkable points of the curve; for, having introduced the conditions, we may obtain the equations for determining the point in question, and find the equation to the locus by eliminating the arbitrary constant between them.

Ex. 1. *A conic is described, touching two straight lines at given distances (a and b) from their point of intersection; find the locus of its centre.*

Taking the two lines as axes, the equation to the conic is, by Art. 303,

$$\left(\frac{x}{a}+\frac{y}{b}-1\right)^2 + kxy = 0;$$

hence, the equations for the centre are (Art. 141)

$$\frac{2}{a}\left(\frac{x}{a}+\frac{y}{b}-1\right)+ky=0,$$

$$\frac{2}{b}\left(\frac{x}{a}+\frac{y}{b}-1\right)+kx=0.$$

Eliminating k, we have for the required locus,

$$ay - bx = 0,$$

a straight line bisecting the chord of contact.

Ex. 2. *To find the locus of the centre of a conic which passes through four points.*

Equation (4), Art. 312 may be written

$$bb'x^2+kxy+aa'y^2-bb'(a+a')x-aa'(b+b')y+aa'bb'=0,$$

where k is undetermined. The equations for the centre are

$$2bb'x+ky-bb'(a+a')=0,$$
$$2aa'y+kx-aa'(b+b')=0.$$

Eliminating k, we obtain for the required equation

$$2bb'x^2-2aa'y^2-bb'(a+a')x+aa'(b+b')y=0.$$

This is a conic passing through the origin, that is, through the intersection of one of the three pairs of straight lines which can be drawn through the four points. By similar reasoning it will pass through the intersection of every pair, as it ought; for each pair is a conic of the system, and the intersection its centre. It cuts the axis of x again, where $x=\frac{a+a'}{2}$; that is, it passes through the middle point of the line joining two of the points. Hence it bisects all the six lines which join the points. It will be an hyperbola if aa' and bb' have the same sign, and an ellipse in the contrary case.

Ex. 3. *If the four points are on a circle, the locus of the centre is a rectangular hyperbola.*

See Euc. III. 35, 36, and Art. 186, Cor. 3.

Ex. 4. *If* $aa'=-bb'$ (*Ex.* 2), *and the axes are rectangular, all the conics drawn through the four points are rectangular hyperbolas, and the locus of the centre is a circle.*

It should be noted, as a particular case of the above, that the three pairs of straight lines which can be drawn through the four points are (Art. 67) at right angles. See Examples XII., 13, 14, 15, and Art. 327, Ex. 3, 4.

*315. *Similar conics.*

DEF. Two curves are said to be similar and similarly placed, when, any point O being taken in the plane of one curve, another point O' can be found in the plane of the other, such that parallel radii, drawn from O and O' to the first and second curve respectively, are to one another in a fixed ratio. They are said to be similar when they can be made to fulfil the above condition, by turning one of them round a fixed point.

The points O, O' are called Centres of Similarity.

*316. *If, in the planes of two curves, one such pair of points, as O and O′, can be found, an infinity of other pairs can be found.*

For suppose OB, $O'B'$ to be parallel radii in the fixed

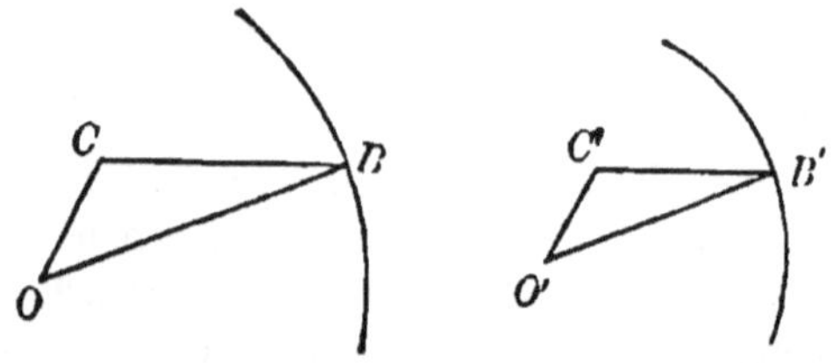

ratio of $1 : k$; and take in the first figure any point C, and draw $O'C'$ parallel to OC, so that

$$O'C' : OC = O'B' : OB = k : 1;$$

and join CB, $C'B'$; then the two triangles OCB, $O'C'B'$ are similar (Euc. VI. 6), and therefore $C'B'$ is parallel to CB; also

$$C'B' : CB = O'B' : OB = k : 1;$$

therefore, since OB, $O'B'$ are any parallel radii, C and C' are centres of similarity. Hence for every point in the first figure there is a point in the second, such that the pair are centres of similarity.

*317. If C be the *centre* of the first curve, C' must be the centre of the second; for every diameter BCA is bisected in C, therefore every parallel diameter $B'C'A'$ through C' is bisected in C'; for otherwise the ratio $CB : C'B'$ would not be equal to the ratio $CA : C'A'$.

Hence, *if two central conics are similar, the centres of the curves are centres of similarity.*

*318. *All conics, whose eccentricity is the same, are similar figures.*

Let two of such conics, which are evidently (Art. 174) of

the same class, be placed with their axes parallel, and a focus in one coincident with the corresponding focus in the other; then, the common focus being pole, their equations are (Art. 304)

$$\frac{l}{\rho} = 1 - e\cos\theta,\quad \frac{l'}{\rho} = 1 - e\cos\theta;$$

hence, when θ is the same in each, we have the parallel radii vectores in the proportion of the latera recta. The two conics then are similar, and the common focus is a centre of similarity.

Cor. All parabolas are similar figures; and they will be similarly placed, if their axes are parallel. This will be the case (Art. 156, Cor.), if the conditions obtained in the next article are satisfied.

*319. *To find the condition that two central conics should be similar and similarly placed.*

By Art. 317 their centres must be centres of similarity. Now we may, by transferring the origin to the centre of each, the axes in each case remaining parallel to their original direction, reduce the equations (Art. 141) to the form

$$Ax^2 + 2Hxy + By^2 + C = 0 \ldots\ldots\ldots\ldots (1),$$

$$A'x^2 + 2H'xy + B'y^2 + C' = 0 \ldots\ldots\ldots\ldots (2),$$

where the first three terms in each remain unchanged. Let the equation to any straight line through the centre of (1) be

$$\frac{x}{c} = \frac{y}{s} = l \ldots\ldots\ldots\ldots\ldots\ldots\ldots\ldots (3),$$

where l is the distance of the centre from (xy); then, for the distance of the intersection of (1) and (3) from the centre, we have

$$l^2(Ac^2 + 2Hcs + Bs^2) + C = 0.$$

Similarly with equation (2) we should have

$$l'^2 (A'c^2 + 2H'cs + B's^2) + C' = 0;$$

hence, writing m for $\frac{s}{c}$, we have

$$\frac{l^2}{l'^2} = \frac{C(A' + 2H'm + B'm^2)}{C'(A + 2Hm + Bm^2)} \quad \text{.................. (4).}$$

If the ratio $l^2 : l'^2$ is constant for all values of m, that is, whenever the direction of the line (3) is the same in each case, we must have

$$\frac{A + 2Hm + Bm^2}{A' + 2H'm + B'm^2} = \text{a constant} = \mu \text{ say,}$$

or $$A - \mu A' + 2(H - \mu H')m + (B - \mu B')m^2 = 0,$$

for *all* values of m; but this can be only the case, when

$$A - \mu A' = 0, \quad H - \mu H' = 0, \quad B - \mu B' = 0,$$

or $$\frac{A}{A'} = \frac{H}{H'} = \frac{B}{B'} \quad \text{........................ (5).}$$

Hence, if two central conics are similar and similarly placed, their asymptotes are (Art. 186) parallel, and the coefficients of the highest powers of the variables are the same in both, or differ by a constant multiplier only.

*320. Condition (5) is *necessary*, in order that the curves should be similar and similarly placed. We proceed to enquire whether it is *sufficient*. Suppose it fulfilled; then the equations to the curves, referred to the centre of each, as before, may evidently be written in the form

$$Ax^2 + 2Hxy + By^2 + C = 0,$$

$$Ax^2 + 2Hxy + By^2 + C' = 0;$$

and we have, from Art. 319, equation (4), for the constant ratio of the corresponding radii,

$$\frac{l^2}{l'^2} = \frac{C}{C'}.$$

If the ratio $l : l'$ is *real* and *finite*, the centres of the curves are centres of similarity, and therefore (Art. 316) the curves are similar and similarly placed. But, if C and C' be of different signs, the ratio $l^2 : l'^2$ will be negative, so that the fixed ratio of the radii is imaginary, and the condition is fulfilled algebraically only. If the curves be of the Ellipse class, we see, by Art. 150, that in this case one of them represents an impossible locus. If they be of the Hyperbola class, they will (Art. 242) represent two hyperbolas with parallel asymptotes, but not lying in the same angles of the asymptotes.

Again, one or both of the quantities C, C' might $=0$, in which case the fixed ratio would be zero, infinite, or indeterminate; but in this case one or both of the equations would (Art. 150) represent straight lines which intersect, imaginary in the case of the ellipse, and real in the case of the hyperbola.

Ex. The three equations

$$4x^2+6xy-4y^2+6x-8y+1=0,$$
$$2x^2+3xy-2y^2+7x-y+3=0,$$
$$6x^2+9xy-6y^2+12x+9y+4=0,$$

fulfil the conditions of Art. 319. If the origin is transferred to the centre of each, the equations become

$$2x^2+3xy-2y^2+\frac{5}{2}=0 \qquad (1),$$
$$2x^2+3xy-2y^2 \quad =0 \qquad (2),$$
$$2x^2+3xy-2y^2-\frac{2}{3}=0 \qquad (3).$$

For (1) and (2) the ratio $l^2 : l'^2$ is infinite: for (2) and (3) it is zero; and for (3) and (1) it is negative. By writing the equations in the form

$$(2x-y)(x+2y)=-\frac{5}{2}, \quad (2x-y)(x+2y)=0, \quad (2x-y)(x+2y)=\frac{2}{3},$$

it is seen (Art. 242) that (1) and (3) represent hyperbolas with parallel asymptotes, but lying in supplemental angles; and (2) represents a pair of straight lines parallel to these asymptotes.

*321. If two conics are similar and not similarly placed, they can, by definition, be made to fulfil the conditions of Art. 319, by turning one of them round a fixed point.

Let equations (1) and (2) of Art. 319, transformed to rectangular axes, if necessary, represent two conics, similar but not similarly placed, and let conic (1) be turned about the origin through such an angle α, that it may be similarly placed with (2). This will, of course, as far as regards the change in its equation, be the same as if the axes were turned through an angle α, and we shall have a new equation,

$$ax^2 + 2hxy + by^2 + C = 0 \ldots\ldots\ldots\ldots\ldots\ldots (3),$$

and, from Art. 76 (ii), we have

$$h^2 - ab = H^2 - AB, \quad a + b = A + B \ldots\ldots\ldots\ldots (4).$$

But, since (3) is similar and similarly placed to (2), we have

$$\frac{a}{A'} = \frac{h}{H'} = \frac{b}{B'} = \mu \text{ say;}$$

from which equations we have

$$h^2 - ab = \mu^2 (H'^2 - A'B'), \quad a + b = \mu (A' + B').$$

Hence $$\frac{H'^2 - A'B'}{(A' + B')^2} = \frac{h^2 - ab}{(a+b)^2},$$

therefore $$\frac{H'^2 - A'B'}{(A' + B')^2} = \frac{H^2 - AB}{(A + B)^2},$$

is, from (4), the required condition. If this is satisfied, the conics will be similar, with the same exceptions as in Art. 320. This is simply the condition (Art. 186, Cor. 3) that the asymptotes should contain equal angles, or that (Art. 294) the eccentricities should be determined by the same equation.

In like manner the condition of similarity for oblique axes is

$$\frac{H'^2 - A'B'}{(A' + B' - 2H' \cos \omega)^2} = \frac{H^2 - AB}{(A + B - 2H \cos \omega)^2}.$$

322. *Sections of the Cone.*

The surface described by an indefinite straight line, which is carried round the perimeter of a given circle, always passing through a given point, is called a cone. The circle is called the base of the cone, the fixed point the vertex, and the line joining the vertex and the centre of the base is called the axis. A cone is said to be *right*, if the axis is perpendicular to the plane of the base, and *oblique*, if the axis is inclined at any other angle to that plane. As the generating line is not limited, the surface of the cone consists of two portions or *sheets* (fig. Art. 324), perfectly similar, situated on opposite sides of the vertex, and of indefinite extent. It is evident, from the method in which the cone is generated, that every plane parallel to the base will cut the cone in a circle, and that every plane through the axis will cut it in two straight lines, in both which cases the section will be represented by an equation of the second degree. We shall now shew that the same is the case, in whatever manner the plane cuts the cone. We shall content ourselves with proving this property in the case of right cones only, since a full investigation of this part of the subject will be found in most Geometrical Treatises.

323 (i). *Every section of a right cone by a plane is a curve of the second degree.*

Obs. The generating line, in a right cone, will always make the same angle with the axis.

Let $HRKL$ be a plane; AB a fixed line, the axis of a cone, inclined at an angle α to the plane; AC a perpendicular from A, the vertex, on the plane; AP the generating line,

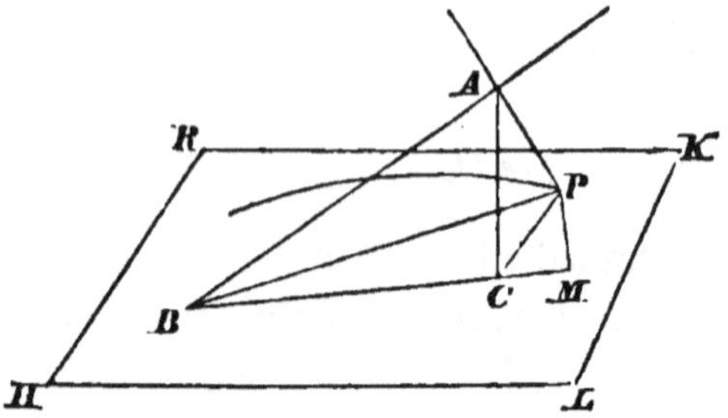

revolving round AB, inclined to it at a constant angle β. Then P, the extremity of AP, will evidently trace out some section of the cone on the plane, which is supposed to be intercepted between the vertex and the circle, round the perimeter of which the generating line is carried.

Draw PM perpendicular to BC produced; join BP and CP. Take C as origin, CM as axis of x, and a perpendicular to it in the plane $HRKL$, as axis of y; let P be the point (xy) and $AB = a$.

$$\text{Then } BP^2 = PM^2 + BM^2$$
$$= y^2 + (x + a\cos\alpha)^2 \ldots\ldots\ldots\ldots\ldots\ldots(1),$$

since $\angle ABC = \alpha$; also

$$BP^2 = a^2 + AP^2 - 2a\,AP\cos\beta,$$
$$= a^2 + (a^2\sin^2\alpha + y^2 + x^2) - 2a\cos\beta\sqrt{a^2\sin^2\alpha + y^2 + x^2} \ldots(2),$$

since $AP^2 = AC^2 + CP^2$.

Equating (1) and (2), we have for the locus of P

$$a^2\cos^2\alpha + 2ax\cos\alpha = a^2 + a^2\sin^2\alpha - 2a\cos\beta\sqrt{a^2\sin^2\alpha + x^2 + y^2},$$

or $$\cos\beta\sqrt{a^2\sin^2\alpha + x^2 + y^2} = a\sin^2\alpha - x\cos\alpha,$$

a curve of the second degree.

323 (ii). Comparing this equation with the general equation of the second degree,

$$Ax^2 + 2Hxy + By^2 + 2Gx + 2Fy + C = 0,$$

we see that in this case

$$A = \cos^2\beta - \cos^2\alpha, \quad H = 0, \quad B = \cos^2\beta.$$

Now (Arts. 150, 153), the curve is an Hyperbola, Parabola, or Ellipse, according as $H^2 - AB > = < 0$, or as

$$-\cos^2\beta\,(\cos^2\beta - \cos^2\alpha) > = < 0,$$

as $$\cos^2\alpha - \cos^2\beta > = < 0,$$

as $$\sin(\beta+\alpha)\sin(\beta-\alpha) > = < 0,$$

as $$\sin(\beta-\alpha) > = < 0,$$

or as $$\beta > = < \alpha,$$

since $\beta + \alpha$ is by construction less than π, and therefore $\sin(\beta+\alpha)$ always positive.

324. We may easily identify the above results with the forms of the curves that we have already discovered; for, let SAR be a cone, AO the axis, B the point where the cutting plane EE' cuts the axis; draw BP parallel to AR, then we have by our assumptions

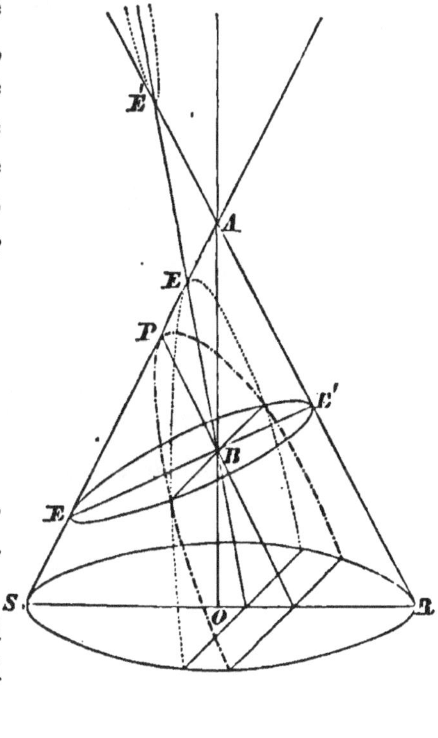

$$AB = a,\ \angle EAB = \beta,$$
$$\angle EBA = \alpha,\ \angle SAR = 2\beta,$$
$$\angle AEB = \pi - \beta - \alpha,$$

and hence

$$\angle SAR + \angle AEB = \pi + \beta - \alpha.$$

If $\beta < \alpha$, these two angles are less than two right angles, and the point E will lie below the point P, and the section will evidently be limited in every direction, and be an Ellipse.

If $\beta = a$, the two angles = two right angles; the point E coincides with P, and the cutting plane is parallel to AR. The section will evidently be limited at P, and unlimited in the direction PB, and be a Parabola.

If $\beta > a$, the two angles are greater than two right angles; the point E lies above P, and the cutting plane meets both sheets of the cone. The section will be unlimited in every direction, and be an Hyperbola.

325. The following subjects are dealt with as examples in the next chapter.

Intersection of hyperbolas, Art. 327. *Circle of Curvature*, Art. 330. *Envelopes*, Arts. 337, 338. *Equation to a pair of tangents to the conic* $\phi(xy) = 0$, Art. 339. *Equation to the director circle*, Art. 339. *Equations to determine the foci*, Art. 341 (ii).

EXAMPLES XII.

1. Employ the method of this chapter to verify the results obtained in Examples VIII.

2. What must be the value of k, in order that the equation

$$2x^2 + kxy - y^2 - 3x + 6y - 9 = 0,$$

may represent a pair of straight lines?

3. Shew that the equation

$$6x^2 - 4xy + y^2 - 6x + 2y - 11 = 0$$

represents an ellipse, lying wholly between the lines $x = 3$ and $x = -2$.

4. Shew that the equation

$$2x^2 - 4xy + y^2 - 2x + 2y + 13 = 0$$

represents an hyperbola, no part of which lies between the lines $x = 3$ and $x = -2$.

5. Shew that the equation

$$x^2 + 2xy + y^2 + 2y + 4 = 0$$

represents a parabola, extending indefinitely in the positive direction of the axis of x.

6. Shew that the equation

$$x^2 - 2xy - y^2 + 2x - 2y + 5 = 0$$

represents an hyperbola, whose ordinates are always real.

7. Shew that the equation

$$3x^2 + 2xy + y^2 + 6x + 2y + 7 = 0$$

represents an imaginary locus.

8. Find the asymptotes and the eccentricity of each of the hyperbolas

$$3x^2 - 4xy + 2ay = 0,\quad 4x^2 - y^2 + 4x - 2y + 9 = 0,$$

the axes being rectangular.

9. Trace the curve $x^2 + y^2 = (ax + by + c)^2$, and determine the focus and directrix.

10. Find the axes of the hyperbolas

$$xy - ax + by = 0 \ldots\ldots(1), \qquad xy - ax + a^2 = 0 \ldots\ldots(2),$$

the axes of co-ordinates being inclined at an angle α.

11. If the equation to a conic is not altered when x and y change places, the bisector of the angle between the positive directions of the axes is an axis of the curve; if it is not altered, when x is changed to $-y$ and y to $-x$, the bisector of the supplementary angle is an axis.

12. Shew from Art. 284, or otherwise, that, if the axes of co-ordinates are parallel to conjugate diameters, $B=0$ in the general equation.

13. If A, B, C, D be four points, situated as in Art. 314, Ex. 4, and A, B, C be any three of the points; then D is the orthocentre of the triangle ABC.

14. The circle which bisects the sides of a triangle passes through the feet of the perpendiculars from the angles on the opposite sides, and bisects the distances of the angular points from the orthocentre.

15. If a conic is drawn through four points, shew that the locus of the centre is a conic whose asymptotes are parallel to the axes of the two parabolas drawn through the four points.

*16. Find the equation to the tangents at the points where the conic whose equation is

$$(ax+by-1)(ax+b'y-1)+cxy=0$$

meets the axis of y; and the equation to the straight line joining their point of intersection with the point where the conic touches the axis of x.

*17. From Art. 301 find the direction of the axes of the conic, the axes of co-ordinates being supposed rectangular; (2) find the equations of the centre, and shew that they represent diameters bisecting chords parallel to the axes of co-ordinates; (3) if these diameters are parallel, shew that the locus is a parabola.

18. In the conic $\phi(xy)=0$, shew that $Gx+Fy=0$ is a chord bisected at the origin.

19. Given the length of a focal chord PSp in a conic, and the rectangle contained by the segments SP, Sp; find the latus rectum.

20. A circle is described concentric with a given conic section; find the equation to their common diameters, and the condition that the curves should touch.

*21. If two hyperbolas have the same asymptotes, their eccentricities are equal, or are connected by the equation

$$\frac{1}{e^2}+\frac{1}{e'^2}=1.$$

22. Two concentric conics have in general one and only one pair of common conjugate diameters.

*23. If any straight line OR be drawn through a point O, and P, the pole of that line, be joined to O, then the straight lines OP, OR will form an harmonic pencil with the tangents from O.

24. Find the locus of the centres of the conics, obtained by varying each of the constants in the general equation of the second degree.

25. If $\phi(xy)=0$ is a parabola, and the axes rectangular, find the latus rectum.

26. How many rectangular hyperbolas can be drawn through four points A, B, C, D, which are not all in the same straight line?

27. A conic section is cut in four points by a circle, and two straight lines, each passing through two of the points of intersection, are taken as axes of co-ordinates; shew that the equation to the conic may be written in the form

$$x^2+2hxy+y^2+2gx+2fy+c=0.$$

28. From a point O two tangents OH, OK are drawn to a conic section; a straight line MQR is drawn, parallel to OK, to intersect OH in M, and the curve in Q, R. Shew that

$$MH^2 : MQ \,.\, MR = OH^2 : OK^2.$$

If QR meets the chord HK in T, then $MQ \,.\, MR = MT^2$.

29. From the equation to the tangent in Art. 305, find the angle that it makes with the axis of x.

30. In the parabola, if SP, SQ be two radii vectores, and PT, TQ tangents at P and Q, then $SP \,.\, SQ = ST^2$.

31. If a chord of a conic section subtends a constant angle at the focus, shew that the locus of the foot of the perpendicular upon it from the focus is in general a circle, but in one case is a straight line.

32. PSP', QSQ' are two focal chords of a conic; shew that a straight line through S bisecting the angle PSQ', will intersect the chord QP produced in the directrix.

33. The angular co-ordinates of the extremities of two chords of a conic are, as in Art. 305, $\alpha - \beta$, $\alpha + \beta$ and $\gamma - \beta$, $\gamma + \beta$; find the locus of their intersection, if $\gamma - \alpha = \epsilon$, where β and ϵ are constants.

34. Having given two points, through which a conic section is to pass, and the directrix; shew that the locus of the corresponding focus is a circle whose centre is on the line joining the two points.

35. A small arc of a conic being traced upon a plane, shew how to determine the nature of the conic to which it belongs.

36. If in a conic section a series of right-angled triangles is described, having the right angles at a given point in the curve, and tangents be drawn at the extremities of the hypotenuses, the points of intersection of these tangents lie in one straight line.

37. A series of triangles is inscribed in a conic section, having a common angular point, the angle at which is bisected by the normal to the curve at the angular point; shew that the sides, opposite the common angular point, will all meet in the point of intersection of the tangents drawn at the two ends of the normal.

38. How must Art. 307 Ex. be enunciated, if the conic is a rectangular hyperbola?

39. If the equation to a conic is $\phi(xy)=0$, interpret the equations

$$Ax^2+2Hxy+By^2=0\ldots(1),\quad 2Gx+2Fy+C=0\ldots(2),$$
$$Ax+2Hy+2G=0\ldots(3),\quad By+2Hx+2F=0\ldots(4).$$

40. Shew that the locus of the centre of a rectangular hyperbola described about a given equilateral triangle is the circle inscribed in the triangle.

41. Shew that two parabolas cannot touch each other in more than one point.

42. If the parallel chords of a conic are divided in any proportion, the locus of the dividing point is a conic.

43. Find the equation to a circle referred to two tangents of length a, containing an angle ω.

*44. Shew that two conics of the same class, whose axes are proportional in magnitude and parallel in direction, cannot have more than two real finite points in common.

*45. A pair of tangents are drawn from a point to a conic, and a straight line bisecting the angle between them cuts the chord of contact in O; shew that the polar of O is the external bisector of the angle between the tangents.

46. If two tangents be taken for axes, the co-ordinates of the centre of a conic are proportional to the lengths of the tangents; hence shew that, if straight lines are drawn from the centre parallel to the tangents, so as to form a parallelogram, the chord joining the points where they cut the conic is parallel to the diagonal.

47. In the sides AB, AC of a given triangle ABC take two points M, N, and in the line joining them take a point P, such that

$$BM : MA = AN : NC = MP : PN;$$

find the locus of P.

48. If a conic be inscribed in a triangle, and touch two sides at the point of their bisection, shew that the straight line joining the centre and the third point of contact will pass through the opposite angular point.

*49. Shew that the equation (axes rectangular) to the equi-conjugate diameters of the conic

$$5x^2 + 2xy + 5y^2 - 12x - 12y = 0$$

is

$$x^2 + 10xy + y^2 - 12x - 12y + 12 = 0.$$

*50. If the equation to an ellipse is $Ax^2 + 2Hxy + By^2 + C = 0$, shew that the equation to an hyperbola upon the same axes is of the form

$$(A + B)(Ax^2 + 2Hxy + By^2) + 2(H^2 - AB)(x^2 + y^2) + C' = 0,$$

and find the value of C' (axes rectangular).

51. Shew that the equation (axes rectangular)

$$(x^2 - a^2)^2 + (y^2 - a^2)^2 = a^4$$

represents two ellipses. Find the principal axes, and shew that the semi-diameter, drawn to the point where their common tangent meets either of them, is equal to $\sqrt{3}a$.

52. If the general equation of the second degree represents a parabola, and the axes (rectangular) are tangents, find the equation to the axis and directrix, and the co-ordinates of the focus.

*53. Shew that the asymptotes and any pair of conjugate diameters of an hyperbola form an harmonic pencil. Hence, if

$$Ax^2 + 2Hxy + By^2 + C = 0, \quad ax^2 + 2hxy + by^2 = 0,$$

are the equations of the curve and the diameters, shew that

$$Ab + Ba = 2Hh.$$

*CHAPTER XIII.

Abridged Notation and Trilinear Co-ordinates.

326. By reasoning precisely similar to that made use of in the case of straight lines, we see that, if

$$S = 0, \qquad S' = 0 \dots\dots\dots\dots\dots\dots (1),$$

be two conic sections (S and S' being of two dimensions in x and y), then

$$S + kS' = 0 \dots\dots\dots\dots\dots\dots (2),$$

where k is some arbitrary constant, since it is also of two dimensions, is also a conic section; and, since the co-ordinates of the points of intersection of equations (1) evidently satisfy equation (2), therefore (2) represents a system of conic sections, passing through the four points of intersection of (1).

327. We say *four*, because the elimination of one of the variables between two equations of the second degree produces, generally (Appendix), an equation of the fourth degree. If the resulting equation should in particular cases fall below the fourth degree, in consequence of the coefficients of one or more of the higher powers vanishing, the curves may still be said to intersect in four points, one or more of these points being infinitely distant. If therefore we take into account points which are infinitely distant or imaginary, or both, we may say, that *two Conic Sections always intersect in four points.* In considering the infinitely distant points of inter-

section of two conics, we must bear in mind (Arts. 42, 179, 253) that, when algebraic results lead to the conclusion that loci meet at infinity, the geometrical interpretation is, that the loci tend to have the same direction, or, in other words, to become parallel straight lines.

Ex. 1. *If two hyperbolas have an asymptote of one parallel to an asymptote of the other, they will meet in three finite points and one infinitely distant point.*

Ex. 2. *If two hyperbolas have a common asymptote, or have the asymptotes of one parallel to the asymptotes of the other, they will meet in two finite and in two infinitely distant points.*

These propositions may be easily proved by Arts. 240, 242. They may be illustrated with every variety of figure, by drawing on thin paper two hyperbolas with their asymptotes, and placing them one over the other, against the light.

Ex. 3. *Every conic, which passes through the intersections of two rectangular hyperbolas, is also a rectangular hyperbola.*

For, if the $A+B$ of $S=0$ and the $A+B$ of $S'=0$ both vanish (the axes being rectangular), so does the $A+B$ of $S+kS'=0$.

The above includes (Art. 67) pairs of straight lines at right angles, which must be considered rectangular hyperbolas. So that every pair of straight lines drawn through the four points of intersection, is one of the conics of the system, and the lines are at right angles.

Ex. 4. *If a rectangular hyperbola circumscribe a triangle, it passes through the point of intersection of the perpendiculars from the angles on the opposite sides.*

Let ABC be the triangle, and let a perpendicular from A on BC meet the curve again in D; then A, B, C, D are the intersections of two rectangular hyperbolas; and therefore BD, CA and CD, AB are at right angles, which proves the proposition.

328. Since three pairs of straight lines (Art. 104, Ex. 1) can be drawn through four points, two conics will have six chords of intersection. The reader who is acquainted with the Theory of Equations will observe, that the imaginary points of intersection of two conics must occur in pairs similar to those of Art. 127; and consequently the chord of

contact, joining a pair of such points, is real. Thus, if two conics intersect in four imaginary points, and the pairs be A, B and C, D, the chords AB and CD are real. The rest are imaginary; for, if AC, for example, were real, C would be the intersection of two real straight lines, and would therefore be real. Again, if the conics intersect in two real and two imaginary points, the chord joining the two real points, and that joining the two imaginary points, will be real, and the rest imaginary.

Ex. *Shew that for three values of k the equation $S+kS'=0$ will represent a pair of straight lines.*

Let the equations, for which $S=0$, $S'=0$ stand, be

$$ax^2+2hxy+by^2+2gx+2fy+c=0,$$
$$a'x^2+2h'xy+b'y^2+2g'x+2f'y+c'=0;$$

then the equation $S+kS'=0$ will be

$$Ax^2+2Hy+By^2+2Gx+2Fy+C=0,$$

where $A=a+ka'$, $H=h+kh'$, &c. The condition that this equation should represent two straight lines (Art. 62) is

$$ABC+2FGH-AF^2-BG^2-CH^2=0;$$

and, if A, H,...&c. are replaced by their values $a+ka'$, $h+kh'$,...&c. we shall obtain an equation of the third degree in k. The roots of this equation, substituted in $S+kS'=0$, would give us the equations to three pairs of straight lines. These are the three pairs of chords, real or imaginary, which can be drawn through the four points of intersection of (S) and (S'). One root of a cubic equation, and therefore one value of k, must always be real.

320. Suppose that $S'=LM$, where L and M are of one dimension in x and y; then $S'=0$ represents two straight lines and

$$S+kLM=0 \ldots\ldots\ldots\ldots\ldots\ldots\ldots (1),$$

will represent a conic, two of whose chords of intersection with the conic (S) are

$$L=0, \qquad M=0.$$

This can be instantly deduced from the previous article; for, since the conic (S') has become two straight lines, the conic (1), which passes through the intersections of (S) and these straight lines, must have the lines for two of its six chords of intersection with (S).

It can be also shewn independently, as follows. Since the co-ordinates of the points of intersection of (L) and (S) make these expressions vanish simultaneously, they also make the expression $S+kLM$ vanish, and therefore satisfy equation (1). Hence the conic (1), passing through the intersections of (L) and (S), has (L) for its chord of intersection with (S). Similarly (M) may be shewn to be a chord of intersection of (1) and (S).

Ex. *If L, M are of the forms $Ax+By+C$, $Ax-By+C'$; then, if the axes of co-ordinates are parallel to the axes of (S), they will also be parallel to the axes of $(S+kLM)$.*

330. The conics (S) and $(S+kLM)$ will touch one another, if two of their points of intersection coincide. This will be the case, if either (L) or (M) touches (S), or if (L) and (M) intersect in a point on (S). Hence, if $T=0$ is the equation to the tangent to (S) at a given point $(x'y')$, then

$$S+kT(Ax+By+C)=0 \ldots\ldots\ldots\ldots\ldots (1)$$

is the equation to a conic touching (S) at the point $(x'y')$.

If the straight line $(Ax+By+C=0)$ pass through the point $(x'y')$, which gives $C=-Ax'-By'$, three of the points of intersection will coincide; and the equation to a conic, having such a contact with (S), is

$$S+kT(Ax+By-Ax'-By')=0 \ldots\ldots\ldots\ldots (2).$$

If we introduce the conditions (Arts. 106, 110) that equation (2) should represent a circle, we shall have two equations, by which A and B may be determined, and we may thus obtain the equation (Art. 296) to the circle of curvature at the point $(x'y')$.

If the line $(Ax+By+C=0)$ coincides with (T), the conics will have four points coincident, and the equation is of the form $S+kT^2=0$.

The above equations may be written without the multiplier k; thus $S+kLM=0$ may be written $S+NM=0$, where $N=kL$, and consequently (L) and (N) represent the same straight line.

Ex. 1. *To find the equation (axes rectangular) to the circle of curvature at the origin of the conic*

$$3x^2+2xy+4y^2+x=0.$$

The tangent $T=0$ is (Art. 297, Cor.) $x=0$, and the chord $A(x-x')+B(y-y')=0$ becomes $Ax+By=0$. Then the required equation is

$$3x^2+2xy+4y^2+x+x(Ax+By)=0,$$

where we omit the multiplier k, since kA, kB are not more general than A, B; then, from the conditions for a circle, $3+A=4$, and $2+B=0$, and the equation becomes

$$4x^2+4y^2+x=0.$$

Ex. 2. *To find the equation (axes rectangular) to the circle of curvature at the point* (μ) *of the parabola* $y^2=4dx$.

The equation to the tangent (Art. 250) is

$$\mu y-x-d\mu^2=0;$$

and since (Art. 296, Cor. 4) the tangent and the chord make equal angles with the axis of the parabola, the equation to the chord is

$$\frac{y-2d\mu}{x-d\mu^2}=-\frac{1}{\mu} \text{ or } \mu y+x-3d\mu^2=0.$$

Hence the required equation is

$$y^2-4dx+k(\mu y-x-d\mu^2)(\mu y+x-3d\mu^2)=0,$$

which becomes after multiplication

$$(1+k\mu^2)y^2-kx^2-4kd\mu^3y+2d(k\mu^2-2)x+3kd^2\mu^4=0.$$

The condition of Art. 296 has made (Art. 329, Ex.) the term containing xy vanish from the equation; the condition that the coefficients of x^2 and y^2 should be equal, gives $k=-\dfrac{1}{1+\mu^2}$; and the equation becomes after substitution

$$x^2+y^2-2d(3\mu^2+2)x+4d\mu^3y-3d^2\mu^4=0.$$

Ex. 3. *To find the radius and the co-ordinates of the centre of the circle of curvature at any point* $(x'y')$ *of a parabola.*

From the result of Ex. 2 we have

$$r=\frac{2(d+x')^{\frac{3}{2}}}{\sqrt{d}},\quad x=2d+3x',\quad y=-\frac{2x'^{\frac{3}{2}}}{\sqrt{d}}.$$

Ex. 4. *To find the equation to the circle of curvature at the point of an ellipse, whose eccentric angle is ϕ.*

The ellipse being referred to its axes, the equation to the tangent (Art. 164) is

$$\frac{x}{a}\cos\phi+\frac{y}{b}\sin\phi=1;$$

and, as in Ex. 2, the equation to the chord is

$$\frac{y-b\sin\phi}{x-a\cos\phi}=\frac{b\cos\phi}{a\sin\phi} \text{ or } \frac{x}{a}\cos\phi-\frac{y}{b}\sin\phi=\cos 2\phi.$$

Hence the required equation is

$$\frac{x^2}{a^2}+\frac{y^2}{b^2}-1+k\left(\frac{x}{a}\cos\phi+\frac{y}{b}\sin\phi-1\right)\left(\frac{x}{a}\cos\phi-\frac{y}{b}\sin\phi-\cos 2\phi\right)=0,$$

or $$\frac{x^2}{a^2}+\frac{y^2}{b^2}-1+k\left(\frac{x^2}{a^2}\cos^2\phi-\frac{y^2}{b^2}\sin^2\phi-2\frac{x}{a}\cos^3\phi+2\frac{y}{b}\sin^3\phi+\cos 2\phi\right)=0;$$

whence, as in Ex. 2, $$k=\frac{a^2-b^2}{a^2\sin^2\phi+b^2\cos^2\phi}.$$

The equation after substitution becomes

$$x^2+y^2-2(a^2-b^2)\left(\frac{x}{a}\cos^3\phi-\frac{y}{b}\sin^3\phi\right)+a^2(3\cos^2\phi-1)+b^2(3\sin^2\phi-1)=0.$$

If the point (ϕ) be $(x'y')$, the equation is

$$x^2+y^2-\frac{2(a^2-b^2)x'^3}{a^4}x+\frac{2(a^2-b^2)y'^3}{b^4}y+a'^2-2b'^2=0,$$

where a' is the semi-diameter through $(x'y')$, and b' the semi-conjugate.

331. If $L=M$, then S' becomes of the form L^2, and the equation $S'=0$, or $L^2=0$, represents two straight lines which coincide. Then, by following out the same train of reasoning as that used above, we see that, since the lines (L) and (M) coincide, the conic $(S+kL^2)$ will pass through the two pairs of coincident points of intersection of (S) and (L^2), and will therefore touch (S) and have (L) for a common chord of contact. The conic $(S+kL^2)$ is said to have a *double contact* with (S) along the line (L). The annexed figures will shew the position of the conics of Arts. 329, 331.

It is evident from an inspection of the figure, that two out of the three pairs of chords common to $(S+kLM)$ and (S) have now come into coincidence with the chord of contact

(L), and that the remaining pair have become tangents at its extremities. If we had made use of these chords instead

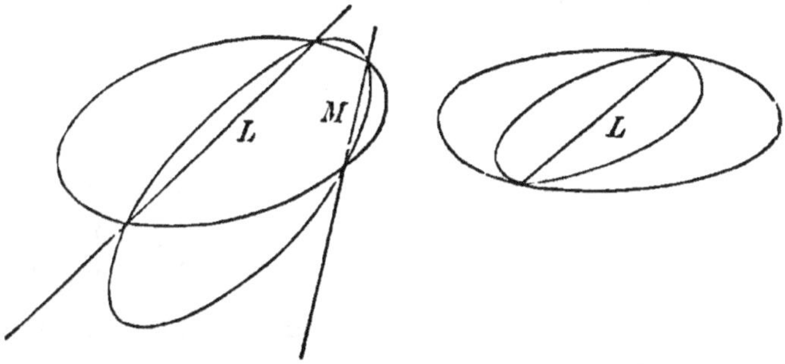

of (L) and (M) in the equation, we should have proved that the equation to a conic, having a double contact with (S) at two given points, may be written in the form $S + kTT' = 0$, where (T) and (T') represent the tangents at these points. These results will be seen to agree with Art. 335.

332. The equation $S + kL = 0$ may be considered as a limiting form of the equation $S + kLM = 0$; for it may be written

$$S + (0 \,.\, x + 0 \,.\, y + k)\, L = 0,$$

which indicates that one of the chords of contact has (Art. 99) become indefinitely distant. It will represent a series of conics, having (L) for their common chord, similar (Art. 319) to (S), and similarly placed. Hence two similar and similarly placed conics have two finite and two infinitely distant points of intersection, a result which may easily be verified by the equations of Arts. 319, 320.

If the curves are hyperbolas, their asymptotes are (Art. 186) parallel, and the infinite points of intersection are explained by the tendency of the curves to merge into two pairs of parallel straight lines. If the curves are ellipses, as the chord moves off, its intersections with (S) become necessarily imaginary, and we can attach no geometrical meaning to the infinitely distant points of intersection. If the curves are parabolas, their axes are parallel, and the curves tend to become four straight lines parallel to their axes, and infinitely distant from them.

333. The equation $S+k=0$ may be written

$$S+k(0.x+0.y+1)^2=0,$$

and is therefore a particular case of the equation $S+kL^2=0$. It represents a conic, having a double contact with (S), the chord of contact being infinitely distant. Now $S+k=0$ differs from $S=0$ in the constant term only; hence the conics are not only similar and similarly placed, but they are, (Art. 141, Cor., 145, Cor.), if central conics, concentric, and their axes and asymptotes are coincident. If the curves are hyperbolas, they have their asymptotes as common tangents at infinity, and therefore touch one another, and have a common chord, at infinity. If they are ellipses, we can attach no geometrical meaning to the infinitely distant, but imaginary, points of contact. If they are parabolas, they are equal, and their axes are (Art. 156, Cor.) coincident; and it may be seen that two such curves approach indefinitely near to one another, or touch, at infinity, by writing the equations to the curves, $y^2=4dx$ and $y^2=4d(x+h)$, and using the method of Art. 171.

334. If S also, as well as S', can be split up into two linear factors, then each of the two conics becomes two straight lines, and our equation $S+kS'=0$ becomes of the form

$$LN+kMR=0,$$

where L, M, N, R are all of one dimension in x and y; and this represents a conic section passing through the four points of intersection of $S=0$ and $S'=0$, or, in other words, circumscribing the quadrilateral formed by these four straight lines. We shall, however, prove this proposition independently; it may be stated thus:

If $L=0$, $M=0$, $N=0$, $R=0$, *be the equations to the four sides of a quadrilateral* taken in order, *then the equation*

$$LN+kMR=0 \ldots\ldots\ldots\ldots\ldots\ldots(1)$$

represents a system of conics circumscribing the quadrilateral.

For the co-ordinates of the point (L, M) make L and M simultaneously vanish, and therefore make the expression $LN+kMR$ vanish, and therefore satisfy equation (1). This corner of the quadrilateral lies therefore in (1). Similarly it may be shewn that the points (M, N), (N, R), (R, L), or the other three corners, lie in (1), and therefore that (1) (which being of two dimensions in x and y must be *some* conic) is one of a system of conics circumscribing the quadrilateral. We have noticed an equation of this form in Art. 312, Ex.

Since the expressions L, M, N, R are proportional to the perpendiculars from the point (xy) on the four sides of the quadrilateral, we have the following geometrical interpretation of equation (1). If any quadrilateral figure be inscribed in a conic, the product of the perpendiculars, drawn from any point on the curve to two opposite sides, is in a constant ratio to the product of the perpendiculars on the other two sides.

335. If $M=R$ in the equation $LN+kMR=0$, it becomes

$$LN+kM^2=0,$$

and we see that, owing to the two opposite sides of the quadrilateral approaching to and ultimately coinciding with each other, instead of two chords of intersection, (M) and (R), of the circumscribing conic, we have only one chord, which will be one of contact; for the lines (L), (N) will then each pass through two consecutive points in the circumscribing conic, and will therefore be tangents to it. We have noticed in Art. 308 the particular case, where the tangents are themselves the axes of co-ordinates.

Ex. 1. The general equation of the second degree may be written

$$Ax^2+2Hxy+By^2+(2Gx+2Fy+C)(0.x+0.y+1)=0,$$

and is therefore of the form $LN+kMR=0$. The first three terms represent two straight lines through the origin, parallel (Art. 186) to the asymptotes, each meeting the curve in one finite and one infinite point. The line $(2Gx+2Fy+C=0)$ is the chord joining the two finite points, and the line at infinity joins the two infinite points.

If $C=0$, or the origin is on the curve, the two finite points coincide, and $(Gx+Fy=0)$ is the tangent at the origin, as in Art. 297, Cor.

Ex. 2. The ordinary equation to central conics may be written

$$(x-a)(x+a)+\frac{a^2}{b^2}y^2=0\text{..............................}(1),$$

where $x-a=0$ and $x+a=0$ are two tangents, and $y=0$ is their chord of contact.

Ex. 3. If the vertex is origin, the equation of Ex. 1 is

$$(x-2a)x+\frac{a^2}{b^2}y^2=0,$$

or, as in Art. 240,

$$\left(2d-\frac{d^2}{a}\right)\left(\frac{x}{a}-2\right)x+y^2=0.$$

Now, when a becomes infinite, or the curve becomes a parabola, this equation becomes

$$2d(0.x-2)x+y^2=0,$$

where one tangent $(0.x-2=0)$ has (Art. 99) become infinitely distant.

Ex. 4. The equation to an hyperbola referred to its asymptotes may be written

$$xy=(0.x+0.y+k)^2,$$

the lines $x=0$, $y=0$ being tangents, whose chord of contact is infinitely distant.

Similarly the ordinary equation to the hyperbola may be written

$$\left(\frac{x}{a}+\frac{y}{b}\right)\left(\frac{x}{a}-\frac{y}{b}\right)=(0.x+0.y+1)^2,$$

with a like interpretation.

Ex. 5. The general equation to the parabola may be written

$$(ax+by)^2+(2Gx+2Fy+C)(0.x+0.y+1)=0,$$

where (Art. 275) the line $(ax+by=0)$ is a diameter, and $(2Gx+2Fy+C=0)$ the tangent at its vertex, the tangent at the other end of the diameter being infinitely distant.

336. If we write the equation of Art. 335 in the form $LM = R^2$, which is equally general, any point on the locus may be defined by the equations

$$\frac{L}{1} = \frac{R}{\mu} = \frac{M}{\mu^2},$$

the equations $R = \mu L,\quad M = \mu R,$

where μ is arbitrary, representing a pair of straight lines, drawn through (L, R) and (M, R) and intersecting at a point on the curve, which we may call 'the point μ.'

337. *To find the equation to the chord joining two points on the conic $LM = R^2$, and to deduce the equation to the tangent.*

Let μ_1, μ_2 (Art. 336) be the two points; then at these points we have

$$\frac{L}{1} = \frac{R}{\mu_1} = \frac{M}{\mu_1^2} \dots\dots\dots (1),\quad \frac{L}{1} = \frac{R}{\mu_2} = \frac{M}{\mu_2^2} \dots\dots\dots (2).$$

Suppose the equation of the chord to be

$$lL + rR + mM = 0 \dots\dots\dots\dots\dots\dots\dots\dots (3),$$

then, since μ_1, μ_2 both lie on this line, we have, from (1) and (2), the equations

$$l + r\mu_1 + m\mu_1^2 = 0 \dots\dots (4),\quad l + r\mu_2 + m\mu_2^2 = 0 \dots\dots (5);$$

hence, obtaining the values of $\frac{m}{l}$, $\frac{r}{l}$ from (4) and (5), and substituting them in (3), we have for the equation to the chord,

$$\mu_1\mu_2 L - (\mu_1 + \mu_2)\ R + M = 0 \dots\dots\dots\dots\dots (6).$$

If we make $\mu_1 = \mu_2$ in equation (6), we obtain, for the equation to the tangent at μ_1,

$$\mu_1^2 L - 2\mu_1 R + M = 0 \dots\dots\dots\dots\dots\dots (7).$$

338. Conversely, if a linear equation contains an arbitrary constant of the second degree, the straight line which it represents will touch a fixed conic. For its most general form is

$$\mu^2L - 2\mu R + M = 0 \ldots\ldots\ldots\ldots\ldots\ldots (1),$$

where μ is arbitrary, and L, M, R are linear; and this represents a tangent to the conic $LM = R^2$. The conic is called the *Envelope* of the system of lines represented by (1).

It is plain that only two lines of the system, that is two tangents to the conic, can be drawn through a given point, namely those answering to the values of μ determined by the equation

$$\mu^2L' - 2\mu R' + M' = 0 \ldots\ldots\ldots\ldots\ldots\ldots (2),$$

where L', M', R' are the results of substituting the co-ordinates of the given point in L, M, R. If the given point is *on* the curve, the two tangents coincide, and the roots of (2) are equal. The condition that this should be the case, is the equation to the envelope. Cf. Arts. 121, 188, 258.

Ex. 1. *To find the envelope of the line*

$$Ax \cos\phi + By \sin\phi = C.$$

Since $\cos\phi = \frac{1-\mu^2}{1+\mu^2}$, $\sin\phi = \frac{2\mu}{1+\mu^2}$, where $\tan\frac{\phi}{2} = \mu$,

the equation becomes

$$(C + Ax)\mu^2 - 2By\mu + C - Ax = 0,$$

and the equation to the envelope is

$$A^2x^2 + B^2y^2 = C^2.$$

Ex. 2. *Given the vertical angle and the sum of the sides of a triangle, to find the envelope of the base.*

If the sides be taken for axes, the equation to the base is

$$\frac{x}{a} + \frac{y}{b} = 1,$$

where $a + b = k$, a constant. Then

$$\frac{x}{a} + \frac{y}{k-a} = 1, \text{ or } a^2 - (x - y + k)a + kx = 0;$$

and the equation to the envelope is

$$(x-y+k)^2=4kx,$$

a parabola touching the sides.

Ex. 3. *Find the envelope of a line such that the sum of the squares of the perpendiculars on it from two fixed points may be constant.*

Take for axes the line joining the two fixed points, and a perpendicular through its middle point, so that the co-ordinates of the fixed points may be $y=0$, $x=\pm c$: then, if the sum of the squares $=k$, the equation is

$$\frac{2x^2}{k-2c^2}+\frac{2y^2}{k}=1.$$

Ex. 4. *If the difference of the squares be given in Ex. 3, to shew that the envelope is a parabola.*

Ex. 5. *The sides BC, CA, AB of a triangle being divided at the points D, E, F, so that*

$$\frac{BD}{DC}=\frac{CE}{EA}=\frac{AF}{FB}=n,$$

to find the envelope of FE.

If AC and AB are taken for axes of x and y, the equation to FE is

$$\frac{x}{\frac{b}{n+1}}+\frac{y}{\frac{nc}{n+1}}=1, \text{ or } n(n+1)\frac{x}{b}+(n+1)\frac{y}{c}=n,$$

which may be written

$$n^2\frac{x}{b}-2n\frac{1-\frac{x}{b}-\frac{y}{c}}{2}+\frac{y}{c}=0,$$

and FE is therefore a tangent to the conic

$$\frac{4xy}{bc}=\left(\frac{x}{b}+\frac{y}{c}-1\right)^2,$$

which (Art. 308) is a parabola touching AB, AC, at the points B, C.

339. It will be observed, that we have considered in the present chapter equations representing conics passing through some four fixed points. Each equation has involved an undetermined constant k, which may receive different values, distinguishing the different conics which can be drawn through the same four points. By giving a suitable value to

k, we may make the equation represent *any conic whatsoever* passing through the four points. For let $\Sigma = 0$ represent any such conic, and let $(x'y')$ be any other point in it; then, by substituting x', y' in the equation, we shall obtain the appropriate value of k, when the locus passes through $(x'y')$. If this value of k is substituted in the equation, the conic represented by it will pass through the same five points as $\Sigma = 0$, and therefore (Art. 312) be identical with it.

Ex. 1. *Find the equation to the conic section that passes through the intersection of the circles*

$$x^2 + y^2 - 4x - 8y = 28 \quad \text{.............................(1)},$$
$$x^2 + y^2 = 4 \quad \text{.............................(2)},$$

and through the centre of (1).

Equation (1) may be written

$$(x-2)^2 + (y-4)^2 = 48$$

and its centre is therefore the point (2, 4).

The equation to the required conic will be of the form

$$(x-2)^2 + (y-4)^2 - 48 + k(x^2 + y^2 - 4) = 0,$$

and, since it passes through the point (2, 4), we have,

$$-48 + k \,.\, 16 = 0,$$

therefore $$k = 3,$$

and the equation becomes

$$(x-2)^2 + (y-4)^2 - 48 + 3(x^2 + y^2 - 4) = 0,$$

or $$4x^2 + 4y^2 - 4x - 8y = 40,$$

or $$x^2 + y^2 - x - 2y = 10;$$

a circle, the co-ordinates of whose centre are $(\frac{1}{2}, 1)$, and whose radius is $\frac{3}{2}\sqrt{5}$.

Ex. 2. *To find the equation to a pair of tangents drawn from the point* $(x'y')$ *to the conic* $\phi(xy) = 0$.

The tangents are a conic having a double contact with $\phi(xy) = 0$ along the polar (say $L = 0$) of $(x'y')$. The equation therefore (Art. 331) is $\phi(xy) + kL^2 = 0$; and the locus passes through the two pairs of coincident points of intersection of $\phi(xy) = 0$ and $L^2 = 0$. Written at length (Art. 298) the equation is

$$\phi(xy) + k\{(Ax' + Hy' + G)x + (Hx' + By' + F)y + Gx' + Fy' + C\}^2 = 0,$$

and, since the locus passes through $(x'y')$, we have from this equation

$$\phi(x'y') + k\{\phi(x'y')\}^2 = 0, \text{ or } k = -\frac{1}{\phi(x'y')}.$$

Hence the only conic answering the conditions is

$$\phi(xy)\,\phi(x'y') = \{(Ax' + Hy' + G)\,x + (Hx' + By' + F)\,y + Gx' + Fy' + C\}^2,$$

which must therefore be the tangents.

Ex. 3. *To find the equation to the director circle of the conic $\phi(xy) = 0$, the axes being rectangular.*

The pair of tangents in Ex. 2 are at right angles, if (Art. 67) the sum of the coefficients of x^2 and y^2 in their equation vanishes; and this is the case, if

$$(Ax' + Hy' + G)^2 + (Hx' + By' + F)^2 = (A + B)\,\phi(x'y');$$

but, if the tangents are at right angles, $(x'y')$ must (Art. 189) be on the director circle; and the required equation is therefore, after expansion,

$$(H^2 - AB)(x^2 + y^2) + 2(HF - BG)\,x + 2(HG - AF)\,y - C(A + B) + F^2 + G^2 = 0.$$

In the case of the parabola, the terms of highest dimensions vanish, and the equation represents the directrix.

340. By means of Art. 335 we may interpret the equation

$$l^2L^2 - m^2M^2 - n^2N^2 = 0;$$

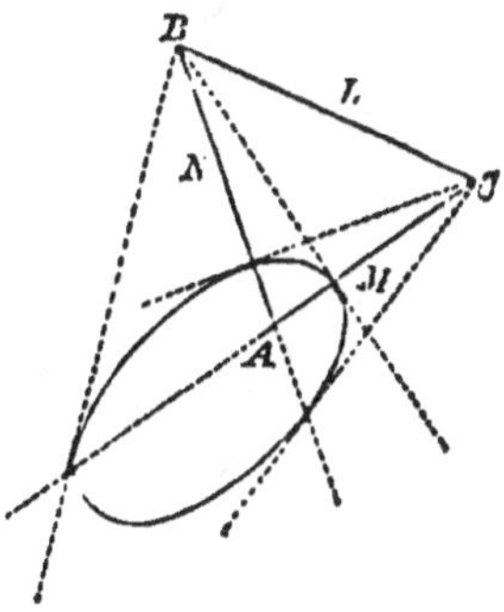

for it may be written in either of the forms

$$(lL + mM)(lL - mM) - n^2N^2 = 0,$$
$$(lL + nN)(lL - nN) - m^2M^2 = 0,$$

and therefore represents a conic, so situated, that AB (N) is the chord of contact of $(lL + mM)$, $(lL - mM)$, the pair of tangents through C (L, M), and CA (M) is the chord of contact of $(lL + nN)$, $(lL - nN)$, the pair of tangents through B (L, N). In other words C (L, M) is the pole of AB (N), and B (L, N) is the pole of CA (M); and consequently (Art. 130) A (M, N) is the pole of BC (L). It will be seen from the equation that, although A (M, N) is the pole of BC (L), the tangents from A are imaginary.

The points of intersection of the lines (L), (M), (N) form (Art. 299) a *conjugate triad*, and the triangle formed by (L), (M), (N) is (Art. 130) *self-conjugate*, or *self-polar*.

341 (i). Suppose the equation to the conic of the preceding article to be (Art. 88)

$$\alpha^2 + \beta^2 = e^2\gamma^2 \quad \ldots\ldots\ldots\ldots\ldots\ldots\ldots (1),$$

and that the lines (α), (β) are at right angles. Then this equation asserts, that the square of the distance of a point on the locus from the point (α, β) has a constant ratio to the square of its distance from the line (γ); hence the locus is a conic with (α, β) for focus, (γ) for directrix, and e for eccentricity. Now (1) may be written

$$(e\gamma - \alpha)(e\gamma + \alpha) = \beta^2;$$

hence (Art. 335) the lines ($e\gamma - \alpha$), ($e\gamma + \alpha$) are tangents, and the focal chord (β) is their chord of contact. But ($e\gamma - \alpha$), ($e\gamma + \alpha$) meet in (γ); hence *tangents at the extremities of any focal chord intersect in the directrix.* Also they meet in (α), which is perpendicular to (β); hence *every focal chord is perpendicular to the straight line joining the pole with the focus.* The lines (α), (β), (γ) occupy the positions of HK, PQ, KX in the fig. of Art. 205.

341 (ii). If we take the focus (α, β) for origin and the lines β, α for rectangular axes of x and y, the equation $\alpha^2 + \beta^2 = e^2\gamma^2$ becomes that obtained in Art. 208,

$$x^2 + y^2 = e^2(x\cos\gamma + y\sin\gamma - p)^2,$$

or $$(x + \sqrt{-1}y)(x - \sqrt{-1}y) = e^2(x\cos\gamma + y\sin\gamma - p)^2.$$

From the form of this equation, we see that the two imaginary straight lines

$$(x + \sqrt{-1}y)(x - \sqrt{-1}y) = 0, \text{ or } x^2 + y^2 = 0,$$

are tangents to the conic, and

$$x \cos \gamma + y \sin \gamma - p = 0$$

is their chord of contact, or the polar of the focus (Arts. 127, 195). But $x^2 + y^2 = 0$ also represents a circle (Art. 107) of indefinitely small radius, whose centre is the origin. That is, if the focus is origin and the axes rectangular, the equation to a pair of tangents from the origin satisfies the conditions for a circle; and the same will be true, if the origin is transferred to any point; for then the equation $x^2 + y^2 = 0$ will become

$$x^2 + y^2 + 2Gx + 2Fy + C = 0.$$

Ex. *To find equations to determine the foci of* $\phi(xy) = 0$.

The equation to a pair of tangents to $\phi(xy)$ from a point $(x'y')$ is (Art. 339, Ex. 2),

$$\phi(xy)\,\phi(x'y') = \{(Ax' + Hy' + G)\,x + (Hx' + By' + F)\,y + Gx' + Fy' + C\}^2;$$

and, if this satisfies the conditions for a circle, i.e. if $(x'y')$ is the focus,

$$(A - B)\,\phi(x'y') = (Ax' + Hy' + G)^2 - (Hx' + By' + F)^2,$$
$$H\phi(x'y') = (Ax' + Hy' + G)(Hx' + By' + F).$$

Hence the foci are given by the equations,

$$\phi(xy) = \frac{(Ax + Hy + G)^2 - (Hx + By + F)^2}{A - B} = \frac{(Ax + Hy + G)(Hx + By + F)}{H}.$$

These equations for the foci may be seen by Arts. 67, 186 to represent rectangular hyperbolas, or straight lines at right angles. The equation

$$\frac{(Ax + Hy + G)^2 - (Hx + By + F)^2}{A - B} = \frac{(Ax + Hy + G)(Hx + By + F)}{H}$$

represents the axes, as proved in Art. 303, Ex. 1.

342. *Pascal's Theorem. The three pairs of opposite sides of a hexagon inscribed in a conic intersect in points which all lie in one straight line.*

Let $L = 0$, $M = 0$, $N = 0$, $R = 0$, $S = 0$, $T = 0$, be the equations to the six sides of the hexagon, and let $Q = 0$ be the equation to the diagonal joining (L, T) and (N, R); then, since the conic is circumscribed about the quadrilateral whose sides are (L), (M), (N), (Q), its equation (Art. 334) can be written in the form

$$LN + kMQ = 0 \quad \text{.....................} \quad (1);$$

also, since it is circumscribed about the quadrilateral whose sides are (R), (S), (T), (Q), its equation can be written in the form

$$RT + k'SQ = 0 \quad \text{.....................} \quad (2);$$

then, since (1) and (2) represent the same locus, one of them must be derived from the other by the introduction of some constant factor λ; that is,

$$LN + kMQ = \lambda(RT + k'SQ)$$

identically; hence

$$LN - \lambda RT = (\lambda k'S - kM)\,Q$$

identically; but $LN - \lambda RT = 0$ represents a locus passing through the points (L, R), (L, T), (N, R), (N, T), hence $(\lambda k'S - kM)\,Q = 0$, which evidently represents two straight lines, represents the same locus; but $Q = 0$ passes through the points (L, T), (N, R), hence $\lambda k'S - kM = 0$ must pass through the other two points (L, R), (N, T); but it evidently passes through (M, S), hence (L, R), (M, S), (N, T) are in one straight line.

343. *Brianchon's Theorem. If tangents be drawn at the six angular points A, B, C, D, E, F, of a hexagon inscribed in a conic, so that the tangents at A, B meet in c, those at B, C in d, &c.; then the three straight lines cf, da, eb meet in a point.*

This follows at once from Pascal's Theorem, or *vice versâ.* For c is the pole of AB and f is the pole of DE; hence (Art. 130) the intersection of AB, DE is the pole of cf.

Similarly .... BC, EF.................. da,

and .... CD, FA.................. eb.

But these three intersections lie on one straight line; therefore the three polars meet in a point. Conversely, if the three polars meet in a point, the three poles will lie on one straight line.

344. *Trilinear Co-ordinates.*

Since (Art. 96) every equation with these co-ordinates can be made homogeneous, the general equation of the second degree (Art. 22) is

$$A\alpha^2 + B\beta^2 + C\gamma^2 + 2F\beta\gamma + 2G\gamma\alpha + 2H\alpha\beta = 0.$$

We shall call the equation $\phi(\alpha\beta\gamma) = 0$.

Precisely as in Art. 97, it may be seen that the Cartesian equation to any conic can be transformed into an equation of the above form; that is to say, the general equation of the second degree in Trilinear Co-ordinates can be made to represent *any* conic, by giving suitable values to the constants.

Cor. If $L = 0$, $M = 0$, $N = 0$ be the equations to the sides of the triangle of reference, it is evident, as in Arts. 92, 98 Cor., that the above statements are true, if we write the abbreviations L, M, N, instead of the trilinear co-ordinates α, β, γ. The same remark will apply to Arts. 345, 347.

345. *To find the equation to a conic section which circumscribes the triangle of reference.*

Let the equation to any such conic be

$$A\alpha^2 + B\beta^2 + C\gamma^2 + 2F\beta\gamma + 2G\gamma\alpha + 2H\alpha\beta = 0;$$

then, since the conic passes through the point $C\left(0, 0, \frac{2\Delta}{c}\right)$, the equation must be satisfied by these values; but this gives

$$C\frac{4\Delta^2}{c^2} = 0;$$

hence we must have $C=0$. Similarly it may be shewn that $A=0$, and $B=0$, and the equation becomes

$$F\beta\gamma + G\gamma\alpha + H\alpha\beta = 0.$$

This equation may be written in the form

$$\frac{l}{\alpha} + \frac{m}{\beta} + \frac{n}{\gamma} = 0,$$

and evidently involves two independent arbitrary constants only, the conic having already fulfilled three conditions.

Ex. *To find the equations to the tangents at A, B, C of the conic circumscribing the triangle ABC.*

By writing the equation $l\beta\gamma + m\gamma\alpha + n\alpha\beta = 0$ in the form

$$l\beta\gamma + \alpha(m\gamma + n\beta) = 0 \text{..........................(1)},$$

we see that it is reduced to $l\beta\gamma = 0$, if we put $m\gamma + n\beta = 0$. Hence the line $m\gamma + n\beta = 0$ meets the conic in the points in which it meets the lines $\beta = 0$, $\gamma = 0$; but these two points coincide, since the line $m\gamma + n\beta = 0$ passes through the intersection of $\beta = 0$, $\gamma = 0$; hence the straight line and the conic meet one another in two coincident points, that is, they touch one another at the point A. We may obtain this result by putting $\beta' = 0$, $\gamma' = 0$ in equation (2) of Art. 351, Ex. 1. Hence the tangents at A, B, C are

$$\frac{\beta}{m} + \frac{\gamma}{n} = 0, \quad \frac{\gamma}{n} + \frac{\alpha}{l} = 0, \quad \frac{\alpha}{l} + \frac{\beta}{m} = 0.$$

346. *To find the condition that the conic circumscribing the triangle of reference should be a circle.*

Let DE be the tangent at the point A; then the angles DAB, EAC are equal to the angles C and B respectively.

Also, since DE lies in the $+-$ and $-+$ compartments formed by the lines (β), (γ), we have for its equation (Art. 57),

$$\frac{\beta}{\gamma} = -\frac{\sin EAC}{\sin DAB} \text{ or } \frac{\beta}{\sin B} + \frac{\gamma}{\sin C} = 0;$$

but (Art. 345, Ex.) its equation is $\frac{\beta}{m} + \frac{\gamma}{n} = 0$; hence

$$\frac{m}{n} = \frac{\sin B}{\sin C}. \quad \text{Similarly } \frac{l}{m} = \frac{\sin A}{\sin B},$$

and the equation to the circle required is

$$\beta\gamma \sin A + \gamma\alpha \sin B + \alpha\beta \sin C = 0,$$

or
$$a\beta\gamma + b\gamma\alpha + c\alpha\beta = 0.$$

Ex. *If one focus of a conic which touches the three sides of the triangle of reference lies on a fixed straight line, shew that the other focus will lie on a conic circumscribing the triangle; and hence deduce Art.* 306, *Ex.* 4.

If $(\alpha\beta\gamma)$, $(\alpha'\beta'\gamma')$ be the foci, we have (Art. 204),

$$\alpha\alpha' = \beta\beta' = \gamma\gamma' = b^2.$$

If therefore the co-ordinates of one focus of the conic satisfy any homogeneous equation $f(\alpha\beta\gamma) = 0$, the co-ordinates of the other focus will satisfy the equation

$$f\left(\frac{1}{\alpha}\frac{1}{\beta}\frac{1}{\gamma}\right) = 0.$$

Thus, if one focus of a conic lies on the line

$$l\alpha + m\beta + n\gamma = 0 \quad \text{..............................(1)},$$

the other focus will lie on the locus

$$\frac{l}{\alpha} + \frac{m}{\beta} + \frac{n}{\gamma} = 0 \quad \text{..............................(2)},$$

that is, on a conic circumscribing the triangle of reference.

If the ratio of $l : m : n$ approaches indefinitely near to the ratio $a : b : c$, the line (1), and consequently the focus, become indefinitely distant, and the conic is a parabola. In this case (2) becomes the circle circumscribing the triangle.

347. *To find the equation to a conic, which touches the three sides of the triangle of reference.*

Let the side BC (α) be a tangent to the conic represented by the equation $\phi(\alpha\beta\gamma)=0$; then, making $\alpha=0$, we have

$$B\beta^2+C\gamma^2+2F\beta\gamma=0,$$

which is the equation to a locus passing through the intersection of (α) and the conic; but, as in Art. 63 (i), we may see that it represents two straight lines

$$\beta-m\gamma=0,\quad \beta-m'\gamma=0,$$

which pass through the angle A (β, γ), and therefore join (β, γ) with the intersection of (α) and the conic. If these two straight lines coincide, that is, if $F^2=BC$, (α) will pass through two coincident points, and be a tangent. Hence, if (α), (β), (γ) are all tangents, we have

$$F^2=BC,\quad G^2=CA,\quad H^2=AB.$$

These conditions shew that A, B, C have the same sign, since the product of any two of them is positive. We shall suppose that the general equation has been so written, as to make A, B, C positive, and shall assume, therefore, that

$$A=l^2,\quad B=m^2,\quad C=n^2;$$

therefore $\quad F=\pm mn,\quad G=\pm nl,\quad H=\pm lm,$

and the equation $\phi(\alpha\beta\gamma)=0$ becomes

$$l^2\alpha^2+m^2\beta^2+n^2\gamma^2\pm 2mn\beta\gamma\pm 2nl\gamma\alpha\pm 2lm\alpha\beta=0.$$

The signs of F, G, H may be taken in eight different ways, thus

$$+++,\ --+,\ -+-,\ +--,$$
$$---,\ ++-,\ +-+,\ -++.$$

The upper line, where all are positive, or only one, will give the equations

$$(l\alpha+m\beta+n\gamma)^2=0,\ (l\alpha+m\beta-n\gamma)^2=0,\ \&c.,$$

and will represent coincident straight lines, which fulfil the conditions of the problem, in that they meet (α), (β), (γ) each in two coincident points; but they cannot be said in any geometrical sense to touch them.

The lower line, where all are negative or only one, will give four forms of the equation, which represent *curves;* they are equivalent to

$$\sqrt{l\alpha}+\sqrt{m\beta}+\sqrt{n\gamma}=0\ldots(1),\quad \sqrt{l\alpha}+\sqrt{m\beta}+\sqrt{-n\gamma}=0\ldots(2),$$

$$\sqrt{l\alpha}+\sqrt{-m\beta}+\sqrt{n\gamma}=0\ldots(3),\quad \sqrt{-l\alpha}+\sqrt{m\beta}+\sqrt{n\gamma}=0\ldots(4),$$

as may be seen by bringing these equations to a rational form. The first of these forms, which corresponds to

$$l^2\alpha^2+m^2\beta^2+n^2\gamma^2-2mn\beta\gamma-2nl\gamma\alpha-2lm\alpha\beta=0\ldots(5),$$

may be considered to include the other three, since these may be obtained from (1) and (5) by changing the sign of one of the quantities l, m, or n. Thus, if we suppose that l, m, or n may denote a negative as well as a positive quantity, we may use (1) or (5) as the general equation to a conic which touches the three straight lines (α), (β), (γ).

Ex. 1. *If a conic is inscribed in a triangle, shew that the three straight lines, joining the angles with the points of contact on the opposite sides, meet in a point.*

Equation (5) may be written in the form

$$n\gamma(n\gamma-2l\alpha-2m\beta)+(l\alpha-m\beta)^2=0,$$

which is (Art. 335) the equation to a conic referred to two tangents (γ) and $(n\gamma-2l\alpha-2m\beta)$, which have $(l\alpha-m\beta)$ for their chord of contact; but $(l\alpha-m\beta)$ passes through (α, β), and is therefore the straight line joining C with the point where the conic touches (γ) or AB. Hence the three straight lines which join the angles with the points of contact on the opposite sides, are

$$l\alpha-m\beta=0\ldots\ldots(1),\quad m\beta-n\gamma=0\ldots\ldots(2),\quad n\gamma-l\alpha=0\ldots\ldots(3),$$

and these lines (Art. 93) meet in a point.

Ex. 2. *To find the equation to the straight line on which* (*Art.* 130) *the poles of* (1), (2), (3) *lie.*

The three tangents to the conic, at the points where (1), (2), (3) meet it again, are

$$n\gamma-2l\alpha-2m\beta=0,\quad l\alpha-2m\beta-2n\gamma=0,\quad m\beta-2n\gamma-2l\alpha=0.$$

For the points where these tangents intersect the opposite sides (γ), (α), (β), respectively, which points are the poles of (1), (2), (3), respectively, we have

$$\gamma=0,\ l\alpha+m\beta=0;\quad \alpha=0,\ m\beta+n\gamma=0;\quad \beta=0,\ n\gamma+l\alpha=0;$$

all which points lie on the line

$$l\alpha+m\beta+n\gamma=0.$$

348. *To find the equations to the inscribed and escribed circles of the triangle of reference.*

The equations must be of the form

$$\sqrt{l\alpha}+\sqrt{m\beta}+\sqrt{n\gamma}=0 \text{ .... (1).}$$

Then, at the point A', where BC meets the inscribed circle, we have $\alpha=0$, and therefore from (1)

$$\sqrt{m\beta}+\sqrt{n\gamma}=0, \text{ or } m\beta=n\gamma \text{ .... (2).}$$

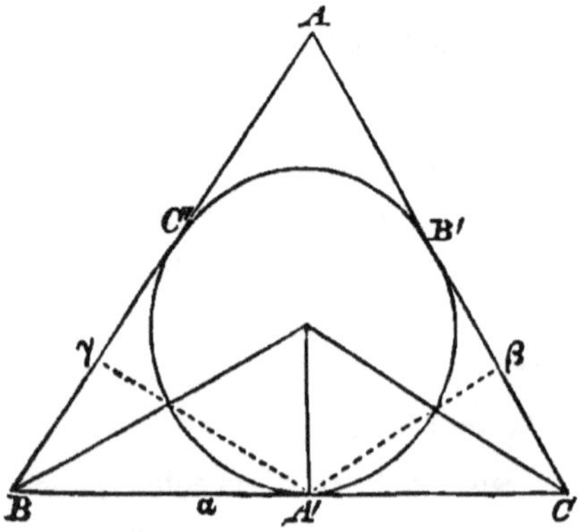

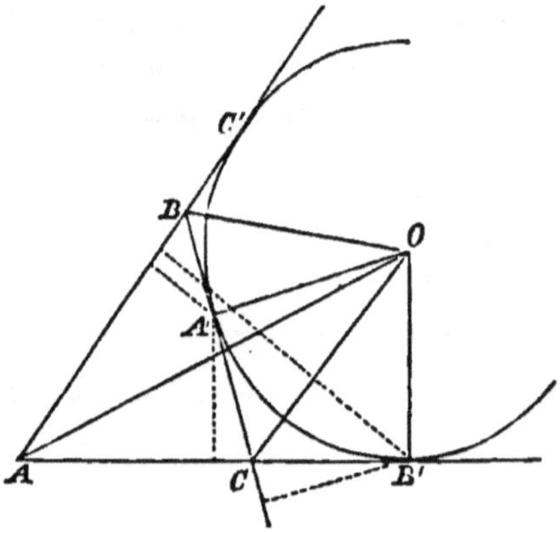

From (2), and from the figure, if the radius $=r$, we have at the point A'

$$\frac{m}{n}=\frac{\gamma}{\beta}=\frac{A'B\sin B}{A'C\sin C}=\frac{r\cot\frac{B}{2}\sin B}{r\cot\frac{C}{2}\sin C}=\frac{\cos^2\frac{B}{2}}{\cos^2\frac{C}{2}};$$

and similar equations may be obtained for the points B', C'; hence

$$l : m : n = \cos^2\frac{A}{2} : \cos^2\frac{B}{2} : \cos^2\frac{C}{2},$$

and the required equation is

$$\cos\frac{A}{2}\sqrt{\alpha} + \cos\frac{B}{2}\sqrt{\beta} + \cos\frac{C}{2}\sqrt{\gamma} = 0.$$

Next let the circle touch BC, and AB, AC produced; then, as before, we have at the point A', $m\beta = n\gamma$; and from this relation and the figure, we have

$$\frac{m}{n} = \frac{\gamma}{\beta} = \frac{A'B\sin B}{A'C\sin C} = \frac{r\cot\frac{\pi - B}{2}\sin B}{r\cot\frac{\pi - C}{2}\sin C} = \frac{\sin^2\frac{B}{2}}{\sin^2\frac{C}{2}}.$$

Also, at the point B' we have $l\alpha = n\gamma$; and from this relation and the figure, since α is negative at B' (Art. 94),

$$\frac{n}{-l} = \frac{-\alpha}{\gamma} = \frac{B'C\sin C}{B'A\sin A} = \frac{r\cot\frac{\pi - C}{2}\sin C}{r\cot\frac{A}{2}\sin A} = \frac{\sin^2\frac{C}{2}}{\cos^2\frac{A}{2}};$$

therefore $l : m : n = -\cos^2\frac{A}{2} : \sin^2\frac{B}{2} : \sin^2\frac{C}{2}$,

and the required equation is

$$\cos\frac{A}{2}\sqrt{-\alpha} + \sin\frac{B}{2}\sqrt{\beta} + \sin\frac{C}{2}\sqrt{\gamma} = 0.$$

Similarly the equations to the other escribed circles may be written down. It may be observed, that, as α is negative for every point of the circle, $\sqrt{-\alpha}$ is an impossible quantity in appearance only.

349. *To find the length of a straight line drawn from the point* $(\alpha'\beta'\gamma')$ *to meet the conic*

$$A\alpha^2 + B\beta^2 + C\gamma^2 + 2F\beta\gamma + 2G\gamma\alpha + 2H\alpha\beta = 0 \ldots\ldots (1).$$

Let the equations to the straight line be (Art. 103)

$$\frac{\alpha-\alpha'}{\lambda}=\frac{\beta-\beta'}{\mu}=\frac{\gamma-\gamma'}{\nu}=\rho\text{..............}(2);$$

then, for the distance of $(\alpha'\beta'\gamma')$ from the curve, we must write in (1)

$$\alpha=\lambda\rho+\alpha',\quad \beta=\mu\rho+\beta',\quad \gamma=\nu\rho+\gamma'.$$

Making this substitution, and arranging, we have

$$\left.\begin{array}{r}\phi(\lambda\mu\nu)\rho^2+2(A\lambda+H\mu+G\nu)\alpha'\\ +2(H\lambda+B\mu+F\nu)\beta'\\ +2(G\lambda+F\mu+C\nu)\gamma'\end{array}\right|\rho+\phi(\alpha'\beta'\gamma')=0\text{....}(3).$$

To this equation the remarks of Art. 176 may be applied without alteration.

350. *To find the equation to the tangent to the conic* $\phi(\alpha\beta\gamma)=0$ *at the point* $(\alpha'\beta'\gamma')$.

Let equations (2) of the last article represent the tangent; then since $(\alpha'\beta'\gamma')$ is on the curve, $\phi(\alpha'\beta'\gamma')=0$, and from equation (3) we obtain for the condition of tangency, as in Art. 117, &c.

$$\left.\begin{array}{r}A\lambda\\ +H\mu\\ +G\nu\end{array}\right|\alpha'+\left.\begin{array}{r}H\lambda\\ +B\mu\\ +F\nu\end{array}\right|\beta'+\left.\begin{array}{r}G\lambda\\ +F\mu\\ +C\nu\end{array}\right|\gamma'=0\text{............}(4),$$

or

$$\left.\begin{array}{r}A\alpha'\\ +H\beta'\\ +G\gamma'\end{array}\right|\lambda+\left.\begin{array}{r}H\alpha'\\ +B\beta'\\ +F\gamma'\end{array}\right|\mu+\left.\begin{array}{r}G\alpha'\\ +F\beta'\\ +C\gamma'\end{array}\right|\nu=0\text{............}(5);$$

then, since λ, μ, ν are, from (2), proportional to $\alpha-\alpha'$, $\beta-\beta'$, $\gamma-\gamma'$ for every point in the tangent, we may substitute these values for λ, μ, ν in (5); hence we obtain

$$\left.\begin{array}{r}A\alpha'\\ +H\beta'\\ +G\gamma'\end{array}\right|\alpha+\left.\begin{array}{r}H\alpha'\\ +B\beta'\\ +F\gamma'\end{array}\right|\beta+\left.\begin{array}{r}G\alpha'\\ +F\beta'\\ +C\gamma'\end{array}\right|\gamma$$

$$=A\alpha'^2+B\beta'^2+C\gamma'^2+2F\beta'\gamma'+2G\gamma'\alpha'+2H\alpha'\beta'=0,$$

since $(\alpha'\beta'\gamma')$ is on the curve; hence the equation to the tangent at $(\alpha'\beta'\gamma')$ is

$$\left.\begin{matrix} A\alpha' \\ +H\beta' \\ +G\gamma' \end{matrix}\right| \alpha + \left.\begin{matrix} H\alpha' \\ +B\beta' \\ +F\gamma' \end{matrix}\right| \beta + \left.\begin{matrix} G\alpha' \\ +F\beta' \\ +C\gamma' \end{matrix}\right| \gamma = 0.$$

351. The reader who is acquainted with the Differential Calculus will observe that the above equation to the tangent is equivalent to

$$\frac{d\phi}{d\alpha'}\alpha + \frac{d\phi}{d\beta'}\beta + \frac{d\phi}{d\gamma'}\gamma = 0.$$

Those not acquainted with that subject may regard $\frac{d\phi}{d\alpha}$ as a symbol used to denote the result of the following operation: Take those terms only in the equation to the conic, which contain α, multiply each term by the index of α in it, and diminish that index by unity. For $\frac{d\phi}{d\alpha'}$ perform the same operation, and then write α', β', γ' for α, β, γ. The symbols $\frac{d\phi}{d\beta}$, $\frac{d\phi}{d\beta'}$, &c. must be understood in a similar manner. The result, when divided by two, will be the equation to the tangent obtained above. Arts. 350, 351 are true for homogeneous equations only.

Ex. 1. *To find the equation to the tangent at the point $(\alpha'\beta'\gamma')$ of the conic.*

$$l\beta\gamma + m\gamma\alpha + n\alpha\beta = 0 \text{..............................(1)},$$

$$\frac{d\phi}{d\alpha'} = n\beta' + m\gamma', \quad \frac{d\phi}{d\beta'} = l\gamma' + n\alpha', \quad \frac{d\phi}{d\gamma'} = m\alpha' + l\beta',$$

and the equation to the tangent is

$$(n\beta' + m\gamma')\,\alpha + (l\gamma' + n\alpha')\,\beta + (m\alpha' + l\beta')\,\gamma = 0 \text{..............(2)}.$$

This may be written in another form; for, since $(\alpha'\beta'\gamma')$ is on the curve,

$$n\beta' + m\gamma' = -\frac{l\beta'\gamma'}{\alpha'}, \qquad l\gamma' + n\alpha' = -\frac{m\gamma'\alpha'}{\beta'}, \qquad m\alpha' + l\beta' = -\frac{n\alpha'\beta'}{\gamma'};$$

therefore, substituting in (2) and dividing by $\alpha'\beta'\gamma'$, we have

$$\frac{l\alpha}{\alpha'^2}+\frac{m\beta}{\beta'^2}+\frac{n\gamma}{\gamma'^2}=0 \quad\text{.................................(3).}$$

Cor. In order that any straight line

$$\lambda\alpha+\mu\beta+\nu\gamma=0 \quad\text{.................................(4)}$$

should touch the conic, we have, by comparing (3) with (4),

$$\frac{\lambda\alpha'^2}{l}=\frac{\mu\beta'^2}{m}=\frac{\nu\gamma'^2}{n};$$

hence, since α', β', γ' satisfy (1), we have for the condition of tangency

$$\sqrt{l\lambda}+\sqrt{m\mu}+\sqrt{n\nu}=0.$$

Ex. 2. *To find the equation to the tangent at the point* $(\alpha'\beta'\gamma')$ *of the conic.*

$$l^2\alpha^2+m^2\beta^2+n^2\gamma^2-2mn\beta\gamma-2nl\gamma\alpha-2lm\alpha\beta=0 \quad\text{............(1),}$$

or
$$\sqrt{l\alpha}+\sqrt{m\beta}+\sqrt{n\gamma}=0 \quad\text{..........................(2).}$$

Here we may obtain

$$\frac{d\phi}{d\alpha'},\ \frac{d\phi}{d\beta'},\ \frac{d\phi}{d\gamma'}$$

from form (1); then, since $(\alpha'\beta'\gamma')$ is on (2), we have, by transposing and squaring,

$$l\alpha'-m\beta'-n\gamma'=2\sqrt{mn\beta'\gamma'},\quad -l\alpha'+m\beta'-n\gamma'=2\sqrt{nl\gamma'\alpha'},\quad -l\alpha'-m\beta'+n\gamma'=2\sqrt{lm\alpha'\beta'};$$

and the equation to the tangent becomes, after substitution and reduction,

$$\sqrt{\frac{l}{\alpha'}}\,\alpha+\sqrt{\frac{m}{\beta'}}\,\beta+\sqrt{\frac{n}{\gamma'}}\,\gamma=0 \quad\text{.......................(3).}$$

Cor. In order that any straight line

$$\lambda\alpha+\mu\beta+\nu\gamma=0 \quad\text{.....................................(4)}$$

should be a tangent, we have, by comparing (3) with (4), the condition

$$\lambda\sqrt{\frac{\alpha'}{l}}=\mu\sqrt{\frac{\beta'}{m}}=\nu\sqrt{\frac{\gamma'}{n}}.$$

Hence, from (2), the required condition is

$$\frac{l}{\lambda}+\frac{m}{\mu}+\frac{n}{\nu}=0.$$

352. Since the equation to the tangent is not altered when we write α', α for α, α', or β', β for β, β', or γ', γ for γ, γ', we see, as in Art. 131, that the equation to the polar of the point $(\alpha'\beta'\gamma')$ is identical in form with the equation to the tangent.

353. *To find the co-ordinates of the centre.*

Let $(\alpha'\beta'\gamma')$ be the centre; then since all chords through the centre are bisected in the centre, the roots of equation (3), Art. 349, must be equal and of opposite signs, whatever be the direction of the chord; hence (Arts. 350, 351)

$$\lambda\frac{d\phi}{d\alpha'}+\mu\frac{d\phi}{d\beta'}+\nu\frac{d\phi}{d\gamma'}=0 \;\ldots\ldots\ldots\ldots\; (1),$$

for the values of λ, μ, ν in every chord. If we eliminate λ from (1) by means of the relation (Art. 103, Cor.)

$$a\lambda+b\mu+c\nu=0,$$

we have $$\mu\left(a\frac{d\phi}{d\beta'}-b\frac{d\phi}{d\alpha'}\right)+\nu\left(a\frac{d\phi}{d\gamma'}-c\frac{d\phi}{d\alpha'}\right)=0,$$

for the values of μ and ν in every chord. But this requires that

$$\frac{1}{a}\frac{d\phi}{d\alpha'}=\frac{1}{b}\frac{d\phi}{d\beta'}=\frac{1}{c}\frac{d\phi}{d\gamma'},$$

which are therefore the equations for finding the centre. See Art. 95 (ii) and examples.

Cor. 1. Since $\frac{d\phi}{d\alpha'}, \frac{d\phi}{d\beta'}, \frac{d\phi}{d\gamma'}$ are proportional to a, b, c, by substituting in the equation to the polar of $(\alpha'\beta'\gamma')$, we obtain

$$a\alpha+b\beta+c\gamma=0,$$

the equation (Art. 99) to a line at an infinite distance. This we might have inferred from the fact that tangents at the extremities of diameters are parallel, or meet at infinity.

COR. 2. If we begin by assuming that the polar of the centre is the line at infinity, we may at once obtain the equations for the centre $(\alpha'\beta'\gamma')$, by comparing the equation to that line with the equation to the polar of the point $(\alpha'\beta'\gamma')$.

EX. *To find the centre of the conic inscribed in the triangle, and thence to deduce the equations to the inscribed and escribed circles.*

The equations for the centre are

$$\frac{l}{a}(l\alpha' - m\beta' - n\gamma') = \frac{m}{b}(m\beta' - n\gamma' - l\alpha') = \frac{n}{c}(n\gamma' - l\alpha' - m\beta'),$$

or
$$\frac{l\alpha' - m\beta' - n\gamma'}{amn} = \frac{m\beta' - n\gamma' - l\alpha'}{bnl} = \frac{n\gamma' - l\alpha' - m\beta'}{clm};$$

adding numerators and denominators of these fractions two and two, we obtain

$$\frac{\alpha'}{bn+cm} = \frac{\beta'}{cl+an} = \frac{\gamma'}{am+bl}.$$

In order that the centre may be at the point $\alpha' = \beta' = \gamma'$, which is the centre of the inscribed circle, we must have

$$bn + cm = cl + an = am + bl,$$

or
$$\frac{abn+cam}{a} = \frac{bcl+abn}{b} = \frac{cam+bcl}{c};$$

whence
$$\frac{bcl}{b+c-a} = \frac{cam}{c+a-b} = \frac{abn}{a+b-c},$$

or
$$l : \cos^2\frac{A}{2} = m : \cos^2\frac{B}{2} = n : \cos^2\frac{C}{2};$$

and we obtain the equation of Art. 348.

Similarly we may obtain the equations to the escribed circles, whose centres are at the points

$$-\alpha' = \beta' = \gamma', \quad \alpha' = -\beta' = \gamma', \quad \alpha' = \beta' = -\gamma'.$$

354. *To find the condition that the conic $\phi(\alpha\beta\gamma) = 0$ should be an ellipse, parabola, or hyperbola.*

If we eliminate γ between the equations

$$\phi(\alpha\beta\gamma) = 0, \quad a\alpha + b\beta + c\gamma = 0,$$

we have a homogeneous equation in α and β, representing two straight lines through the angular point C of the triangle of reference, drawn to the points where the conic meets the line at infinity. These will be parallel to the asymptotes of the conic; and the condition that these lines should be imaginary, real and coincident, or real and different, will be the condition required (Arts. 177, 252).

Ex. *To find the condition that the equations*

$$l\beta\gamma + m\gamma\alpha + n\alpha\beta = 0 \ldots\ldots (1),$$

$$\sqrt{l\alpha} + \sqrt{m\beta} + \sqrt{n\gamma} = 0 \ldots\ldots (2),$$

should represent ellipses, parabolas, or hyperbolas.

Equation (1) after the above substitution is

$$ma\alpha^2 + (la + mb - nc)\,\alpha\beta + lb\beta^2 = 0,$$

from which we get the required condition,

$$l^2a^2 + m^2b^2 + n^2c^2 - 2lmab - 2mnbc - 2nlca <=> 0.$$

Eliminating γ from equation (2), we have

$$\sqrt{lc\alpha} + \sqrt{mc\beta} + \sqrt{-n\,(a\alpha + b\beta)} = 0,$$

which becomes, when brought to a rational form,

$$(na + lc)^2\,\alpha^2 + 2\,\{(na - lc)\,mc + (na + lc)\,nb\}\,\alpha\beta + (mc + nb)^2\,\beta^2 = 0,$$

from which we obtain the required condition,

$$lmn\left(\frac{l}{a} + \frac{m}{b} + \frac{n}{c}\right) >=< 0.$$

355. *To find the condition that the conic $\phi(\alpha\beta\gamma) = 0$ should be a rectangular hyperbola.*

Transform to rectangular Cartesian co-ordinates, as in Art. 100 (ii); then the sum of the coefficients of x^2 and y^2 must $= 0$. The condition will be found to be

$$A + B + C - 2F\cos A - 2G\cos B - 2H\cos C = 0.$$

356. *To find the condition that the conic $\phi(\alpha\beta\gamma) = 0$ should be a circle.*

If $(\alpha'\beta'\gamma')$ be the centre in Art. 349, the coefficient of ρ

in equation (3) must vanish, and we have, to determine the semi-diameters,

$$\phi(\lambda\mu\nu)\rho^2 + \phi(\alpha'\beta'\gamma') = 0.$$

Now suppose the line (2), Art. 349, to be drawn parallel to the side BC of the triangle (fig. Art. 103); then

$$\lambda = \sin\theta = 0;\ \mu = \sin\phi = \sin C;\ \nu = \sin\psi = -\sin B;$$

and we obtain for the semi-diameters parallel to BC

$$\phi(0, \sin C, -\sin B)\, l^2 + \phi(\alpha'\beta'\gamma') = 0,$$

or $$B\sin^2 C + C\sin^2 B - 2F\sin B\sin C = -\frac{\phi(\alpha'\beta'\gamma')}{\rho^2}.$$

Similar expressions may be obtained for the semi-diameters parallel to CA, AB; but, if the curve is a circle, these are all equal, and conversely; hence the required condition is, after substituting for $\sin A$, $\sin B$, $\sin C$ the proportionals a, b, c,

$$Ab^2 + Ba^2 - 2Hab = Bc^2 + Cb^2 - 2Fbc = Ca^2 + Ac^2 - 2Gca.$$

From this may be obtained the results of Arts. 346, 348. Cf. Art. 353, Ex.

357. *To find the condition that the conic $\phi(\alpha\beta\gamma) = 0$ should be a pair of straight lines.*

This is the condition that $\phi(\alpha\beta\gamma)$ should be the product of two linear factors. It is the same as that of Art. 62, since the equation may be made to assume the same form as the Cartesian equation, by dividing by γ^2.

358. *To find the equation to the asymptotes of the conic $\phi(\alpha\beta\gamma) = 0$.*

Since the equations of a conic and its asymptotes differ by a constant only, the equation will be $\phi(\alpha\beta\gamma) + k = 0$, or in a homogeneous form,

$$\phi(\alpha\beta\gamma) + \frac{k}{4\Delta^2}(a\alpha + b\beta + c\gamma)^2 = 0 \ldots\ldots\ldots\ldots(1),$$

a form of equation which reminds us that the asymptotes are a conic having a double contact with $\phi(\alpha\beta\gamma)=0$ along the line $a\alpha+b\beta+c\gamma=0$ at an infinite distance. Also since the asymptotes pass through the centre $(\alpha'\beta'\gamma')$, we have $\phi(\alpha'\beta'\gamma')+k=0$, and the equation becomes

$$\phi(\alpha\beta\gamma)=\phi(\alpha'\beta'\gamma').$$

359. The equation to any two circles, with Cartesian co-ordinates, can be written (Art. 110) so as to have the terms of the second degree the same; that is to say, if $S=0$ represents a circle, *any* other circle can be represented by

$$S+Ax+By+C=0.$$

Hence it follows, as in Art. 344, that, if $S=0$ is the equation to a circle, with Trilinear Co-ordinates,

$$S+l\alpha+m\beta+n\gamma=0$$

will represent any other circle; or we may use (Art. 96) the homogeneous form

$$S+(l\alpha+m\beta+n\gamma)(a\alpha+b\beta+c\gamma)=0.$$

If the two circles are concentric, their equations will differ only by a constant; that is, any circle concentric with $S=0$ will be represented by $S+c=0$, or, in a homogeneous form,

$$S+k(a\alpha+b\beta+c\gamma)^2=0.$$

Cor. 1. If (S) is the circle circumscribing the triangle of reference, the equation to any other circle may be written in the form

$$a\beta\gamma+b\gamma\alpha+c\alpha\beta+(l\alpha+m\beta+n\gamma)(a\alpha+b\beta+c\gamma)=0;$$

and if the general equation

$$A\alpha^2+B\beta^2+C\gamma^2+2F\beta\gamma+2G\gamma\alpha+2H\alpha\beta=0\ldots(2)$$

be the equation to a circle, it must be identical with

$$k(a\beta\gamma+b\gamma\alpha+c\alpha\beta)+\left(\frac{A\alpha}{a}+\frac{B\beta}{b}+\frac{C\gamma}{c}\right)(a\alpha+b\beta+c\gamma)=0\ldots(3).$$

By equating the coefficients of (2) and (3), we may obtain the conditions of Art. 356, that (2) should represent a circle.

COR. 2. The radical axis of two circles, whose equations are written in the above form, will be found by subtracting one equation from the other, which leaves a linear equation, as in Art. 133.

Ex. 1. *To find the equation to the circle, which passes through the feet of the perpendiculars from the angles of the triangle ABC upon the opposite sides.*

Let the equation be (Arts. 359, Cor. 1)

$$a\beta\gamma + b\gamma\alpha + c\alpha\beta = (l\alpha + m\beta + n\gamma)(a\alpha + b\beta + c\gamma);$$

then for the point A', the foot of the perpendicular from A on BC, we have

$$\alpha : \beta : \gamma = 0 : AA' \cos C : AA' \cos B = 0 : \frac{1}{\cos B} : \frac{1}{\cos C}.$$

Substituting (Art. 95 (ii)) the latter proportionals in the equation, we have

$$\frac{a}{\cos B \cos C} = \left(\frac{m}{\cos B} + \frac{n}{\cos C}\right)\left(\frac{b}{\cos B} + \frac{c}{\cos C}\right);$$

or, since $b \cos C + c \cos B = a$,

$$\frac{m}{\cos B} + \frac{n}{\cos C} = 1 \quad \ldots\ldots\ldots\ldots\ldots\ldots (1).$$

Similarly

$$\frac{n}{\cos C} + \frac{l}{\cos A} = 1 \ldots\ldots\ldots (2), \quad \frac{l}{\cos A} + \frac{m}{\cos B} = 1 \ldots\ldots\ldots (3).$$

Adding (2) and (3) and subtracting (1), we obtain $\frac{2l}{\cos A} = 1$, and therefore, from symmetry,

$$\frac{l}{\cos A} = \frac{m}{\cos B} = \frac{n}{\cos C} = \frac{1}{2};$$

hence the required equation is

$$a\beta\gamma + b\gamma\alpha + c\alpha\beta = \frac{1}{2}(\alpha \cos A + \beta \cos B + \gamma \cos C)(a\alpha + b\beta + c\gamma).$$

Ex. 2. *To find the equation to the circle, which passes through the middle points of the sides of the triangle ABC.*

Let the equation be

$$a\beta\gamma+b\gamma\alpha+c\alpha\beta=(l\alpha+m\beta+n\gamma)(a\alpha+b\beta+c\gamma);$$

then, for the middle point of BC, we have

$$\alpha:\beta:\gamma=0:\frac{a}{2}\sin C:\frac{a}{2}\sin B=0:\frac{1}{b}:\frac{1}{c}.$$

Substituting the latter proportionals in the equation, we have

$$\frac{a^2}{abc}=2\left(\frac{m}{b}+\frac{n}{c}\right),$$

and similar equations may be obtained from the other middle points; hence

$$\frac{m}{b}+\frac{n}{c}=\frac{a^2}{2abc},\quad \frac{n}{c}+\frac{l}{a}=\frac{b^2}{2abc},\quad \frac{l}{a}+\frac{m}{b}=\frac{c^2}{2abc};$$

adding the two last equations and subtracting the first, we have

$$\frac{2l}{a}=\frac{b^2+c^2-a^2}{2abc}=\frac{\cos A}{a};$$

therefore, by symmetry,

$$\frac{l}{\cos A}=\frac{m}{\cos B}=\frac{n}{\cos C}=\frac{1}{2},$$

and, substituting these proportionals in the equation to the circle, we have

$$a\beta\gamma+b\gamma\alpha+c\alpha\beta=\frac{1}{2}(\alpha\cos A+\beta\cos B+\gamma\cos C)(a\alpha+b\beta+c\gamma).$$

This equation is the same as that obtained in Ex. 1; hence in every triangle the same circle passes through these six points.

Suppose the three perpendiculars AA', BB', CC', to meet in P; then A', B', C' will be the feet of the perpendiculars dropped from B, A, P, on the sides of the triangle BAP; hence the circle which passes through A', B', C' passes through the middle points of AP and BP, and for a similar reason must pass through the middle point of CP. Hence the bisections of the sides, the feet of the perpendiculars, and the bisections of AP, BP, CP all lie on the same circle. This is called the *Nine Points Circle* of the triangle ABC. See Art. 314, Ex. 4, and Examples XII., 14.

In the following examples we shall find the equations to the inscribed and escribed circles of the triangle of reference, in the form of Art. 359, in order to prove that they are touched by the Nine Points Circle.

Ex. 3. *To find the equation to the circle inscribed in the triangle of reference.*

This equation may (Art. 348) be written

$$\alpha^2 \cos^4 \frac{A}{2} + \beta^2 \cos^4 \frac{B}{2} + \gamma^2 \cos^4 \frac{C}{2} - 2\beta\gamma \cos^2 \frac{B}{2} \cos^2 \frac{C}{2} - \&c. = 0,$$

which must (Art. 359, Cor. 1) be identical with

$$k\,(a\beta\gamma + b\gamma\alpha + c\alpha\beta) + \left(\frac{\alpha}{a} \cos^4 \frac{A}{2} + \frac{\beta}{b} \cos^4 \frac{B}{2} + \frac{\gamma}{c} \cos^4 \frac{C}{2}\right)(a\alpha + b\beta + c\gamma) = 0.$$

Equating the coefficients of $\beta\gamma$ in these two equations, we obtain

$$-abck = c^2 \cos^4 \frac{B}{2} + b^2 \cos^4 \frac{C}{2} + 2bc \cos^2 \frac{B}{2} \cos^2 \frac{C}{2}$$

$$= \left(c \cos^2 \frac{B}{2} + b \cos^2 \frac{C}{2}\right)^2 = s^2,$$

where $2s = a + b + c$. Hence the equation to the circle becomes

$$a\beta\gamma + b\gamma\alpha + c\alpha\beta = \frac{abc}{s^2}\left(\frac{\alpha}{a} \cos^4 \frac{A}{2} + \frac{\beta}{b} \cos^4 \frac{B}{2} + \frac{\gamma}{c} \cos^4 \frac{C}{2}\right)(a\alpha + b\beta + c\gamma)$$

$$= \frac{1}{abc}\{a\alpha\,(s-a)^2 + b\beta\,(s-b)^2 + c\gamma\,(s-c)^2\}\,(a\alpha + b\beta + c\gamma).$$

Ex. 4. *To find the equation to the escribed circle opposite to A.*

As in Ex. 3, it may be shewn that this equation may be written

$$a\beta\gamma + b\gamma\alpha + c\alpha\beta = \frac{abc}{(s-a)^2}\left(\frac{\alpha}{a} \cos^4 \frac{A}{2} + \frac{\beta}{b} \sin^4 \frac{B}{2} + \frac{\gamma}{c} \sin^4 \frac{C}{2}\right)(a\alpha + b\beta + c\gamma),$$

$$= \frac{1}{abc}\{a\alpha s^2 + b\beta\,(s-c)^2 + c\gamma\,(s-b)^2\}\,(a\alpha + b\beta + c\gamma),$$

from which the equations to the other escribed circles may be written down by symmetry.

Ex. 5. *To shew that the Nine Points Circle touches the inscribed and escribed circles.*

By subtracting the equation to the inscribed circle from the equation to the Nine Points Circle, we obtain for their radical axis

$$\alpha\left(\cos A - \frac{2\,(s-a)^2}{bc}\right) + \beta\left(\cos B - \frac{2\,(s-b)^2}{ca}\right) + \gamma\left(\cos C - \frac{2\,(s-c)^2}{ab}\right) = 0,$$

or $$\frac{\alpha}{bc}(c-a)(a-b) + \frac{\beta}{ca}(a-b)(b-c) + \frac{\gamma}{ab}(b-c)(c-a) = 0,$$

or $$\frac{a\alpha}{b-c} + \frac{b\beta}{c-a} + \frac{c\gamma}{a-b} = 0;$$

and (Art. 351, Ex. 2) this is a tangent to the inscribed circle if

$$\frac{\cos^2\frac{A}{2}(b-c)}{a}+\frac{\cos^2\frac{B}{2}(c-a)}{b}+\frac{\cos^2\frac{C}{2}(a-b)}{c}=0,$$

an identity which may easily be proved. Similarly the radical axis of the Nine Points Circle and any of the escribed circles may be shewn to be a tangent to the two circles. Hence the Nine Points Circle touches the inscribed and escribed circles.

360. *Areal Co-ordinates.*

In this system an equation of the second degree will (Art. 104 (ii)) represent a conic section; and, if the same conic be represented in trilinear and areal co-ordinates, respectively, by

$$A\alpha^2+B\beta^2+C\gamma^2+2F\beta\gamma+2G\gamma\alpha+2H\alpha\beta=0,$$

and $$A'x^2+B'y^2+C'z^2+2F'yz+2G'zx+2H'xy=0,$$

we have $$\frac{x}{a\alpha}=\frac{y}{b\beta}=\frac{z}{c\gamma}=\frac{1}{2\Delta},$$

whence $$\frac{A}{A'a^2}=\frac{B}{B'b^2}=\frac{C}{C'c^2}=\frac{F}{F'bc}=\frac{G}{G'ca}=\frac{H}{H'ab};$$

so that we can at once transform any equation or expression from the one system to the other.

Ex. The general equation in areal co-ordinates will (Art. 356) represent a circle, if

$$\frac{B+C-2F}{a^2}=\frac{C+A-2G}{b^2}=\frac{A+B-2H}{c^2}.$$

It will (Art. 355) represent a rectangular hyperbola, if

$$a^2(A+F-G-H)+b^2(B+G-H-F)+c^2(C+H-F-G)=0.$$

*EXAMPLES XIII.

1. If two conics have each a double contact with a third conic, a pair of the chords of intersection of the first two, as well as the chords of contact of the first two with the third, all pass through one point.

2. Find the equation to an ellipse referred to any pair of conjugate diameters, by considering it as circumscribed about the quadrilateral formed by joining the extremities of these diameters.

3. Shew from Art. 345, Ex. that $l\beta\gamma + m\gamma\alpha + n\alpha\beta = 0$ is the equation to a conic circumscribed about a quadrilateral, when two of its angular points coincide.

4. Interpret the equation (Art. 329) $S + kLM = 0$, (i) when one of the asymptotes of (S) is parallel to (L), and (ii) when one is parallel to (L) and the other to (M).

5. If three conics have each a double contact with a fourth, six of their chords of intersection will pass, three by three, through the same points.

6. ABC is a triangle, and P any point, such that the squares of the three areas PAB, PBC, PCA is equal to the square of the triangle ABC; prove that the locus of P is an ellipse.

7. If $L = 0$, $M = 0$, $N = 0$ meet in a point, can *any* conic be represented by $\phi(LMN) = 0$?

8. Interpret the following equations, where L, M, N are linear, and k constant.

(i) $LM + kN = 0$, (ii) $LM + k^2 = 0$, (iii) $L^2 + kM = 0$.

9. The equations to two conics are $\alpha\beta = \mu^2\gamma^2$ and $\alpha\gamma = \lambda^2\beta^2$; shew that $\alpha + 4\lambda\mu(\lambda\beta + \mu\gamma) = 0$ is the equation to their common tangent.

10. AB and AC are two tangents to a conic, and the bisector of the angle BAC meets BC in D; prove that the segments of any chord through D subtend equal angles at A.

11. If the asymptotes of an hyperbola ($xy = \frac{1}{4}m^2$) coincide with the conjugate diameters of an ellipse which has a double contact with the hyperbola, then $4a^2b^2 = m^4$, where a, b are the semi-conjugate diameters.

12. If three conics have a common chord, the other three common chords meet in a point.

13. Find the envelope of a line which cuts off from the axes intercepts whose sum is constant.

14. The bisectors of the angles A, B, C meet the conic round ABC in D, E, F; shew that the equations to BD, CD, DE are, with the notation of Art. 345,

$$\gamma + \frac{m+n}{l}\alpha = 0, \qquad \frac{m+n}{l}\alpha + \beta = 0,$$

$$(m+n)\alpha + (n+l)\beta - n\gamma = 0.$$

Hence shew that DE, EF, FD will meet AB, BC, CA, respectively, in points which lie on one straight line.

15. In Art. 338, Ex. 5, if FD and DE meet CA and AB respectively in Q and R, shew that the envelope of QR is the same parabola.

16. Shew that a pair of tangents through any angle of a self-conjugate triangle form an harmonic pencil with the sides which meet in that angle.

17. From the equation to the circle circumscribed about a triangle shew that the feet of the perpendiculars dropped upon the sides of the triangle from any point in the circumference lie in one straight line.

18. With the angular points of a triangle ABC as centres, and the sides as asymptotes, three hyperbolas are described, so that the perpendicular distance of the vertex of each from its

asymptotes is the same; shew that the intersection of each pair lies on the axis of the third.

19. The asymptotes to an hyperbola are tangents to an ellipse; shew that the chords which join the points of intersection of the two curves are parallel.

20. If $L=0$, $M=0$, $N=0$ are parallel straight lines, the equation $\phi(LMN)=0$ can only represent parabolas and parallel straight lines.

21. Shew that the tangents to a conic, drawn at the angular points of an inscribed triangle, will meet the opposite sides in points which all lie in one straight line.

22. A conic is inscribed in the triangle of reference; shew that the equations to the straight lines joining the points of contact are

$$l\alpha+m\beta-n\gamma=0, \quad n\gamma+l\alpha-m\beta=0, \quad m\beta+n\gamma-l\alpha=0.$$

23. If ABC be a triangle, and AP, CP be so drawn, that

$$\sin PAC : \sin PAB = \sin PCB : \sin PCA,$$

shew that the locus of P is a conic section, to which two sides of the triangle are tangents.

24. A fixed conic circumscribes a triangle ABC, and is touched at B and C by a variable conic, which meets CA, AB again in D and E; shew that BD, CE intersect in a fixed line passing through A, and that DE passes through a fixed point.

25. Three circles, which touch each other mutually, are described with the angular points of a triangle as centres; shew that the three straight lines, which join the centre of one circle to the point of contact of the other two, meet in a point.

26. If the tangents to a circumscribed conic at the angular points of the triangle of reference are parallel to the opposite sides, the conic is an ellipse, whose equation is

$$\frac{\beta\gamma}{a}+\frac{\gamma\alpha}{b}+\frac{\alpha\beta}{c}=0.$$

27. The equation to the circle which passes through the centres of the three escribed circles is

$$a\beta\gamma + b\gamma\alpha + c\alpha\beta + (\alpha + \beta + \gamma)(a\alpha + b\beta + c\gamma) = 0.$$

28. The equation to the circle, which passes through the centres of the inscribed circle and two escribed circles which touch CA and AB, is

$$a\beta\gamma + b\gamma\alpha + c\alpha\beta + (\alpha - \beta - \gamma)(a\alpha + b\beta + c\gamma) = 0.$$

29. A conic circumscribes a triangle; and any conic is described having a double contact with this, and such that the bisector of the angle C is the chord of contact. Prove that the straight line joining the points, in which the latter conic cuts CB and CA, meets AB in a fixed point.

30. Interpret the equations

$$\sqrt{\alpha \sin A} + \sqrt{\beta \sin B} + \sqrt{\gamma \sin C} = 0,$$

$$\sqrt{\alpha \cos A} + \sqrt{\beta \cos B} + \sqrt{\gamma \cos C} = 0,$$

$$\sqrt{\alpha} + \sqrt{\beta} + \sqrt{\gamma} = 0.$$

31. If a conic touch the sides of the triangle of reference in A', B', C', then (i) any one of the lines AA', BB', CC', drawn from the opposite angles, passes through the intersection of tangents drawn to the conic at the points where the other two cut it; and (ii) if p, q, r be perpendiculars drawn from the point where these three lines meet (Art. 347, Ex. 1), the equation to the conic is

$$\sqrt{\frac{\alpha}{p}} + \sqrt{\frac{\beta}{q}} + \sqrt{\frac{\gamma}{r}} = 0.$$

32. Find the diameter of a circle described about a semi-ellipse bounded by its axis minor.

33. If a conic be inscribed in a quadrilateral, the line joining its points of contact with two opposite sides passes through the intersection of the diagonals.

34. From the equation to a conic circumscribing a quadrilateral, shew that, if the conic be a circle, the opposite angles of the quadrilateral are supplementary.

35. Show that the three conics

$$\frac{\alpha\beta}{lm} - \frac{\gamma^2}{n^2} = 0, \quad \frac{\beta\gamma}{mn} - \frac{\alpha^2}{l^2} = 0, \quad \frac{\gamma\alpha}{nl} - \frac{\beta^2}{m^2} = 0,$$

meet in a single point, and that, if the constants l, m, n are connected by the equation $\frac{\lambda}{l} + \frac{\mu}{m} + \frac{\nu}{n} = 0$, where λ, μ, ν are fixed quantities, the locus of that point is a conic circumscribing the triangle of reference. Explain this by a geometrical figure.

36. The curve $l\alpha^2 + m\beta^2 + n\gamma^2 = 0$ will be a circle, if

$$\frac{l}{\sin 2A} = \frac{m}{\sin 2B} = \frac{n}{\sin 2C}.$$

37. Find the condition that the equation $l\beta\gamma + m\gamma\alpha + n\alpha\beta = 0$ may represent a rectangular hyperbola; hence prove that every rectangular hyperbola described about a triangle passes through the point of intersection of the perpendiculars from the angles on the opposite sides.

38. Shew that $l\alpha^2 + m\beta^2 + n\gamma^2 = 0$ will represent a rectangular hyperbola, if $l + m + n = 0$.

39. An equilateral hyperbola is described, with regard to which a given triangle is self-conjugate; shew that the curve passes through the centres of the inscribed and escribed circles of the triangle.

40. Find the locus of a point (i) such that the square of the tangent from it to a fixed circle is in a constant ratio to the product of its distances from two fixed lines; and (ii) such that the tangent from it to a fixed circle is in a constant ratio to its distance from a fixed line.

PROPERTIES OF QUADRATIC EQUATIONS.

I. THE most general form of a quadratic is

$$ax^2 + bx + c = 0 \ldots\ldots\ldots\ldots\ldots\ldots (1).$$

Solving this in the ordinary manner, we obtain

$$x = \frac{-b \pm \sqrt{b^2 - 4ac}}{2a} \ldots\ldots\ldots\ldots\ldots\ldots (2).$$

The roots of (1) are, therefore,

$$\frac{-b + \sqrt{b^2 - 4ac}}{2a}, \quad \frac{-b - \sqrt{b^2 - 4ac}}{2a} \ldots\ldots\ldots (3);$$

hence, (i) if $b^2 > 4ac$, we shall have $b^2 - 4ac$ a *positive* quantity, and therefore $\sqrt{b^2 - 4ac}$ a *possible* quantity, and the two roots will be *real and different.* (ii) If $b^2 = 4ac$, $\sqrt{b^2 - 4ac} = 0$, and therefore the two roots will be *real and equal.* (iii) If $b^2 < 4ac$, $b^2 - 4ac$ is a *negative* quantity, and $\sqrt{b^2 - 4ac}$ is imaginary, and therefore the two roots are *imaginary.*

Hence, the roots of equation (1) are *real and different, real and equal,* or *imaginary,* according as

$$b^2 > = < 4ac.$$

II. If α, β represent the roots of $ax^2 + bx + c = 0$, given in (3), we have

$$\alpha + \beta = -\frac{b}{a}, \quad \alpha\beta = \frac{b^2 - (b^2 - 4ac)}{4a^2} = \frac{c}{a}.$$

This property is expressed by saying that, if any quadratic be so written, that the coefficient of x^2 is unity, then the coefficient of x is equal to *minus the sum* of the roots, and the last term is equal to the *product* of the roots.

If $\frac{c}{a}$ is positive, the roots have the *same* sign; if $\frac{c}{a}$ is negative, they have *different* signs.

III. If we write equation (1) in the form

$$c\frac{1}{x^2}+b\frac{1}{x}+a=0 \text{.......................} (4),$$

that is, as a quadratic in $\frac{1}{x}$, so that the roots are $\frac{1}{\alpha}$ and $\frac{1}{\beta}$, we have in the same way

$$\frac{1}{\alpha}+\frac{1}{\beta}=-\frac{b}{c},\quad \frac{1}{\alpha\beta}=\frac{a}{c}.$$

IV. (i) If $c=0$ in (1), the quadratic is divisible by x, so that one root is $x=0$, and the other $x=-\frac{b}{a}$. If $b=0$ as well as $c=0$, the quadratic is divisible by x^2, so that each of the roots is $=0$. This is also evident from (3).

(ii) If b only $=0$, the sum of the roots $=0$; that is, the roots are equal and of opposite signs.

(iii) If $a=0$, then one of the roots is $x=\infty$; for, if we write the equation in the form (4), the roots are by (i) $\frac{1}{x}=0$, $\frac{1}{x}=-\frac{b}{c}$, which gives $x=\infty$, $x=-\frac{c}{b}$. This may be seen from the form of solution given above, for, if we multiply the numerator of (2) by $-b\mp\sqrt{b^2-4ac}$, we obtain the roots in the form

$$\frac{2c}{-b-\sqrt{b^2-4ac}},\quad \frac{2c}{-b+\sqrt{b^2-4ac}} \text{............} (5),$$

and, by putting $a=0$ in (5), we obtain the same result as above. If $b=0$ as well as $a=0$, it will be seen in the same way that each root $=\infty$.

V. *If there be two equations given of the second degree between two unknown quantities, the elimination of either of them will produce, generally, an equation of the fourth degree.*

Since the general equation of the second degree may be written

$$x^2 + 2\frac{Hy + G}{A}x + \frac{By^2 + 2Fy + C}{A} = 0,$$

two equations of this form may be written

$$x^2 + Qx + R = 0, \qquad x^2 + Q'x + R' = 0 \ldots\ldots\ldots(1).$$

By subtraction we obtain

$$(Q - Q')x + R - R' = 0 \ldots\ldots\ldots\ldots\ldots\ldots(2),$$

and, eliminating x between (2) and either of equations (1), we have

$$(R - R')^2 + (QR' - RQ')(Q - Q') = 0 \ldots\ldots\ldots\ldots(3).$$

A consideration of the values of Q, R, &c., will make it evident that (3) will contain y^4, but no higher power of y.

ANSWERS TO THE EXAMPLES.

I.

2. $\left(1, -\frac{3}{2}\right)$, $\left(-\frac{1}{2}, 0\right)$, $\left(\frac{1}{2}, -\frac{3}{2}\right)$. 3. $x = -\frac{9}{4}, y = 1$.

4. 5, $\sqrt{37}$. 5. $3\sqrt{2}$, $3\sqrt{3}$. 6. $\frac{17}{7}, \frac{24}{7}$.

7. $PQ = \sqrt{21}$. 8. (1) $\rho = 2, \theta = 30^\circ$; (2) $\rho = 2, \theta = 150^\circ$; (3) $\rho = \sqrt{2}, \theta = 135^\circ$. 9. (1) $x = \frac{5}{\sqrt{2}} = y$; (2) $x = -\frac{5}{2}$, $y = -\frac{5\sqrt{3}}{2}$; (3) $x = \frac{5}{\sqrt{2}}, y = -\frac{5}{\sqrt{2}}$. 10. $\rho \cos(\theta - a) = a$, $\rho^2 (2 + \sin 2\theta) = 2b^2$. 11. $(x^2 + y^2)^2 = a^2 (x^2 - y^2)$.

12. $\frac{2}{3}, 1; -\frac{2}{3}, -1$. 13. (i) $\frac{c}{2}, \frac{b}{2}$; (ii) $\frac{bc \cos C}{a}$, $\frac{bc \cos B}{a}$; (iii) $x = y = \frac{bc}{b + c}$. 14. (i) $\frac{c + b \cos A}{2}, \frac{b \sin A}{2}$; (ii) $c \sin^2 B$, $c \cos B \sin B$; (iii) $y = \frac{bc \sin A}{b + c}$, $x = y \cot \frac{A}{2}$.

II.

2. $y - 6x + 7 = 0$. 3. $(2y - y_2 - y_1)(x_3 - x_2) = (2x - x_1 - x_2)(y_3 - y_2)$ is one equation. 4. $y - x - 7 = 0$.

5. (i) $y + x - 6\sqrt{2} = 0$; (ii) $y + x + 6\sqrt{2} = 0$.

6. $x = \frac{3}{7}, y = \frac{1}{7}$. 7. $3x + 4y - 5a = 0$.

8. $y = mx + c$. 9. $4(x + y) - 5a = 0$.

10. $(a^2 + b^2)^{\frac{1}{2}}, \frac{(a^2 + b^2)^{\frac{3}{2}}}{ab}$. 11. $y - x - 1 = 0$.

12. $x + y = a + b$. 13. $(a' - a) y - (b' - b) x = a'b - ab'$, $(a - a') y + (b - b') x = ab - a'b'$. 14. $x = \sqrt{3} y$.

15. $x=2,\ y=3,\ x=3,\ y=4$. 16. $-\frac{6}{\sqrt{13}}$.

17. $y+2x=0$. 18. $x+y=2a$, or $=4a$, according to the side on which the line is drawn. 20. $m=3$.

21, 22. Use Art. 43. 24. $\cos^2 a=\frac{(A-B\sqrt{2})^2}{2(A^2-AB\sqrt{2}+B^2)}$.

25. $\frac{m}{n}$. 26. $x=y=\frac{a}{4}\sqrt{6}$.

28. Co-ordinates of vertex being $o,\ y'$, the intersection is $o,\ \frac{1}{3}y'$. 29. $\rho=2a,\ \theta=\frac{\pi}{2}$; the angle $=\frac{\pi}{3}$.

31. Take the general polar equation (Art. 44), and proceed as in Art. 30. One triangle is equal in area to the sum of two others.

III.

1. $11y-(8\pm5\sqrt{3})x-5(5\mp\sqrt{3})=0$. 2. (i) $x=c$, $y=0$; (ii) 135°. 3. $(a\pm b)y=(b\mp a)(x-c)$.

4. $5y-8x-40=0$. 5. $\frac{25}{\sqrt{629}}$. 6. 90°.

7. $y=(1\pm\sqrt{2})(x+2)$. 9. $99x-27y=79$, $21x+77y=1$. 13. Any angle. 14. $y=0,\ y+x=a$.

15. 45° or 135°. 16. $2x=y,\ 2y=x$.

17. $\frac{62}{11\sqrt{26}}$. 18. $\pm\frac{C-C'}{\sqrt{A^2+B^2}}$.

19. (1, 2) on the origin side; (3, -4) on the side remote from the origin. 20. The side remote from the origin.

22. (i) and (ii) Straight lines inclined at an angle a to the initial line. (iii) A circle whose radius $=a$; a circle whose equation is $\rho+a\cos\theta=0$; and a straight line $\rho\cos\theta-a=0$. 23. 45°.

24. (i) The initial line and a perpendicular to it through the pole. (ii) A perpendicular to the initial line through the pole, and two straight lines drawn through the pole, making angles of

30° and 150° with the initial line. (iii) Three straight lines, real or imaginary, passing through the pole. 25. (i) $\frac{ab}{2}$; (ii) $4a^2$; (iii) $\pm\frac{1}{2}(x'y - xy')$. 26. $\frac{9}{14}$.

IV.

1. $x^2 + y^2 = c^2$. 2. $x^2 \cos 2a - xy \sin 2a = a^2$.
3. $\sqrt{2}x = c$, $xy = 0$. 5. $y^2 \sin^2 a = 4ax$.
6. $27y^2 - x^2 = 12$. 7. $3x^2 + 10y^2 - 7\sqrt{3}xy = 6$.

V.

4. Take Ox, OC as axes. 10. A straight line through the intersection of the perpendicular lines. 11. A straight line perpendicular to the base.

VI.

1. $b\beta + c\gamma = 0$. 2. $-a\alpha + b\beta + c\gamma = 0$.
3. $l(\alpha - \alpha') + m(\beta - \beta') + n(\gamma - \gamma') = 0$.
4. $\alpha \cos A + \beta \cos B - \gamma \cos C = 0$. 6. $\frac{\alpha}{mn' - m'n} = \frac{\beta}{nl' - n'l} = \frac{\gamma}{lm' - l'm} = \frac{2\Delta}{a(mn' - m'n) + \&c.}$; see Art. 95, Ex.
7. $\alpha'(\beta''\gamma''' - \beta'''\gamma'') + \beta'(\gamma''\alpha''' - \gamma'''\alpha'') + \gamma'(\alpha''\beta''' - \alpha'''\beta'') = 0$.
10. $lmn = -1$.

11. It is parallel to $(\beta \cos B - \gamma \cos C)$, Art. 93, Ex. 4; hence, using Art. 86 and the condition that it passes through the bisection of BC, where $\beta = \frac{a}{2}\sin C$, $\gamma = \frac{a}{2}\sin B$, we obtain

$$\beta \cos B - \gamma \cos C = \frac{a}{2}\sin(C - B).$$

13. $\beta + \alpha \cos C = 0$. 14. $al + bm + cn = 0$.
16. $(b + c)l = (m + n)a$. 17. $AM = \frac{bcl}{lb - ma}$, $AN = \frac{bcl}{lc - na}$.

20. A straight line passing through the points where the external bisectors of the angles of the triangle meet the opposite sides. The three sides of a triangle formed by joining the points where the bisectors of the angles meet the opposite sides.

VII.

1. Co-ordinates of centre are 3 and -2; rad. $= 3$.

2. A circle whose rad. $= r$, and a tangent to it. 3. $x + y = 3$.

4. $x + y + 1 = 0$. 5. $(1 + m^2)(x^2 + y^2) - 2r(x + my) = 0$.

6. $x^2 - ax + y^2 = r^2 - \dfrac{a^2}{2}$. 7. An imaginary locus.

8. (i) $G^2 = AC$, (ii) $F^2 = AC$. 9. See Art. 121.

10. $\dfrac{1}{r^2} = \dfrac{1}{a^2} + \dfrac{1}{b^2}$. 11. $x^2 + y^2 = \dfrac{9}{5}$. 12. $\dfrac{r}{\cos a}, \dfrac{r}{\sin a}$.

13. $x \cos(a' + a'') + y \sin(a' + a'') = r \cos(a'' - a')$.

14. Use Art. 129 and Cor.

16. $x^2 + y^2 + xy + x + y - 1 = 0$. 17. $x = a + c$.

18. A circle, whose diameter = the radius of the given circle.

20. The segment of a circle on the base, which contains the given angle. 22. $\dfrac{h(x' - x'') + k(y' - y'')}{\{(x' - x'')^2 + (y' - y'')^2\}^{\frac{1}{2}}}$. 23. $x \pm 3y = 10$.

24. $y - b = m(x - a) \pm r(1 + m^2)^{\frac{1}{2}}$. 25. $4y - 3x = 0$.

26. $\{4r^2 - 2(a - b)^2\}^{\frac{1}{2}}$. 27. Area $= \pi$.

28. $2\left(r^2 - \dfrac{a^2b^2}{a^2 + b^2}\right)^{\frac{1}{2}}$. 29. Take CS as axis of x; the locus is a straight line. 31. $r^2 = \dfrac{a^2 - 2ab\cos\omega + b^2}{4\sin^2\omega}$.

33. $x^2 + y^2 + 8ax - 6ay = 0$. 34. Take the given point as pole, and a straight line through the centre as initial line; then the radius vector of the locus is half the sum of the roots in the polar equation to the circle. The result is a circle described on the line joining the point and the centre, as diameter.

35. A circle whose centre is on the line joining the given point with the given circle. 41. The triangle must be isosceles, and the ratio one of equality (Art. 106, Cor.).

42. The triangle must be equilateral. 43. A circle.

44. The centre is on the base.

45. $(x - a \pm b' \mp b)^2 + (y - b - b')^2 = (b' - b)^2$. 47. $y = x$.

48. $x^2 + y^2 - 2r(x + y) + r^2 = 0$. 49. Take the bounding radii as axes, and use the equation of Ex. 48; then the circle passes through (hk) on the quadrant, where $h^2 + k^2 = R^2$, and the roots of the equation obtained from this condition are r and r' &c.

50. The tangent to the first circle is $x \cos \alpha + y \sin \alpha = r$, and to the second $(x - a) \cos \beta + y \sin \beta = r'$; if these represent the same line, the equation is $(r \pm r') x \pm \{a^2 - (r \pm r')^2\}^{\frac{1}{2}} y = ar$.

51. Write $r = \dfrac{a + b \pm \sqrt{a^2 + b^2}}{2}$ in the result of Ex. 48. See Art. 119, Cor. 1. 52. $x^2 + y^2 - 2\{a + b \pm (2ab)^{\frac{1}{2}}\}(x + y) + \{a + b \pm (2ab)^{\frac{1}{2}}\}^2 = 0$. See Ex. 49. 53. If the circle touches BC in L, then $BL = s - c$, &c.; take two sides as axes. 54. If AE be initial line, $AP = \rho$, $PAE = \theta$, the polar equation to the locus is $\rho \sin \theta = a \sin (\theta + \alpha)$, where $a = AE$, $\alpha = \angle EAB$.

55. $\dfrac{x}{y} = \dfrac{OB - OB'}{OA - OA'}$. 56. The polar co-ordinates of the centre are $\rho = 2$, $\theta = \dfrac{\pi}{3}$, and rad. $= 3$. 57. A circle whose centre is the intersection of the diagonals. 58. Take A as pole and the diameter through A as initial line; if $CP = n \,.\, AC$, the locus is a circle whose radius $= (n + 1) r$.

59. If α be the given angle, centre origin, the equation is

$$y = \tan\left(\alpha - \frac{\pi}{4}\right) x + \frac{r}{\sqrt{2}} \sec\left(\alpha - \frac{\pi}{4}\right).$$

60. A circle whose centre is the fixed extremity of one of the lines. 61. Take the common tangent and common diameter as axes, and use Art. 43. 62. A straight line perpendicular to the line joining the fixed point and the centre.

63. Take AB, AC as axes; then the equation to the circle is (Art. 110) $x^2 + 2xy \cos \omega + y^2 + 2Gx + 2Fy = 0$; then assume

$BM : MA = AN : NC = n : 1$, and thus find AM and AN; D and E may be found from the condition of the circle passing through M and N. For the fixed point $\frac{x}{b} = \frac{y}{c} = \frac{bc}{b^2 + c^2 + 2bc \cos A}$.

VIII.

14. $y - x = 0$; $y - x = 0$, $y + x - 2 = 0$. 15. Use Art. 73.

IX.

4. $\cos\theta = \frac{5\sqrt{5} - 3\sqrt{3}}{4\sqrt{5}}$. 7. $\cos\theta = \pm\frac{5\sqrt{17}}{27}$.

8. (i) $\cos^{-1} e$; (ii) $\tan^{-1}\frac{a}{b}$. 11. In any triangle ABC, $\tan\frac{A}{2}\tan\frac{B}{2} = \frac{s-c}{s}$, &c.

14. $\frac{a^2}{m^2} - \frac{b^2}{n^2} = 1$. 17. The extremities of the latus rectum; $\tan^{-1} \pm e$. 18. The equation $\pm 3y \pm \sqrt{5}x - 9a = 0$ will represent the four tangents. 21. $r = eb$.

22. The extremity of the latus rectum. 25. PQ is a tangent to the inner ellipse at $(x'y')$, and the polar of R (hk) with regard to the outer; by comparing these equations we get x', y' in terms of h, k, &c. 28 and 29. Use Art. 197.

30. $(x^2 + y^2)^2 = a^2x^2 - b^2y^2$. 31. $a\frac{\cos\frac{\theta+\phi}{2}}{\cos\frac{\theta-\phi}{2}}$; $b\frac{\sin\frac{\theta+\phi}{2}}{\cos\frac{\theta-\phi}{2}}$.

33. $\tan^{-1}\frac{\sqrt{5}}{2}$. 34. Use equation Art. 34, P being $(x'y')$, and combine with equation to asymptotes as one locus, &c.

35. See Art. 188.

37. $\frac{1}{\sqrt{2}}$, 3. 38. $y = 3x$, $3y = x$; $\tan^{-1} 2\sqrt{6}$. See Arts. 186, 67. 39. See Art. 192. Four, if the points on

the major and minor axes are not further from the centre than $\frac{a^2-b^2}{a}$ and $\frac{a^2-b^2}{b}$ respectively; from other points, two.

41. $\left(\frac{bm}{y}\right)^2+\left(\frac{an}{x}\right)^2=(m+n)^2$. 42. An ellipse whose equation is $\frac{x^2}{a^2}+\frac{b^2y^2}{a^4}=1$. 44. Use the polar equation of Art. 209; the eccentricity is $\sqrt{\frac{2e}{1+e}}$. 45. See Art. 187.

46. The equations to the ellipses are $\frac{x^2}{a^2}+\frac{y^2}{b^2}=1$, $\frac{x^2}{a^2}+\frac{y^2}{\beta^2}=1$, where $a^2-b^2=\alpha^2-\beta^2$; take equations to tangents in form of Art. 182. 47. The centre (xy), $P(x'y')$; then $\frac{y'^2}{(ae+x)^2}$ $=\tan^2\frac{PSH}{2}=\frac{1-\cos PSH}{1+\cos PSH}$, which may be expressed in terms of x'. Similarly $\frac{y'^2}{(ae-x)^2}=\tan^2\frac{PHS}{2}$ &c. Dividing the resulting equations, we obtain $x=a$ for the equation.

X.

10. Use equation of Art. 34, and so obtain a polar equation to the locus with the fixed point as pole. 14. See Art. 163.

15. $\frac{\sqrt{6}}{3}$. 17. Use Art. 223 to form the equations to CD and PF. The equation to the locus is $\frac{b^2}{y^2}+\frac{a^2}{x^2}=\left(\frac{a^2-b^2}{x^2+y^2}\right)^2$.

18. Use Art. 223. 20. Art. 155 will suggest the solution. 21. The perpendicular from the centre $(x', 0)$ on the asymptote $=a-x'$, &c. 22. Use the method of Arts. 214—216 with the equation of Art. 240. 23. Draw a diameter by Art. 214, and the axes by Art. 234; then $BS=a$, hence the focus and latus rectum; draw a tangent at the extremity of the latus rectum by Art. 214, Cor.; this gives the directrix, Examples IX. 17.

24. $\tan^{-1}\frac{1}{b}$. 26. $x=3$, $y=2$; $x=0$, $y+2x=0$.

27. Use Art. 223 to form the equations to the tangents at P and D. The equation is $a^2y^2 + b^2x^2 = 2a^2b^2$. 28. If $2p$ be the length of the chord, the equation is $\frac{x^2}{a^2} + \frac{y^2}{b^2} + p^2\frac{a^2y^2+b^2x^2}{a^4y^2+b^4x^2} = 1$.

32. See Arts. 218, 225, 230; $a^2 = 2c^2 \cos^2 \frac{a}{2}$, $b^2 = 2c^2 \sin^2 \frac{a}{2}$, $e^2 = \frac{2 \cos a}{1 + \cos a}$. 34. Use equation of Art. 34, and Art. 219.

37. If $(x'y')$ is the point of contact, and m, c constants, we have $-\frac{b^2x'}{a^2y'} = m$, $a^2 - b^2 = c$; from these and the equation to the ellipse we must eliminate a and b.

39. Use Arts. 132, 182; the equation is $(x^2 + y^2)^2 = a^2x^2 + b^2y^2$.

40. Use Art. 223; the normal at D is perpendicular to CP.

41. A rectangular hyperbola, of which AB is an asymptote.

42. Take the sides as axes; then, when $x = 0$, the difference of the roots = one of the lines, and similarly when $y = 0$, &c.

43. An hyperbola whose asymptotes are parallel to the lines containing the given angle. 49. Use equation of Art. 34 for any one of the chords, $(x'y')$ being middle point, and combine it with the equations to the two hyperbolas; then, if λ satisfies one equation, $-\lambda$ will satisfy the other; substituting and eliminating λ, we have for the required equation

$$\left(\frac{c^2}{a^2} - \frac{s^2}{b^2}\right) + 4\left(\frac{cx}{a^2} - \frac{sy}{b^2}\right)^2\left(\frac{x^2}{a^2} - \frac{y^2}{b^2}\right) = 0.$$

50. Prove the ordinates of the points proportional to the conjugate axes, &c.

XI.

1. $4d(2 \pm \sqrt{3})$. 2. $(0, 0)$, $(4, -8)$.

3. $y_1 = d(\sqrt{5} - 2)^{\frac{1}{2}}$, $y_2 = d(\sqrt{5} + 2)^{\frac{1}{2}}$. 5. Art. 261. Cor.

8. Taking the centre of the circle as origin, and the given diameter as axis of x, the latus rectum = twice the abscissa of the point of contact. 11. $x = \frac{ab' - a'b}{b - b'}$, $y = \frac{1}{2}(b + b')$.

12. A parabola having its axis parallel to that of y, and its vertex at the point $x = \frac{1}{2}$, $y = \frac{1}{4}$. The straight line is a tangent at

the point $x=1$, $y=0$. 13. Take (hk) a point on the outer, and find the points of tangency by Art. 123, Cor. The difference of the ordinates of these points is constant. 14. See Art. 136.

15. $y=\dfrac{2\mu\mu'}{\mu+\mu'}x+\dfrac{d(\mu+\mu')}{2\mu\mu'}$. See Arts. 283, 257.

16. Parabolas whose latera recta = half that of the original curve. Use Art. 34. 17. See Art. 304, Ex.

19. Distance $=2\sqrt{2}d$. 20. See Art. 129, Cor.

21. $y^2=\dfrac{4dx^2}{2d-x}$. 22. See Art. 259.

24. $\sqrt{x}+\sqrt{y}=\sqrt{2d\sqrt{2}}$. 25. $2a^{\frac{3}{2}}(d^{\frac{1}{2}}\sim\delta^{\frac{1}{2}})$. 26. $2d$.

27. Art. 256 for equations to sides. 28. $\dfrac{d\cos^2\phi}{\sin\phi}$, $\dfrac{d}{\sin\phi}$.

29. A parabola whose focus is at the centre of the given circle.

30. Use Art. 256. The equation is $d(y+x)^2=(y^2+x^2)(y-x)$.

31. If the focus be pole, the axis initial line, the polar equation is $\rho=2d\cot\dfrac{\theta}{2}$. 33. $y=mx+r\sqrt{1+m^2}$, where $m=\pm\left(\dfrac{\sqrt{4d^2+r^2}-r}{2r}\right)^{\frac{1}{2}}$. 35. Find a diameter by Art. 276, and draw a tangent at its vertex by Art. 278; draw a focal chord by Art. 269, the diameter of which will intersect the tangent in the directrix; then find the focus by Art. 272. 36. Equal tangents from $(x'y')$ would make equal angles with the polar of $(x'y')$. Use Art. 257. 38. Latus rectum $=4\rho\sin^2\phi$, if ρ and ϕ are the given quantities. 43. $y^2=h(x-h)$, where h is any constant. 44. $(y-x)^2-8dx\sqrt{2}=0$. 45. Take equation Art. 261, Cor., $m^3+\dfrac{2d-x}{d}m+\dfrac{y}{d}=0$, and suppose the roots to be μ, $-\dfrac{1}{\mu}$, μ'; then the sum $=\mu-\dfrac{1}{\mu}+\mu'=0$; the sum of the products, two and two, $=-1+\mu'\left(\mu-\dfrac{1}{\mu}\right)=\dfrac{2d-x}{d}$: the product $=-\mu'=-\dfrac{y}{d}$; eliminating μ and μ', we have $y^2=d(x-3d)$.

46. $x=0$, $x=2(c-2d)$, $y=\pm\left(\dfrac{2d}{c-2d}\right)^{\frac{1}{2}}x$.

49. See Art. 271. 50. Take equation (2) of Art. 251, and divide by l^2, so as to get a quadratic in $\dfrac{1}{l}$; then follow the method of Art. 310. 51. A parabola whose equation is $y^2=d(x-d)$. 53. An extremity of the latus rectum.

55. $a^2=2b^2$. 56. Let the tangent of one of the parabolas at the given point be inclined at an angle θ to the diameter; take the diameter and a perpendicular through the point as axes, and find the co-ordinates of the vertex (Art. 280) in terms of θ and the parameter; then eliminate θ.

57. Take $y^2=4dx$ as the equation to the parabola in one of its positions; then the equations to the rectangular axes are $y=mx+\dfrac{d}{m}$, $y=-\dfrac{x}{m}-md$; the perpendiculars on these lines from the focus or vertex will be the x and y of the locus, if the lines are axes of co-ordinates; writing these equations, and eliminating m, we obtain (i) $x^2y^2=d^2(x^2+y^2)$, (ii) $x^{\frac{2}{3}}y^{\frac{2}{3}}(x^{\frac{2}{3}}+y^{\frac{2}{3}})=d^2$.

XII.

2. $m=1$. 8. $2x-a=0$, $8y-6x-3a=0$, $e=\dfrac{\sqrt{5}}{2}$; $y-2x=0$, $y+2x+2=0$, $e=\dfrac{\sqrt{5}}{2}$. 10. Semi-transverse (1) $2\sqrt{ab}\sin\dfrac{\alpha}{2}$, (2) $2a\sin\dfrac{\alpha}{2}$; Semi-conjugate (1) $2\sqrt{ab}\cos\dfrac{\alpha}{2}$, (2) $2a\cos\dfrac{\alpha}{2}$. See end of Art. 241. 13. Since all pairs of lines drawn through the four points are at right angles, AD, BC; BD, CA; CD, BA are at right angles, and D is orthocentre. Conversely, if D is orthocentre of ABC, and AD, BD, CD meet the sides in A', B', C'; then $DC'=AC'\tan DAC'=AC'\cot B$, and $CC'=b\sin A$; whence $CC'\,.\,DC'=AC'\,.\,BC'$; and, if $C'A$, $C'C$ be axes, we have $aa'=-bb'$, &c. 14. Let D be the orthocentre of ABC; then the locus of the centres of the conics $ABCD$, all of

which are rectangular hyperbolas, will be the circle. The feet of the perpendiculars are the centres of conics of the system; for the bisections see Art. 314, Ex. 2. This is the *Nine Points Circle* of the triangle ABC. See Art. 359, Ex. 1—5. 15. See Arts. 312 and 313, Ex. 1. 16. Transfer the origin to the point of tangency; then use Art. 297, Cor., and retransfer. The result is

$$(by-1)(b'-b)+x\{c+a(b'-b)\}=0;\ (b'y-1)(b'-b)-x\{c-a(b'-b)\}=0;$$
$$2ax+(b+b')y=2.$$

19. $\dfrac{4SP \, . \, Sp}{PSp}$. 20. $(A-k)x^2+2Hxy+(B-k)y^2=0$; $H^2=(A-k)(B-k)$; where the conic and circle are

$$Ax^2+2Hxy+By^2=1, \text{ and } k(x^2+y^2)=1.$$

21. See Art. 294 (1). 22. Use Art. 301. Cor.
23. See Art. 310. 24. For A or G, $Hx+By+F=0$; for H, $By^2+Fy=Ax^2+Gx$; for B or F, $Ax+Hy+G=0$; for C, the conics are concentric. 25. $\dfrac{2(F\sqrt{A}-G\sqrt{B})}{(A+B)^{\frac{3}{2}}}$. See Art. 156.

26. Generally one; but an infinite number if the perpendiculars from A on BC, B on CA, C on AB meet in D, where A, B, C are any three of the points. 27. Use Art. 296, Cor. 3.

28. Use Arts. 296, 308. 29. $\tan\theta=\dfrac{e-\cos\alpha}{\sin\alpha}$.

30—33. Use Art. 305. 33. The locus is a conic, whose eccentricity is $e\cos\beta\sec\dfrac{\epsilon}{2}$. 35. Draw two diameters, &c.

36. See Art. 307, Ex. 38. The hypotenuse is parallel to the normal. 39. (1) Two straight lines through the origin, parallel to the asymptotes. (2) The straight line joining the points at a finite distance, in which (1) meets the curve. (3) If lines be drawn parallel to Ox, from the points of intersection of the curve with Oy, they meet the curve in two points, the line joining which is represented by (3). (4) Similarly for Oy and Ox. 41. Use Art. 308. 42. Use Art. 295.

43. $(x+y-a)^2=4xy\sin^2\dfrac{\omega}{2}$. 44. See Arts. 318—320.

46. Art. 314, Ex. 1, and Art. 296, Cor. 2. 47. A parabola to which AB, AC are tangents. 48. Take the two sides as axes, and use Art. 314, Ex. 1. 50. For the first part: See Arts. 218, 294 (II), 186, Cor. 1. To find C': Let the equations to the ellipse and hyperbola, referred to their axes, be $A'x^2+B'y^2+C=0$ and $A''x^2+B''y^2+C'=0$; then the latter equation must be equivalent to $A'x^2-B'y^2\pm C=0$. Comparing these forms and using Art. 147, we find $C'=\pm C\sqrt{(A-B)^2+4H^2}$.

51. The ellipses are equal, and if $2a$, 2β are their axes, $a^2=a^2\sqrt{2}(\sqrt{2}+1)$, $\beta^2=a^2\sqrt{2}(\sqrt{2}-1)$. 52. The directrix is $Gy+Fx=0$; for the focus, $\frac{x}{G}=\frac{y}{F}=-\frac{C}{G^2+F^2}$; the axis is $(G^2+F^2)(Gx-Fy)+(G^2-F^2)C=0$. See Arts. 259, 263, 272, 283. 53. See Arts. 92, 233. Let the asymptotes be $y=ax$, $y=\beta x$, and the conjugate diameters $y=a'x$, $y=\beta'x$; then (Art. 92) the two latter equations are equivalent to

$$y-ax+k(y-\beta x)=0,\quad y-ax-k(y-\beta x)=0.$$

Comparing these equations and eliminating k, we have

$$(a+\beta)(a'+\beta')=2(a\beta+a'\beta');$$

but $a+\beta=-\frac{2H}{B}$, $a\beta=\frac{A}{B}$; $a'+\beta'=-\frac{2h}{b}$, $a'\beta'=\frac{a}{b}$, &c.

XIII.

4. (i) Conics having three points of intersection finite, and one infinite. (ii) Conics having two points of intersection finite, and two infinite. 7. See Art. 86. 8. (i) Similar and similarly placed hyperbolas, having that portion of (N) which is intercepted between (L) and (M) as common chord. (ii) Hyperbolas having (L) and (M) for asymptotes. (iii) Parabolas, of which (L) is a diameter, and (M) the tangent at its vertex.

9. Compare the equation $l\alpha+m\beta+n\gamma=0$ with the equation to the tangent (Art. 337) of each curve, and eliminate μ_1.

14. Use equation $\beta\gamma+k\alpha^2=0$; then the chord through D is $\alpha+k'(\beta-\gamma)=0$; eliminate α, and use Art. 89. 11. From the symmetry of the figs. the common chord must pass through the

centre. 12. See Art. 329. 13. A parabola.

17. The equation asserts that two triangles are together equal in area to a third. 18. The three hyperbolas will be $\beta\gamma = l^2$, $\gamma\alpha = m^2$, $\alpha\beta = n^2$, and their axes $\beta - \gamma = 0$, $\gamma - \alpha = 0$, $\alpha - \beta = 0$; then the condition gives $l = m = n$; hence the first may be written $\gamma\alpha - l^2 - (\alpha - \beta)\gamma = 0$, &c. 19. Take the asymptotes as two sides of the triangle of reference. 27, 28. See Art. 359. 30. Conics touching the sides of the triangle of reference; (i) at their middle points; (ii) at the feet of the perpendiculars from the angles; (iii) where the bisectors of the angles meet the sides.

31. See Art. 347, Ex. 1. 32. The equation to a circle, having two common chords (L), (M) with the ellipse (S), would be included in the equation $S + kLM = 0$ (Art. 329); hence the equation is

$$\frac{x^2}{a^2} + \frac{y^2}{b^2} - 1 + kx(x - a) = 0,$$

where the chord $x - a = 0$ has become a tangent, &c. then use Art. 106, Cor. 33. Take equation (Art. 348) to the conic touching three sides; then use Art. 351, Ex. 2, Cor. for the fourth side. 34. Take the equation $\alpha\gamma + k\beta\delta = 0$, and write for α, $p - x\cos\alpha - y\sin\alpha$, &c.; then use Art. 106, Cor.

35. The point is defined by $\frac{\alpha}{l} = \frac{\beta}{m} = \frac{\gamma}{n}$. 36. See Art. 356.

37. See Art. 355. The condition is $l\cos A + m\cos B + n\cos C = 0$.

39. See Art. 340.

40. (i) A conic passing through the four points in which the fixed lines intersect the circle. (ii) A conic touching the circle at the two points where the fixed line meets it. See Arts. 134, 334, and suppose (S) to be a circle in Arts. 329, 331.

CAMBRIDGE: PRINTED BY C. J. CLAY, M.A. & SON, AT THE UNIVERSITY PRESS.

www.ingramcontent.com/pod-product-compliance
Lightning Source LLC
Chambersburg PA
CBHW021356210326
41599CB00011B/899